# ADVANCED
# PROCESS
# IDENTIFICATION
# AND CONTROL

# CONTROL ENGINEERING

*A Series of Reference Books and Textbooks*

Editor

## NEIL MUNRO, Ph.D., D.Sc.

Professor
Applied Control Engineering
University of Manchester Institute of Science and Technology
Manchester, United Kingdom

*Additional Volumes in Preparation*

# ADVANCED PROCESS IDENTIFICATION AND CONTROL

## Enso Ikonen

*University of Oulu*
*Oulu, Finland*

## Kaddour Najim

*Institut National Polytechnique de Toulouse*
*Toulouse, France*

CRC Press
Taylor & Francis Group
Boca Raton London New York

CRC Press is an imprint of the
Taylor & Francis Group, an **informa** business

CRC Press
Taylor & Francis Group
6000 Broken Sound Parkway NW, Suite 300
Boca Raton, FL 33487-2742

First issued in paperback 2019

ISBN-13: 978-0-8247-0648-7 (hbk)
ISBN-13: 978-0-367-39688-6 (pbk)

**Visit the Taylor & Francis Web site at**
**http://www.taylorandfrancis.com**

**and the CRC Press Web site at**
**http://www.crcpress.com**

# Series Introduction

Many textbooks have been written on control engineering, describing new techniques for controlling systems, or new and better ways of mathematically formulating existing methods to solve the ever-increasing complex problems faced by practicing engineers. However, few of these books fully address the applications aspects of control engineering. It is the intention of this new series to redress this situation.

The series will stress applications issues, and not just the mathematics of control engineering. It will provide texts that present not only both new and well-established techniques, but also detailed examples of the application of these methods to the solution of real-world problems. The authors will be drawn from both the academic world and the relevant applications sectors.

There are already many exciting examples of the application of control techniques in the established fields of electrical, mechanical (including aerospace), and chemical engineering. We have only to look around in today's highly automated society to see the use of advanced robotics techniques in the manufacturing industries; the use of automated control and navigation systems in air and surface transport systems; the increasing use of intelligent control systems in the many artifacts available to the domestic consumer market; and the reliable supply of water, gas, and electrical power to the domestic consumer and to industry. However, there are currently many challenging problems that could benefit from wider exposure to the applicability of control methodologies, and the systematic systems-oriented basis inherent in the application of control techniques.

This series presents books that draw on expertise from both the academic world and the applications domains, and will be useful not only as academically recommended course texts but also as handbooks for practitioners in many applications domains. *Advanced Process Identification and Control* is another outstanding entry to Dekker's Control Engineering series.

*Neil Munro*

# Preface

The study of control systems has gained momentum in both theory and applications. Identification and control techniques have emerged as powerful techniques to analyze, understand and improve the performance of industrial processes. The application of modeling, identification and control techniques is an extremely wide field. Process identification and control methods play an increasingly important role in the solution of many engineering problems.

There is extensive literature concerning the field of systems identification and control. Far too often, an engineer faced with the identification and control of a given process cannot identify it in this vast literature, which looks like the cavern of Ali Baba. This book will introduce the basic concepts of advanced identification, prediction and control for engineers. We have selected recent ideas and results in areas of growing importance in systems identification, parameter estimation, prediction and process control. This book is intended for advanced undergraduate students of process engineering (chemical, mechanical, electrical, etc.), or can serve as a textbook of an introductory course for postgraduate students. Practicing engineers will find this book especially useful. The level of mathematical competence expected of the reader is that covered by most basic control courses.

This book consists of nine chapters, two appendices, a bibliography and an index. A detailed table of contents provides a general idea of the scope of the book. The main techniques detailed in this book are given in the form of algorithms, in order to emphasize the main tools and facilitate their implementation. In most books it is important to read all chapters in consecutive order. This is not necessarily the only way to read this book.

Modeling is an essential part of advanced control methods. Models are extensively used in the design of advanced controllers, and the success of the methods relies on the accuracy modeling of relevant features of the process to be controlled. Therefore the first part (Chapters 1–6) of the book is dedicated to process identification—the experimental approach to process modeling.

v

Linear models, considered in Chapters 1–3, are by far the most common in industrial practice. They are simple to identify and allow analytical solutions for many problems in identification and control. For many real-world problems, however, sufficient accuracy can be obtained only by using non-linear system descriptions. In Chapter 4, a number of structures for the identification of non-linear systems are considered: power series, neural networks, fuzzy systems, and so on. Dynamic non-linear structures are considered in Chapter 5, with a special focus on Wiener and Hammerstein systems. These systems consist of a combination of linear dynamic and non-linear static structures. Practical methods of parameter estimation in non-linear and constrained systems are briefly introduced in Chapter 6, including both gradient-based and random search techniques.

Chapters 7–9 constitute the second part of the book. This part focuses on advanced control methods, the predictive control methods in particular. The basic ideas behind the predictive control technique, as well as the generalized predictive controller (GPC), are presented in Chapter 7, together with an application example.

Chapter 8 is devoted to the control of multivariable systems. The control of MIMO systems can be handled by two approaches, i.e., the implementation of either global multi-input–multi-output controllers or distributed controllers (a set of SISO controllers for the considered MIMO system). To achieve the design of a distributed controller it is necessary to select the best input–output pairing. We present a well-known and efficient technique, the relative gain array method. As an example of decoupling methods, a multivariable PI-controller based on decoupling at both low and high frequencies is presented. The design of a multivariable GPC based on a state-space representation ends this chapter.

Finally, in order to solve complex problems faced by practicing engineers, Chapter 9 deals with the development of predictive controllers for non-linear systems (adaptive control, Hammerstein and Wiener control, neural control, etc.). Predictive controllers can be used to design both fixed parameter and adaptive strategies, to solve unconstrained and constrained control problems.

Application of the control techniques presented in this book are illustrated by several examples: fluidized-bed combustor, valve, binary distillation column, two-tank system, pH neutralization, fermenter, tubular chemical reactor. The techniques presented are general and can be easily applied to many processes. Because the example concerning fluidized bed combustion (FBC) is repeatedly used in several sections of the book, an appendix is included on the modeling of the FBC process. An ample bibliography is given at the end of the book to allow readers to pursue their interests further.

Any book on advanced methods is predetermined to be incomplete. We have selected a set of methods and approaches based on our own preferences, reflected by our experience—and, undoubtedly, lack of experience—with many of the modern approaches. In particular, we concentrate on the discrete time approaches, largely omitting the issues related to sampling, such as multi-rate sampling, handling of missing data, etc. In parameter estimation, sub-space methods have drawn much interest during the past years. We strongly suggest that the reader pursue a solid understanding of the bias-variance dilemma and its implications in the estimation of non-linear functions. Concerning the identification of non-linear dynamic systems, we only scratch the surface of Wiener and Hammerstein systems, not to mention the multiplicity of the other paradigms available. Process control can hardly be considered a mere numerical optimization problem, yet we have largely omitted all frequency domain considerations so invaluable for any designer of automatic feedback control. Many of our colleagues would certainly have preferred to include robust control in a cookbook of advanced methods. Many issues in adaptive and learning control would have deserved inspection, such as identification in closed-loop, input–output linearization, or iterative control. Despite all this, we believe we have put together a solid package of material on the relevant methods of advanced process control, valuable to students in process, mechanical, or electrical engineering, as well as to engineers solving control problems in the real world.

We would like to thank Professor M. M'Saad, Professor U. Kortela, and M.Sc. H. Aaltonen for providing valuable comments on the manuscript. Financial support from the Academy of Finland (Projects 45925 and 48545) is gratefully acknowledged.

**Enso Ikonen**

**Kaddour Najim**

# Contents

## II  Control

# III  Appendices

# Part I

# Identification

# Chapter 1

# Introduction to Identification

Identification is the experimental approach to process modeling [5]. In the following chapters, an introductory overview to some important topics in process modeling is given. The emphasis is on methods based on the use of measurements from the process. In general, these types of methods do not require detailed knowledge of the underlying process; the chemical and physical phenomena need not be fully understood. Instead, good measurements of the plant behavior need to be available.

In this chapter, the role of identification in process engineering is discussed, and the steps of identification are briefly outlined. Various methods, techniques and algorithms are considered in detail in the chapters to follow.

## 1.1 Where are models needed?

An engineer who is faced with the characterization or the prediction of the plant behavior, has to model the considered process. A modeling effort always reflects the intended use of the model. The needs for process models arise from various requirements:

- In *process design*, one wants to formalize the knowledge of the chemical and physical phenomena taking place in the process, in order to understand and develop the process. Because of safety and/or financial reasons, it might be difficult or even impossible to perform experiments on the real process. If a proper model is available, experimenting can be conducted using the model instead. Process models can also help to scale-up the process, or integrate a given system in a larger production scheme.

- In *process control*, the short-term behavior and dynamics of the process may need to be predicted. The better one is able to predict the output of a system, the better one is able to control it. A poor control system may lead to a loss of production time and valuable raw materials.

- In *plant optimization*, an optimal process operating strategy is sought. This can be accomplished by using a model of the plant for simulating the process behavior under different conditions, or using the model as a part of a numerical optimization procedure. The models can also be used in an operator *decision support* system, or in *training* the plant personnel.

- In *fault detection*, anomalies in different parts of the process are monitored by comparing models of known behavior with the measured behavior. In *process monitoring*, we are interested in physical states (concentrations, temperatures, *etc.*) which must be monitored but that are not directly (or reliably) available through measurements. Therefore, we try to deduce their values by using a model. *Intelligent sensors* are used, *e.g.*, for inferring process outputs that are subject to long measurement delays, by using other measurements which may be available more rapidly.

## 1.2   What kinds of models are there?

Several approaches and techniques are available for deriving the desired process model. Standard modeling approaches include two main streams:

- the first-principle (white-box) approach and

- the identification of a parameterized black-box model.

The first-principle approach (white-box models)  denotes models based on the physical laws and relationships (mass and energy balances, *etc.*) that are supposed to govern the system's behavior. In these models, the structure reflects all physical insight about the process, and all the variables and the parameters all have direct physical interpretations (heat transfer coefficients, chemical reaction constants, etc.)

**Example 1 (Conservation principle)** A typical first-principle law is the general conservation principle:

$$\text{Accumulation} = \text{Input} - \text{Output} + \text{Internal production} \qquad (1.1)$$

The fundamental quantities that are being conserved in all cases are either mass, momentum, or energy, or combinations thereof.

**Example 2 (Bioreactor)** Many biotechnological processes consist of fermentation, oxidation and/or reduction of feedstuff (substrate) by microorganisms such as yeasts and bacteria. Let us consider a continuous-flow fermentation process. Mass balance considerations lead to the following model:

$$\frac{dx}{dt} = (\mu - u) x \qquad (1.2)$$

$$\frac{ds}{dt} = -\frac{1}{R}\mu x + u (s_{in} - s) \qquad (1.3)$$

where $x$ is the biomass concentration, $s$ is the substrate concentration, $u$ is the dilution rate, $s_{in}$ is the influent substrate concentration, $R$ is the yield coefficient and $\mu$ is the specific growth rate.

The specific growth rate $\mu$ is known to be a complex function of several parameters (concentrations of biomass, $x$, and substrate, $s$, pH, *etc.*) Many analytical formulae for the specific growth rate have been proposed in the literature [1] [60]. The Monod equation is frequently used as the kinetic description for growth of micro-organisms and the formation of metabolic products:

$$\mu = \mu_{\max}\frac{s}{K_M + s} \qquad (1.4)$$

where $\mu_{\max}$ is the maximum growth rate and $K_M$ is the Michaelis-Menton parameter.

Often, such a direct modeling may not be possible. One may say that:

> The physical models are as different from the world as a geographic map is from the surface of the earth (Brillouin).

The reason may be that the

- knowledge of the system's mechanisms is incomplete, or the

- properties exhibited by the system may change in an unpredictable manner. Furthermore,

- modeling may be time-consuming and

- may lead to models that are unnecessarily complex.

In such cases, variables characterizing the behavior of the considered system can be measured and used to construct a model. This procedure is usually called identification [55]. Identification governs many types of methods. The models used in identification are referred to as black-box models (or experimental models), since the parameters are obtained through identification from experimental data.

Between the two extremes of white-box and black-box models lay the semiphysical grey-box models. They utilize physical insight about the underlying process, but not to the extent that a formal first-principle model is constructed.

**Example 3 (Heating system)** If we are dealing with the modeling of an electric heating system, it is preferable to use the electric power $V^2$ as a control variable, rather than the voltage, $V$. In fact, the heater power, rather than the voltage, causes the temperature to change. Even if the heating system is non-linear, a linear relationship between the power and the temperature will lead to a good representation of the behavior of this system.

**Example 4 (Tank outflow)** Let us consider a laboratory-scale tank system [53]. The purpose is to model how the water level $y(t)$ changes with the inflow that is generated by the voltage $u(t)$ applied to the pump. Several experiments were carried out, and they showed that the best linear black-box model is the following

$$y(t) = a_1 y(t-1) + a_2 u(t-1) \tag{1.5}$$

Simulated outputs from this model were compared to real tank measurements. They showed that the fit was not bad, yet the model output was physically impossible since the tank level was negative at certain time intervals. As a matter of fact, all linear models tested showed this kind of behavior.

Observe that the outflow can be approximated by Bernoulli's law which states that the outflow is proportional to square root of the level $y(t)$. Combining these facts, it is straightforward to arrive at the following non-linear model structure

$$y(t) = a_1 y(t-1) + a_2 u(t-1) + a_3 \sqrt{y(t-1)} \tag{1.6}$$

This is a grey box model. The simulation behavior of this model was found better than that of the previous one (with linear black-box model), as the constraint on the origin of the output (level) was no longer violated.

Modeling always involves approximations since all real systems are, to some extent, non-linear, time-varying, and distributed. Thus it is highly improbable that any set of models will contain the 'true' system structure. All that can be hoped for is a model which provides an acceptable level of approximation, as measured by the use to which the model will be dedicated.

Another problem is that we are striving to build models not just for the fun of it, but to use the model for analysis, whose outcome will affect our decision in the future. Therefore we are always faced with the problem of having model 'accurate enough,' *i.e.*, reflecting enough of the important aspects of the problem. The question of what is 'accurate enough' can only, eventually, be settled by real-world experiments.

In this book, emphasis will be on the discrete time approaches. Most processes encountered in process engineering are continuous time in nature. However, the development of discrete-time models arises frequently in practical situations where system measurements (observations) are made, and control policies are implemented at discrete time instants on computer systems. Discrete time systems (discrete event systems) exist also, such as found from manufacturing systems and assembly lines, for example. In general, for a digital controller it is convenient to use discrete time models. Several techniques are also available to transform continuous time models to a time discrete form.

## 1.2.1 Identification vs. first-principle modeling

Provided that adequate theoretical knowledge is available, it may seem obvious that the first-principle modeling approach should be preferred. The model is justified by the underlying laws and principles, and can be easily transferred and used in any other context bearing similar assumptions.

However, these assumptions may become very limiting. This can be due to the complexity of the process itself, which forces the designer to use strong simplifications and/or to fix the model components too tightly. Also, advances in process design together with different local conditions often result in that no two plants are identical.

**Example 5 (Power plant constructions)** Power plant constructions are usually strongly tailored to match the local conditions of each individual site. The construction depends on factors such as the local fuels available, the ratio and amount of thermal and electrical power required, new technological innovations towards better thermal efficiency and emission control, *etc.* To make the existing models suit a new construction, an important amount of redesign and tuning is required.

Solving of the model equations might also pose problems with highly detailed first-principle models. Either cleverness of a mathematician is required from the engineer developing the model, or time-consuming iterative computations need to be performed.

In addition to the technical point of view, first-principle models can be criticized due to their costs. The more complex and *a priori* unknown the various chemical/physical phenomena are to the model developer, or to the scientific community as a whole, the more time and effort the building of these models requires. Although the new information adds to the general knowledge of the considered process, this might not be the target of the model development project. Instead, as in projects concerning plant control and optimization, the final target is in improving the plant behavior and productivity. Just as plants are built and run in order to fabricate a product with a competitive price, the associated development projects are normally assessed against this criterion.

The description of the process phenomena given by the model might also be incomprehensible for users other than the developer, and the obtained knowledge of the underlying phenomena may be wasted. It might turn out to be difficult to train the process operators to use a highly detailed theoretical model, not to mention teaching them to understand the model equations. Furthermore, the intermediate results, describing the sub-phenomena of the process, are more difficult to put to use in a process automation system. Even an advanced modern controller, such as a predictive controller, typically requires only estimates of the future behavior of the controlled variable.

Having accepted these points of view, a semi- or full-parameterized approach seems much more meaningful. This is mainly due to the saved design time, although collecting of valid input-output observations from a process might be time consuming. Note however, that it is very difficult to over-perform the first-principle approach in the case where few measurements are available, or when good understanding of the plant behavior has already been gained. In process design, for example, there are no full-scale measurement data at all (as the plant has not been built yet) and the basic phenomena are (usually) understood. In many cases, however, parameterized experimental models can be justified by the reduced time and effort required in building the models, and their flexibility in real-world modeling problems.

## 1.3    Steps of identification

Identification is the experimental approach to process modeling [5]. Identification is an iterative process of the following components:

- experimental planning (data acquisition),

- selection of the model structure,

- parameter estimation, and

- model validation.

The basis for the identification procedure is experimental planning, where process experiments are designed and conducted so that suitable data for the following three steps is obtained. The purpose is to maximize the information content in the data, within the limits imposed by the process.

- In modeling of dynamic systems, the *sampling period*[1] must be small enough so that significant process information is not lost. A peculiar effect called *aliasing* may also occur if the sampled signal contains frequencies that are higher than half of the sampling frequency: In general, if a process measurement is sampled with a sampling frequency $w_s$, high frequency components of the process variable with a frequency greater than $\frac{w_s}{2}$ appear as low-frequency components in the sampled signal, and may cause problems if they appear in the same frequency range as the normal process variations. The sampling frequency should be, if at all possible, ten times the maximum system bandwidth. For low *signal-to-noise ratios*, a filter should be considered. In some cases, a time-varying sampling period may be useful (related, *e.g.*, to the throughflow of a process).

- The signal must also be *persistently exciting*, such as a pseudo random (binary) sequence, PRBS, which exhibits spectral properties similar to those of the white noise.

Selection of the model structure is referred to as structure estimation, where the model *input-output signals* and the *internal components* of the model are determined. In general, the model structure is derived using prior knowledge.

---

[1]When a digital computer is used for data acquisition, real-valued continuous signals are converted into digital form. The time interval between successive samples is referred to as sampling period (sampling rate). In recursive identification the length of the time interval between two successive measurements can be different from the sampling rate associated with data acquisition (for more details, see *e.g.* [5]).

- Most of the suggested criteria can be seen as a minimization of a loss function (prediction error, Akaike Information Criterion, *etc.*). In dynamic systems, the choice of the *order of the model* is a nontrivial problem. The choice of the model order is a compromise between reducing the unmodelled dynamics and increasing the complexity of the model which can lead to model stabilizability difficulties. In many practical cases, a second order (or even a first order) model is adequate.

Various model structures will be discussed in detail in the following chapters.

In general, conditioning of data is necessary: *scaling* and *normalization* of data (to scale the variables to approximately the same scale), and *filtering* (to remove noise from the measurements).

- Scaling process is commonly used in several aspects of applied physics (heat transfer, fluid mechanics, *etc.*). This process leads to dimensionless parameters (Reynolds number of fluid mechanics, *etc.*) which are used as an aid to understanding similitude and scaling. In [9] a theory of scaling for linear systems using method from Lie theory is described. The scaling of the input and output units has very significant effects for multivariable systems [16]. It affects interaction, design aims, weighting functions, model order reduction, *etc.*

- The unmodeled dynamics result from the use of input-output models to represent complex systems: parts of the process dynamics are neglected and these introduce extra modeling errors which are not necessarily bounded. It is therefore advisable to perform normalization of the input-output data before they are processed by the identification procedure. The normalization procedure based on the norm of the regressor is commonly used [62].

- Data filtering permits to focus the parameter estimator on an appropriate bandwidth. There are two aspects, namely high-pass filtering to eliminate offsets, load disturbances, *etc.*, and low-pass filtering to eliminate irrelevant high frequency components including noise and system response. The rule of thumb governing the design of the filter is that the upper frequency should be about twice the desired system bandwidth and the lower frequency should be about one-tenth the desired bandwidth.

In parameter estimation, the values of the unknown parameters of a parameterized model structure are estimated. The choice of the parameter estimation method depends on the structure of the model, as well as the properties of the data. Parameter estimation techniques will be discussed in detail in the following chapters.

In validation, the goodness of the identified model is assessed. The validation methods depend on the properties that are desired from the model. Usually, *accuracy* and good generalization (*interpolation/extrapolation*) abilities are desired; *transparency* and computational efficiency may also be of interest. Simulations provide a useful tool for model validation. Accuracy and generalization can be tested by cross-validation techniques, where the model is tested on a test data set, previously unseen to the model. Also statistical tests on prediction error may provide useful. With dynamic systems, *stability*, zeros and poles, and the effect of the variation of the poles, are of interest.

- Most model *validation tests* are based on simply the difference between the simulated and measured output. Model validation is really about model falsification. The validation problem deals with demonstrating the confidence in the model. Often prior knowledge concerning the process to be modeled and statistical tests involving confidence limits are used to validate a model.

## 1.4 Outline of the book

In the remaining chapters, various model structures, parameter estimation techniques, and predictive control of different kinds of systems (linear, non-linear, SISO and MIMO) are discussed. In the second chapter, linear regression models and methods for estimating model parameters are presented. The method of least squares (LS) is a very commonly used batch method. It can be written in a recursive form, so that the components of the recursive least squares (RLS) algorithm can be updated with new information as soon as it becomes available. Also the Kalman filter, commonly used both for state estimation as well as for parameter estimation, is presented in Chapter 2. Chapter 3 considers linear dynamic systems. The polynomial time-series representation and stochastic disturbance models are introduced. An *i*-step-ahead predictor for a general linear dynamic system is derived.

Structures for capturing the behavior of non-linear systems are discussed in Chapter 4. A general framework of generalized basis function networks is introduced. As special cases of the basis function network, commonly used non-linear structures such as power series, sigmoid neural networks and Sugeno fuzzy models are obtained. Chapter 5 extends to non-linear dynamical systems. The general non-linear time-series approaches are briefly viewed.

A detailed presentation of Wiener and Hammerstein systems, consisting of linear dynamics coupled with non-linear static systems, is given.

To conclude the chapters on identification, parameter estimation techniques are presented in Chapter 6. Discussion is limited to prediction error methods, as they are sufficient for most practical problems encountered in process engineering. An extension to optimization under constraints is done, to emphasize the practical aspects of identification of industrial processes. A brief introduction to learning automata, and guided random search methods in general, is also given.

The basic ideas behind predictive control are presented in Chapter 7. First, a simple predictive controller is considered. This is followed by an extension including a noise model: the generalized predictive controller (GPC). State space representation is used, and various practical features are illustrated. Appendix A gives some background on state space systems.

Chapter 8 is devoted to the control of multiple-input–multiple-output (MIMO) systems. There are two main approaches to handle the control of MIMO systems: the implementation of a global MIMO controllers, or implementation of a distributed controller (a set of SISO controllers for the considered MIMO system). To achieve the design of a distributed controller it is necessary to be able to select the best input–output pairing. In this chapter we present a well known and efficient technique, the relative gain array (RGA) method. As an example of decoupling methods, a multivariable PI-controller based on decoupling at both low and high frequencies, is presented. Finally, the design of a multivariable GPC based on a state space representation is considered.

In order to solve increasingly complex problems faced by practicing engineers, Chapter 9 deals with the development of predictive controllers for non-linear systems. Various approaches (adaptive control, control based on Hammerstein and Wiener models, or neural networks) are considered to deal with the time-varying and non-linear behavior of systems. Detailed descriptions are provided for predictive control algorithms to use. Using the inverse model of the non-linear part of both Hammerstein and Wiener models, we show that any linear control strategy can be easily implemented in order to achieve the desired performance for non-linear systems.

The applications of the different control techniques presented in this book are illustrated by several examples including: fluidized-bed combustor, valve, binary distillation column, two-tank system, pH neutralization, fermenter, tubular chemical reactor, *etc.* The example concerning the fluidized bed combustion is repeatedly used in several sections of the book. This book ends with Appendix B concerning the description and modeling of a fluidized bed combustion process.

# Chapter 2

# Linear Regression

A major decision in identification is how to parameterize the characteristics and properties of a system using a model of a suitable structure. Linear models usually provide a good starting point in the structure selection of the identification procedure. In general, linear structures are simpler than the non-linear ones and analytical solutions may be found. In this chapter, linear structures and parameter estimation in such structures are considered.

## 2.1  Linear systems

The dominating distinction between linear and non-linear systems is the principle of superposition [19].

**Definition 1 (Principle of superposition)** The following holds only if $a$ is linearly dependent on $b$:

$$\begin{aligned}
&\text{If } a_1 \text{ is the output due to } b_1 \\
&\text{and } a_2 \text{ is the output due to } b_2, \\
&\text{then } \alpha a_1 + \beta a_2 \text{ is the output due to } \alpha b_1 + \beta b_2.
\end{aligned} \tag{2.1}$$

In above, the $\alpha$ and $\beta$ are constant parameters, and $a_i$ and $b_i$ $(i = 1, 2)$ are some values assumed by variables $a$ and $b$.

The characterization of linear time-invariant dynamic systems, in general, is virtually complete because the principle of superposition applies to all such systems. As a consequence, a large body of knowledge concerning the analysis and design of linear time-invariant systems exists. By contrast, the state of non-linear systems analysis is not nearly complete.

13

With parameterized structures $f(\boldsymbol{\varphi}, \boldsymbol{\theta})$, two types of linearities are of importance: Linearity of the model output with respect to model inputs $\boldsymbol{\varphi}$; and linearity of the model output with respect to model parameters $\boldsymbol{\theta}$. The former considers the mapping capabilities of the model, while the latter affects the estimation of the model parameters. If at least one parameter appears non-linearly, models are referred to as non-linear regression models [78].

In this chapter, linear regression models are considered. Consider the following model of the relation between the inputs and output of a system [55]:

$$y(k) = \boldsymbol{\theta}^T \boldsymbol{\varphi}(k) + \xi(k) \tag{2.2}$$

where

$$\boldsymbol{\theta} = \begin{bmatrix} \theta_1 \\ \theta_2 \\ \vdots \\ \theta_i \\ \vdots \\ \theta_I \end{bmatrix} \tag{2.3}$$

and

$$\boldsymbol{\varphi}(k) = \begin{bmatrix} \varphi_1(k) \\ \varphi_2(k) \\ \vdots \\ \varphi_i(k) \\ \vdots \\ \varphi_I(k) \end{bmatrix} \tag{2.4}$$

The model describes the observed variable $y(k)$ as an unknown linear combination of the observed vector $\boldsymbol{\varphi}(k)$ plus noise $\xi(k)$. Such a model is called a linear regression model, and is a very common type of model in control and systems engineering. $\boldsymbol{\varphi}(k)$ is commonly referred to as the **regression vector**; $\boldsymbol{\theta}$ is a vector of constants containing the parameters of the system; $k$ is the sample index. Often, one of the inputs is chosen to be a constant, $\varphi_I \equiv 1$, which enables the modeling of bias.

If the statistical characteristics of the disturbance term are not known, we can think of

$$\hat{y}(k) = \boldsymbol{\theta}^T \boldsymbol{\varphi}(k) \tag{2.5}$$

as a natural prediction of what $y(k)$ will be. The expression (2.5) becomes a prediction in an exact statistical (mean squares) sense, if $\{\xi(k)\}$ is a sequence of independent random variables, independent of the observations $\varphi$, with zero mean and finite variance[1].

In many practical cases, the parameters $\theta$ are not known, and need to be estimated. Let $\widehat{\theta}$ be the estimate of $\theta$

$$\widehat{y}(k) = \widehat{\theta}^T \varphi(k) \tag{2.6}$$

Note, that the output $\widehat{y}(k)$ is linearly dependent on both $\widehat{\theta}$ and $\varphi(k)$.

**Example 6 (Static system)** The structure (2.2) can be used to describe many kinds of systems. Consider a noiseless static system with input variables $u_1$, $u_2$ and $u_3$ and output $y$

$$y(k) = a_1 u_1(k) + a_2 u_2(k) + a_3 u_3(k) + a_4 \tag{2.7}$$

where $a_i$ $(i = 1, 2, 3, 4)$ are constants. It can be presented in the form of (2.2)

---

[1] We are looking for a predictor $\widehat{y}(k)$ which minimizes the mean square error criterion

$$\widehat{y}(k) = \arg\min_{\widehat{y}} E\left\{(y(k) - \widehat{y})^2\right\}$$

Replacing $y(k)$ by its expression $\theta^T \varphi(k) + \xi(k)$ it follows:

$$
\begin{aligned}
E\left\{(y(k) - \widehat{y})^2\right\} &= E\left\{\left(\theta^T \varphi(k) + \xi(k) - \widehat{y}\right)^2\right\} \\
&= E\left\{\left(\theta^T \varphi(k) - \widehat{y}\right)^2 + [\xi(k)]^2 + 2\xi(k)\left(\theta^T \varphi(k) - \widehat{y}\right)\right\} \\
&= E\left\{\left(\theta^T \varphi(k) - \widehat{y}\right)^2\right\} + E\left\{[\xi(k)]^2\right\} + 2E\left\{\xi(k)\left(\theta^T \varphi(k) - \widehat{y}\right)\right\}
\end{aligned}
$$

If the sequence $\{\xi(k)\}$ is independent of the observations $\varphi(k)$, then

$$E\left\{\xi(k)\left(\theta^T \varphi(k) - \widehat{y}\right)\right\} = E\{\xi(k)\} E\left\{\left(\theta^T \varphi(k) - \widehat{y}\right)\right\}$$

In view of the fact that $\{\xi(k)\}$ is a sequence of independent random variables with zero mean value, it follows $E\left\{\xi(k)\left(\theta^T \varphi(k) - \widehat{y}\right)\right\} = 0$. As a consequence,

$$J = E\left\{\left(\theta^T \varphi(k) - \widehat{y}\right)^2\right\} + E\left\{(\xi(k))^2\right\}$$

and the minimum is obtained for (2.5). The minimum value of the criterion is equal to $E\left\{(\xi(k))^2\right\}$, the variance of the noise.

by choosing

$$\boldsymbol{\theta} = \begin{bmatrix} a_1 \\ a_2 \\ a_3 \\ a_4 \end{bmatrix} \tag{2.8}$$

$$\boldsymbol{\varphi}(k) = \begin{bmatrix} u_1(k) \\ u_2(k) \\ u_3(k) \\ 1 \end{bmatrix} \tag{2.9}$$

and we have

$$y(k) = \boldsymbol{\theta}^T \boldsymbol{\varphi}(k) \tag{2.10}$$

**Example 7 (Dynamic system)** Consider a dynamic system with input signals $\{u(k)\}$ and output signals $\{y(k)\}$, sampled at discrete time instants[2] $k = 1, 2, 3, \dots$ . If the values are related through a linear difference equation

$$\begin{aligned} y(k) &+ a_1 y(k-1) + \dots + a_{n_A} y(k - n_A) \\ &= b_0 u(k-d) + \dots + b_{n_B} u(k - d - n_B) + \xi(k) \end{aligned} \tag{2.11}$$

where $a_i$ $(i = 1, \dots, n_A)$ and $b_i$ $(i = 0, \dots, n_B)$ are constants and $d$ is the time delay, we can introduce a parameter vector $\boldsymbol{\theta}$

$$\boldsymbol{\theta} = \begin{bmatrix} -a_1 \\ \vdots \\ -a_{n_A} \\ b_0 \\ \vdots \\ b_{n_B} \end{bmatrix} \tag{2.12}$$

and a vector of lagged input-output data $\boldsymbol{\varphi}(k)$

$$\boldsymbol{\varphi}(k) = \begin{bmatrix} y(k-1) \\ \vdots \\ y(k - n_A) \\ u(k - d) \\ \vdots \\ u(k - d - n_B) \end{bmatrix} \tag{2.13}$$

---

[2]Observed at sampling instant $k$ $(k \in 1, 2, \dots)$ at time $t = kT$, where $T$ is referred to as the sampling interval, or sampling period. Two related terms are used: the sampling frequency $f = \frac{1}{T}$, and the angular sampling frequency, $w = \frac{2\pi}{T}$.

and represent the system in the form of (2.2) as

$$y(k) = \boldsymbol{\theta}^T \boldsymbol{\varphi}(k) + \xi(k) \tag{2.14}$$

The backward shift $d$ is a convenient way to deal with process time delays. Often, there is a noticeable delay between the instant when a change in the process input is implemented and the instant when the effect can be observed from the process output. When a process involves mass or energy transport, a transportation lag (time delay) is associated with the movement. This time delay is equal to the ratio $L/V$ where $L$ represents the length of the process (furnace for example), and $V$ is the velocity (*e.g.*, of the raw material).

In system identification, both the structure and the true parameters $\boldsymbol{\theta}$ of a system may be *a priori* unknown. Linear structures are a very useful starting point in black-box identification, and in most cases provide predictions that are accurate enough. Since the structure is simple, it is also simple to validate the performance of the model. The selection of a model structure is largely based on experience and the information that is available of the process.

Similarly, parameter estimates $\hat{\boldsymbol{\theta}}$ may be based on the available *a priori* information concerning the process (physical laws, phenomenological models, *etc.*). If these are not available, efficient techniques exist for estimating some or all of the unknown parameters using sampled data from the process. In what follows, we shall be concerned with some methods related to the estimation of the parameters in linear systems. These methods assume that a set of input-output data pairs is available, either off-line or on-line, giving examples of the system behavior.

## 2.2  Method of least squares

The method of least squares[3] is essential in systems and control engineering. It provides a simple tool for estimating the parameters of a linear system.

In this section, we deal with linear regression models. Consider the model (2.2):

$$y(k) = \boldsymbol{\theta}^T \boldsymbol{\varphi}(k) + \xi(k) \tag{2.15}$$

where $\boldsymbol{\theta}$ is a column vector of parameters to be estimated from observations $y(k)$, $\boldsymbol{\varphi}(k)$, $k = 1, 2, ..., K$, and where regressor $\boldsymbol{\varphi}(k)$ is independent of $\boldsymbol{\theta}$

---

[3]The least squares method was developed by Karl Gauss. He was interested in the estimation of six parameters characterising the motions of planets and comets, using telescopic measurements.

(linear regression)[4]. $K$ is the number of observations. This type of model is commonly used by engineers to develop correlations between physical quantities. Notice that $\varphi(k)$ may correspond to *a priori* known functions (log, exp, *etc.*) of a measured quantity.

The goal of parameter estimation is to obtain an estimate of the parameters of the model, so that the model fit becomes 'good' in the sense of some criterion. A commonly accepted method for a 'good' fit is to calculate the values of the parameter vector that minimize the *sum of the squared residuals*. Let us consider the following estimation criterion

$$J(\boldsymbol{\theta}) = \frac{1}{K} \sum_{k=1}^{K} \alpha_k \left[ y(k) - \boldsymbol{\theta}^T \varphi(k) \right]^2 \qquad (2.16)$$

This quadratic cost function (to be minimized with respect to $\boldsymbol{\theta}$) expresses the average of the *weighted* squared errors between the $K$ observed outputs, $y(k)$, and the predictions provided by the model, $\boldsymbol{\theta}^T \varphi(k)$. The scalar coefficients $\alpha_k$ allow the weighting of different observations.

The important benefit of having a quadratic cost function is that it can be minimized analytically. Remember that a quadratic function has the shape of a parabola, and thus possesses a single optimum point. The optimum (minimum or maximum) can be solved analytically by setting the derivative to zero and the examination of the second derivative shows whether a minimum or a maximum is in question.

## 2.2.1   Derivation

Let us minimize the cost function $J$ with respect to parameters $\boldsymbol{\theta}$

$$\widehat{\boldsymbol{\theta}} = \arg\min_{\boldsymbol{\theta}} J \qquad (2.17)$$

where $J$ is given by (2.16)

$$J = \frac{1}{K} \sum_{k=1}^{K} \alpha_k \left[ y(k) - \boldsymbol{\theta}^T \varphi(k) \right]^2 \qquad (2.18)$$

---

[4]Note that this poses restrictions on the choice of $\varphi(k)$.

Assuming that $\varphi(k)$ is not a function of $\theta_i$, the partial derivative for the $i$'th term can be calculated, which gives

$$\frac{\partial J}{\partial \theta_i} = \frac{\partial}{\partial \theta_i} \frac{1}{K} \sum_{k=1}^{K} \alpha_k \left[ y(k) - \boldsymbol{\theta}^T \boldsymbol{\varphi}(k) \right]^2 \tag{2.19}$$

$$= \frac{1}{K} \sum_{k=1}^{K} \alpha_k \frac{\partial}{\partial \theta_i} \left\{ \left[ y(k) - \boldsymbol{\theta}^T \boldsymbol{\varphi}(k) \right]^2 \right\} \tag{2.20}$$

$$= \frac{1}{K} \sum_{k=1}^{K} \alpha_k \left\{ 2 \left[ y(k) - \boldsymbol{\theta}^T \boldsymbol{\varphi}(k) \right] \left[ -\varphi_i(k) \right] \right\} \tag{2.21}$$

$$= \frac{2}{K} \left[ \boldsymbol{\theta}^T \sum_{k=1}^{K} \alpha_k \boldsymbol{\varphi}(k) \varphi_i(k) - \sum_{k=1}^{K} \alpha_k y(k) \varphi_i(k) \right] \tag{2.22}$$

For the second derivative we have

$$\frac{\partial^2 J}{\partial \theta_{i,j}^2} = \frac{2}{K} \frac{\partial}{\partial \theta_j} \left[ \boldsymbol{\theta}^T \sum_{k=1}^{K} \alpha_k \boldsymbol{\varphi}(k) \varphi_i(k) - \sum_{k=1}^{K} \alpha_k y(k) \varphi_i(k) \right] \tag{2.23}$$

$$= \frac{2}{K} \sum_{k=1}^{K} \alpha_k \varphi_j(k) \varphi_i(k) \tag{2.24}$$

Collecting all terms, the first derivatives can be written as a row vector:

$$\left( \frac{\partial J}{\partial \boldsymbol{\theta}} \right)^T = \left[ \frac{\partial J}{\partial \theta_i} \right]_i \tag{2.25}$$

$$= \frac{2}{K} \left[ \boldsymbol{\theta}^T \sum_{k=1}^{K} \alpha_k \boldsymbol{\varphi}(k) \varphi_i(k) - \sum_{k=1}^{K} \alpha_k y(k) \varphi_i(k) \right]_i \tag{2.26}$$

$$= \frac{2}{K} \left[ \boldsymbol{\theta}^T \sum_{k=1}^{K} \alpha_k \boldsymbol{\varphi}(k) \boldsymbol{\varphi}^T(k) - \sum_{k=1}^{K} \alpha_k y(k) \boldsymbol{\varphi}^T(k) \right] \tag{2.27}$$

$$= \frac{2}{K} \left[ \left( \left[ \sum_{k=1}^{K} \alpha_k \boldsymbol{\varphi}(k) \boldsymbol{\varphi}^T(k) \right] \boldsymbol{\theta} \right)^T \right. \tag{2.28}$$

$$\left. - \sum_{k=1}^{K} \alpha_k y(k) \boldsymbol{\varphi}^T(k) \right]$$

Taking the transpose gives

$$\frac{\partial J}{\partial \boldsymbol{\theta}} = \frac{2}{K} \left[ \left[ \sum_{k=1}^{K} \alpha_k \boldsymbol{\varphi}(k) \boldsymbol{\varphi}^T(k) \right] \boldsymbol{\theta} - \sum_{k=1}^{K} \alpha_k y(k) \boldsymbol{\varphi}(k) \right] \tag{2.29}$$

The optimum of a quadratic function is found by setting all partial derivatives to zero:

$$\frac{\partial J}{\partial \boldsymbol{\theta}} = 0 \tag{2.30}$$

$$\frac{2}{K}\left[\left[\sum_{k=1}^{K} \alpha_k \boldsymbol{\varphi}(k) \boldsymbol{\varphi}^T(k)\right]\widehat{\boldsymbol{\theta}} - \sum_{k=1}^{K} \alpha_k y(k) \boldsymbol{\varphi}(k)\right] = 0 \tag{2.31}$$

$$\left[\sum_{k=1}^{K} \alpha_k \boldsymbol{\varphi}(k) \boldsymbol{\varphi}^T(k)\right]\widehat{\boldsymbol{\theta}} = \sum_{k=1}^{K} \alpha_k \boldsymbol{\varphi}(k) y(k) \tag{2.32}$$

The second derivative can be collected in a matrix:

$$\frac{\partial^2 J}{\partial \boldsymbol{\theta}^2} = \left[\frac{\partial^2 J}{\partial \theta^2_{i,j}}\right]_{i,j} \tag{2.33}$$

$$= \left[\sum_{k=1}^{K} \alpha_k \varphi_j(k)\varphi_i(k)\right]_{i,j} \tag{2.34}$$

$$= \sum_{k=1}^{K} \alpha_k \boldsymbol{\varphi}(k) \boldsymbol{\varphi}^T(k) \tag{2.35}$$

For the optimum to be a minimum, we require that the matrix is positive definite[5].

Finally, the parameter vector $\widehat{\boldsymbol{\theta}}$ minimizing the cost function $J$ is given by (if the inverse of the matrix exists):

$$\widehat{\boldsymbol{\theta}} = \left[\sum_{k=1}^{K} \alpha_k \boldsymbol{\varphi}(k) \boldsymbol{\varphi}^T(k)\right]^{-1} \sum_{k=1}^{K} \alpha_k \boldsymbol{\varphi}(k) y(k) \tag{2.36}$$

The optimum is a minimum if the second derivative is positive, *i.e.* the matrix $\frac{\partial^2 J}{\partial \boldsymbol{\theta}^2}$ is positive definite.

## 2.2.2  Algorithm

Let us represent the celebrated least squares parameter estimate as an algorithm.

---

[5]Matrix $\mathbf{A}$ is positive definite if $\mathbf{x}^T\mathbf{A}\mathbf{x} \geq 0$ for $\mathbf{x} \neq 0$.

**Algorithm 1 (Least squares method for a fixed data set)** Let a system be given by

$$y(k) = \boldsymbol{\theta}^T \boldsymbol{\varphi}(k) + \xi(k) \tag{2.37}$$

where $y(k)$ is the scalar output of the system; $\boldsymbol{\theta}$ is the true parameter vector of the system of size $I \times 1$; $\boldsymbol{\varphi}(k)$ is the regression vector of size $I \times 1$; and $\xi(k)$ is system noise. The least squares parameter estimate $\widehat{\boldsymbol{\theta}}$ of $\boldsymbol{\theta}$ that minimizes the cost function

$$J = \frac{1}{K} \sum_{k=1}^{K} \alpha_k \left[ y(k) - \boldsymbol{\theta}^T \boldsymbol{\varphi}(k) \right]^2 \tag{2.38}$$

where $\alpha_k$ are scalar weighting factors, is given by

$$\widehat{\boldsymbol{\theta}} = \left[ \sum_{k=1}^{K} \alpha_k \boldsymbol{\varphi}(k) \boldsymbol{\varphi}^T(k) \right]^{-1} \sum_{k=1}^{K} \alpha_k \boldsymbol{\varphi}(k) y(k) \tag{2.39}$$

If

$$\sum_{k=1}^{K} \alpha_k \boldsymbol{\varphi}(k) \boldsymbol{\varphi}^T(k) \tag{2.40}$$

is invertible, then there is a unique solution. The inverse exists if the matrix is positive definite.

Hence, a linear regression model

$$\widehat{y}(k) = \widehat{\boldsymbol{\theta}}^T \boldsymbol{\varphi}(k) \tag{2.41}$$

was identified using sampled measurements of the plant behavior, where $\widehat{y}(k)$ is the output of the model (predicted output of the system) and $\widehat{\boldsymbol{\theta}}$ is a parameter estimate (based on $K$ samples).

### 2.2.3 Matrix representation

Often, it is more convenient to calculate the least squares estimate from a compact matrix form. Let us collect the observations at the input of the model to a $K \times I$ matrix

$$\Phi = \begin{bmatrix} \boldsymbol{\varphi}^T(1) \\ \boldsymbol{\varphi}^T(2) \\ \vdots \\ \boldsymbol{\varphi}^T(K) \end{bmatrix} = \begin{bmatrix} \varphi_1(1) & \varphi_2(1) & \cdots & \varphi_I(1) \\ \varphi_1(2) & \varphi_2(2) & & \varphi_I(2) \\ \vdots & & \ddots & \vdots \\ \varphi_1(K) & \varphi_2(K) & \cdots & \varphi_I(K) \end{bmatrix} \tag{2.42}$$

and observations at the output to a $K \times 1$ vector

$$\mathbf{y} = \begin{bmatrix} y(1) \\ y(2) \\ \vdots \\ y(K) \end{bmatrix} \qquad (2.43)$$

The $K$ equations can be represented by a matrix equation

$$\mathbf{y} = \Phi\theta + \mathbf{E} \qquad (2.44)$$

where $\mathbf{E}$ is a $K \times 1$ column vector of modeling errors. Now the least squares algorithm (assuming $\alpha_k = 1$ for all $k$) that minimizes

$$J = \frac{1}{K}(\mathbf{y} - \Phi\theta)^T(\mathbf{y} - \Phi\theta) \qquad (2.45)$$

can be represented in a more compact form by

$$\widehat{\theta} = \left[\Phi^T\Phi\right]^{-1}\Phi^T\mathbf{y} \qquad (2.46)$$

where

$$\frac{\partial^2 J}{\partial\theta^2} = \Phi^T\Phi \qquad (2.47)$$

must be positive definite.

Consider Example 7 (dynamic system). If the input signal is constant, say $\bar{u}$, the right side of equation (2.11) may be written as follows

$$\bar{u}\sum_{i=0}^{n_B} b_i + \xi(k) \qquad (2.48)$$

It is clear that we can not identify separately the parameters $b_i$ ($i = 0, ..., n_B$). Mathematically, the matrix $\Phi^T\Phi$ is *singular*. From the point of view of process operation, the constant input fails to excite all the dynamics of the system. In order to be able to identify all the model parameters, the input signal must fluctuate enough, *i.e.* it has to be *persistently exciting*.

Let us illustrate singularity by considering the following matrix:

$$\mathbf{A} = \begin{bmatrix} 1 & \varepsilon \\ \frac{1}{\varepsilon} & 1 \end{bmatrix} \qquad (2.49)$$

| $Q_C[\frac{kg}{s}]$ | $P[MW]$ |
|---|---|
| 2.2 | 19.1 |
| 2.3 | 19.3 |
| 2.3 | 19.2 |
| 2.3 | 19.1 |
| 1.6 | 13.1 |
| 1.7 | 15.1 |
| 1.7 | 14.3 |
| 3.1 | 26.0 |
| 3.0 | 27.0 |
| 3.0 | 25.6 |

Table 2.1: Steady-state data from an FBC plant.

which is singular for all $\varepsilon \in \Re$. However, if $\varepsilon$ is very small we can neglect the term $a_{1,2} = \varepsilon$, and obtain

$$\mathbf{A}_1 = \begin{bmatrix} 1 & 0 \\ \frac{1}{\varepsilon} & 1 \end{bmatrix} \tag{2.50}$$

The determinant of $\mathbf{A}_1$ is equal to 1. Thus, the determinant provides no information on the closeness of singularity of a matrix. Recall that the determinant of a matrix is equal to the product of its eigenvalues. We might therefore think that the eigenvalues contain more information. The eigenvalues of the matrix $\mathbf{A}_1$ are both equal to 1, and thus the eigenvalues give no additional information. The *singular values* (the positive square roots of the eigenvalues of the matrix $\mathbf{A}^T\mathbf{A}$) of a matrix represent a good quantitative measure of the near singularity of a matrix. The ratio of the largest to the smallest singular value is called the *condition number* of the considered matrix. It provides a measure of closeness of a given matrix to being singular. Observe that the condition number associated with the matrix $\mathbf{A}_1$ tends to infinity as $\varepsilon \to 0$.

Let us illustrate the least squares method with two examples.

**Example 8 (Effective heat value)** Let us consider a simple application of the least squares method. The following steady state data (Table 2.1) was measured from an FBC plant (see Appendix B). In steady state, the power $P$ is related to the fuel feed by

$$P = HQ_C + h_0 \tag{2.51}$$

where $H$ is the effective heat value $[\frac{MJ}{kg}]$ and $h_0$ is due to losses. Based on the data, let us determine the least squares estimate of the effective heat value of the fuel.

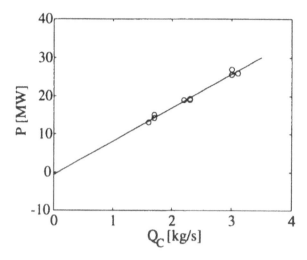

Figure 2.1: Least squares estimate of the heat value.

Substituting $\theta \leftarrow [H, h_0]^T$, $\Phi \leftarrow [\mathbf{Q}_C, 1]$, $\mathbf{y} \leftarrow \mathbf{P}$ we have

$$\Phi = \begin{bmatrix} 2.2 & 1 \\ 2.3 & 1 \\ \vdots & \vdots \\ 3.0 & 1 \end{bmatrix} ; \mathbf{y} = \begin{bmatrix} 19.1 \\ 19.3 \\ \vdots \\ 25.6 \end{bmatrix} \tag{2.52}$$

Using (2.46), or Algorithm 1, we obtain

$$\begin{aligned} \widehat{\theta} &= \left[ \Phi^T \Phi \right]^{-1} \Phi^T \mathbf{y} \\ &= \begin{bmatrix} 8.7997 \\ -0.6453 \end{bmatrix} \end{aligned} \tag{2.53}$$

Thus, $H = 8.7997$ is the least squares estimate of the effective heat value of the fuel. Fig. 2.1 shows the data points (dots) and the estimated linear relation (solid line).

**Example 9 ($O_2$ dynamics)** From an FBC plant (see Appendix B), the fuel feed $Q_C$ $[\frac{kg}{s}]$ and flue gas oxygen content $C_F$ $[\frac{Nm^3}{Nm^3}]$ were measured with a sampling interval of 4 seconds. The data set consisted of 91 noisy measurement patterns from step experiments around a steady-state operating point: fuel feed $\overline{Q}_C = 2.6$ $[\frac{kg}{s}]$, primary air $\overline{F}_1 = 3.6$ $[\frac{Nm^3}{s}]$, secondary air $\overline{F}_2 = 8.4$ $[\frac{Nm^3}{s}]$. Based on the measurements, let us determine the parameters $a$, $b$, and $c$ of the following difference equation:

$$\left[ C_F(k) - \overline{C}_F \right] = -a \left[ C_F(k-1) - \overline{C}_F \right] + b \left[ Q_C(k-6) - \overline{Q}_C \right] + c \tag{2.54}$$

Let us construct the input matrix

$$\Phi = \begin{bmatrix} C_F(6) - \overline{C}_F & Q_C(1) - \overline{Q}_C & 1 \\ C_F(7) - \overline{C}_F & Q_C(2) - \overline{Q}_C & 1 \\ \vdots & \vdots & \vdots \\ C_F(90) - \overline{C}_F & Q_C(85) - \overline{Q}_C & 1 \end{bmatrix} \tag{2.55}$$

and the vector of measured outputs

$$y = \begin{bmatrix} C_F(7) \\ C_F(8) \\ \vdots \\ C_F(91) \end{bmatrix} \tag{2.56}$$

The least squares estimate of $[-a, b, c]^T$ is then calculated by (2.46), resulting in:

$$\widehat{\theta} = \begin{bmatrix} 0.648 \\ -0.0172 \\ -0.0000 \end{bmatrix} \tag{2.57}$$

Thus, the dynamics of the $O_2$ content from fuel feed are described by

$$\begin{aligned} &\left[ C_F(k) - \overline{C}_F \right] \\ &= \quad 0.648 \left[ C_F(k-1) - \overline{C}_F \right] - 0.0172 \left[ Q_C(k-6) - \overline{Q}_C \right] \end{aligned} \tag{2.58}$$

or, equivalently using the backward shift operator, $x(k-1) = q^{-1}x(k)$

$$\left( 1 - 0.648 q^{-1} \right) \left[ C_F(k) - \overline{C}_F \right] = -0.0172 q^{-6} \left[ Q_C(k) - \overline{Q}_C \right] \tag{2.59}$$

The data (dots) and a simulation with the estimated model (solid lines) is illustrated in Fig. 2.2.

## 2.2.4 Properties

Next, we will be concerned with the properties of the least squares estimator $\widehat{\theta}$. Owing to the fact that the measurements are disturbed, the vector parameter estimation $\widehat{\theta}$ is random. An estimator is said to be unbiased if the mathematical expectation of the parameter estimation is equal to the true parameters $\theta$. The least squares estimation is unbiased if the noise $E$ has zero mean and if the noise and the data $\Phi$ are statistically independent. Notice, that the statistical independence of the observations and a zero mean

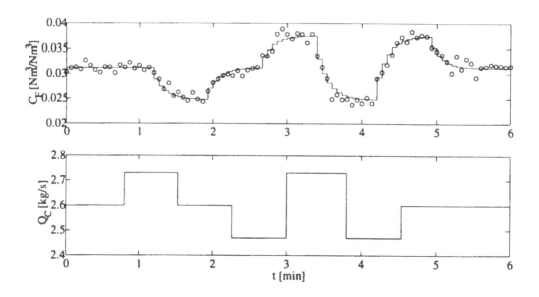

Figure 2.2: Prediction by the estimated model. Upper plot shows the pre-dicted (solid line) and measured (circles) flue gas oxygen content. The lower plot shows the model input, fuel feed.

noise is sufficient but not necessary for carrying out unbiased estimation of the vector parameters [62].

The estimation error is given by

$$\widetilde{\theta} = \theta - \widehat{\theta} \tag{2.60}$$

The mathematical expectation is given by

$$E\left\{\widetilde{\theta}\right\} = E\left\{\theta - \left[\Phi^T\Phi\right]^{-1}\Phi^T\mathbf{y}\right\} \tag{2.61}$$

$$= E\left\{\theta - \left[\Phi^T\Phi\right]^{-1}\Phi^T\left[\Phi\theta + \mathbf{E}\right]\right\} \tag{2.62}$$

$$= -E\left\{\left[\Phi^T\Phi\right]^{-1}\Phi^T\right\}E\left\{\mathbf{E}\right\} \tag{2.63}$$

since $\left[\Phi^T\Phi\right]^{-1}\Phi^T\Phi = \mathbf{I}$, and $\mathbf{E}$ and $\Phi$ are statistically independent. It follows that if $\mathbf{E}$ has zero mean, the LS estimator is unbiased, *i.e.*

$$E\left\{\widetilde{\theta}\right\} = 0 \text{ and } E\left\{\widehat{\theta}\right\} = \theta \tag{2.64}$$

Let us now consider the covariance matrix of the estimation error which represents the dispersion of $\widetilde{\theta}$ about its mean value. The covariance matrix[6]

---

[6]The covariance of a random variable $x$ is defined by $cov\left(x\right) = E\left\{\left[x - E\left\{x\right\}\right]\left[x - E\left\{x\right\}\right]^T\right\}$. If $x$ is zero mean, $E\left\{x\right\} = 0$, then $cov\left(x\right) = E\left\{xx^T\right\}$.

of the estimation error is given by

$$\mathbf{P} = E\left\{\widetilde{\boldsymbol{\theta}}\widetilde{\boldsymbol{\theta}}^T\right\} \tag{2.65}$$

$$= E\left\{\left[\boldsymbol{\theta} - \widehat{\boldsymbol{\theta}}\right]\left[\boldsymbol{\theta} - \widehat{\boldsymbol{\theta}}\right]^T\right\} \tag{2.66}$$

$$= E\left\{\left[\boldsymbol{\theta} - \left[\boldsymbol{\Phi}^T\boldsymbol{\Phi}\right]^{-1}\boldsymbol{\Phi}^T\left[\boldsymbol{\Phi}\boldsymbol{\theta} + \mathbf{E}\right]\right]\right. \tag{2.67}$$

$$\times \left.\left[\boldsymbol{\theta} - \left[\boldsymbol{\Phi}^T\boldsymbol{\Phi}\right]^{-1}\boldsymbol{\Phi}^T\left[\boldsymbol{\Phi}\boldsymbol{\theta} + \mathbf{E}\right]\right]^T\right\}$$

$$= E\left\{\left[\boldsymbol{\Phi}^T\boldsymbol{\Phi}\right]^{-1}\boldsymbol{\Phi}^T\mathbf{E}\mathbf{E}^T\boldsymbol{\Phi}\left[\boldsymbol{\Phi}^T\boldsymbol{\Phi}\right]^{-1}\right\} \tag{2.68}$$

$$= E\left\{\left[\boldsymbol{\Phi}^T\boldsymbol{\Phi}\right]^{-1}\boldsymbol{\Phi}^T\boldsymbol{\Phi}\left[\boldsymbol{\Phi}^T\boldsymbol{\Phi}\right]^{-1}\right\}E\left\{\mathbf{E}\mathbf{E}^T\right\} \tag{2.69}$$

$$= E\left\{\left[\boldsymbol{\Phi}^T\boldsymbol{\Phi}\right]^{-1}\right\}\sigma^2 \tag{2.70}$$

since $\mathbf{E}$ has zero mean and variance $\sigma^2$ (and its components are identically distributed), and $\mathbf{E}$ and $\boldsymbol{\Phi}$ are statistically independent. It is a measure of how well we can estimate the unknown $\boldsymbol{\theta}$. In the least squares approach we operate on given data, $\boldsymbol{\Phi}$ is known. This results in

$$\mathbf{P} = \left[\boldsymbol{\Phi}^T\boldsymbol{\Phi}\right]^{-1}\sigma^2 \tag{2.71}$$

The square root of the diagonal elements of $\mathbf{P}$, $\sqrt{P_{i,i}}$, represents the *standard errors* of each element $\widehat{\theta}_i$ of the estimate $\widehat{\boldsymbol{\theta}}$. The variance can be estimated using the sum of squared errors divided by degrees of freedom

$$\widehat{\sigma}^2 = \frac{\left[\mathbf{y} - \boldsymbol{\Phi}\widehat{\boldsymbol{\theta}}\right]^T\left[\mathbf{y} - \boldsymbol{\Phi}\widehat{\boldsymbol{\theta}}\right]}{K - I} \tag{2.72}$$

where $I$ is the number of parameters to estimate.

**Example 10 (Effective heat value: continued)** Consider Example 8. We have $K = 10$ data points and two parameters, $I = 2$. Using (2.72) we obtain $\widehat{\sigma}^2 = 0.3639$, a standard error of 0.3582 for the estimate of $H$, and 0.8527 for the bias $h_0$.

**Remark 1 (Covariance matrix)** For $\sigma^2 = 1$ we obtain

$$\mathbf{P} = \left[\boldsymbol{\Phi}^T\boldsymbol{\Phi}\right]^{-1} \tag{2.73}$$

Therefore, in the framework of parameter estimation, the matrix $\mathbf{P} = \left[\boldsymbol{\Phi}^T\boldsymbol{\Phi}\right]^{-1}$ is called the error covariance matrix.

## 2.3  Recursive LS method

The least squares method provides an estimate for the model parameters, based on a set of observations. Consider the situation when the observation pairs are obtained one-by-one from the process, and that we would like to update the parameter estimate whenever new information becomes available. This can be done by adding the new observation to the previous set of observations and recomputing (2.39). In what follows, a recursive formulation is derived, however [55]. Instead of recomputing the estimates with all available data, the previous parameter estimates are updated with the new data sample. In order to do this, the least squares estimation formula is written in the form of a recursive algorithm.

**Definition 2 (Recursive algorithm)** A recursive algorithm has the form

$$
\begin{matrix} \text{new} \\ \text{estimate} \end{matrix} = \begin{matrix} \text{old} \\ \text{estimate} \end{matrix} \tag{2.74}
$$

$$
+ \begin{matrix} \text{correction} \\ \text{factor} \end{matrix} \left( \begin{matrix} \text{new} \\ \text{observation} \end{matrix} - \begin{matrix} \text{prediction} \\ \text{with old} \\ \text{estimate} \end{matrix} \right)
$$

### 2.3.1  Derivation

The least squares estimate at sample instant $k - 1$ is given by (2.39)

$$
\widehat{\theta}(k-1) = \left[ \sum_{i=1}^{k-1} \alpha_i \varphi(i)\varphi^T(i) \right]^{-1} \sum_{i=1}^{k-1} \alpha_i \varphi(i)y(i) \tag{2.75}
$$

At sample instant $k$, new information is obtained and the least squares estimate is given by

$$
\widehat{\theta}(k) = \left[ \sum_{i=1}^{k-1} \alpha_i \varphi(i)\varphi^T(i) + \alpha_k \varphi(k)\varphi^T(k) \right]^{-1} \tag{2.76}
$$

$$
\times \left( \sum_{i=1}^{k-1} \alpha_i \varphi(i)y(i) + \alpha_k \varphi(k)y(k) \right)
$$

Define

$$
\mathbf{R}(k) = \sum_{i=1}^{k} \alpha_i \varphi(i)\varphi^T(i) \tag{2.77}
$$

which leads to the following recursive formula for $\mathbf{R}(k)$

$$\mathbf{R}(k) = \mathbf{R}(k-1) + \alpha_k \varphi(k) \varphi^T(k) \tag{2.78}$$

Using (2.77), the least squares estimate (2.76) can be rewritten as

$$\widehat{\theta}(k) = \mathbf{R}^{-1}(k) \left[ \sum_{i=1}^{k-1} \alpha_i \varphi(i) y(i) + \alpha_k \varphi(k) y(k) \right] \tag{2.79}$$

Based on (2.77), the estimate at iteration $k-1$, (2.75), can be rewritten as follows:

$$\widehat{\theta}(k-1) = \mathbf{R}^{-1}(k-1) \sum_{i=1}^{k-1} \alpha_i \varphi(i) y(i) \tag{2.80}$$

which gives

$$\sum_{i=1}^{k-1} \alpha_i \varphi(i) y(i) = \mathbf{R}(k-1) \widehat{\theta}(k-1) \tag{2.81}$$

Substituting this equation into (2.79), we find

$$\widehat{\theta}(k) = \mathbf{R}^{-1}(k) \left[ \mathbf{R}(k-1) \widehat{\theta}(k-1) + \alpha_k \varphi(k) y(k) \right] \tag{2.82}$$

From (2.78), we have a recursive formula for which is substituted in (2.82)

$$\widehat{\theta}(k) = \mathbf{R}^{-1}(k) \left[ \left[ \mathbf{R}(k) - \alpha_k \varphi(k) \varphi^T(k) \right] \widehat{\theta}(k-1) + \alpha_k \varphi(k) y(k) \right] \tag{2.83}$$

Reorganizing gives:

$$\widehat{\theta}(k) = \widehat{\theta}(k-1) + \mathbf{R}^{-1}(k) \alpha_k \varphi(k) \left[ y(k) - \varphi^T(k) \widehat{\theta}(k-1) \right] \tag{2.84}$$

which, together with (2.78), is a recursive formula for the least squares estimate.

In the algorithm given by (2.84), the matrix $\mathbf{R}(k)$ needs to be inverted at each time step. In order to avoid this, introduce

$$\mathbf{P}(k) = \mathbf{R}^{-1}(k) \tag{2.85}$$

The recursion of $\mathbf{R}(k)$, (2.78), now becomes

$$\mathbf{P}^{-1}(k) = \mathbf{P}^{-1}(k-1) + \alpha_k \varphi(k) \varphi^T(k) \tag{2.86}$$

The target is to be able to update $\mathbf{P}(k)$ directly, without needing to do matrix inversion. This can be done by using the matrix inversion lemma.

**Lemma 1 (Matrix inversion lemma)** Let $\mathbf{A}$, $\mathbf{B}$, $\mathbf{C}$ and $\mathbf{D}$ be matrices of compatible dimensions so that $\mathbf{A} + \mathbf{BCD}$ exists. Then

$$[\mathbf{A} + \mathbf{BCD}]^{-1} = \mathbf{A}^{-1} - \mathbf{A}^{-1}\mathbf{B}\left[\mathbf{DA}^{-1}\mathbf{B} + \mathbf{C}^{-1}\right]^{-1}\mathbf{DA}^{-1} \qquad (2.87)$$

The verification of the lemma can be obtained by multiplying the right-hand side by $A + BCD$ from the right, which gives unit matrix (for proof, see [64], p. 64).

Making the following substitutions

$$\mathbf{A} \leftarrow \mathbf{P}^{-1}(k-1) \qquad (2.88)$$
$$\mathbf{B} \leftarrow \varphi(k) \qquad (2.89)$$
$$\mathbf{C} \leftarrow \alpha_k \qquad (2.90)$$
$$\mathbf{D} \leftarrow \varphi^T(k) \qquad (2.91)$$

and applying Lemma 1 to (2.86) gives

$$\mathbf{P}(k) = \left[\mathbf{P}^{-1}(k-1) + \varphi(k)\alpha_k\varphi^T(k)\right]^{-1} \qquad (2.92)$$

$$= \mathbf{P}(k-1) - \frac{\mathbf{P}(k-1)\varphi(k)\varphi^T(k)\mathbf{P}(k-1)}{\frac{1}{\alpha_k} + \varphi^T(k)\mathbf{P}(k-1)\varphi(k)} \qquad (2.93)$$

Thus the inversion of a square matrix of size $\dim \theta$ is replaced by the inversion of a scalar.

The algorithm can be more conveniently expressed by defining a gain vector $\mathbf{L}(k)$

$$\mathbf{L}(k) = \frac{\mathbf{P}(k-1)\varphi(k)}{\frac{1}{\alpha_k} + \varphi^T(k)\mathbf{P}(k-1)\varphi(k)} = \alpha_k\mathbf{P}(k)\varphi(k) \qquad (2.94)$$

where the second equality can be verified by substituting (2.93) for $\mathbf{P}(k)$ and reorganizing.

The recursive algorithm needs some initial values to be started up. In the absence of prior knowledge, one possibility to obtain initial values is to use the least squares method on the first $k_0 > \dim \theta$ samples. Another common choice is to set the initial parameter vector to zero

$$\widehat{\theta}(k_0) = 0 \qquad (2.95)$$

and let the initial error covariance matrix to be a multiple of identity matrix

$$\mathbf{P}(k_0) = C\mathbf{I} \qquad (2.96)$$

where $C$ is some large positive constant. A large value implies that the confidence in $\widehat{\theta}(k_0)$ is poor and ensures a high initial degree of correction (adaptation). Notice that this makes the updating direction coincide with the negative gradient of the least squares criterion.

## 2.3.2 Algorithm

The recursive least squares algorithm can now be given, using (2.94), (2.84)-(2.85), and (2.93)

**Algorithm 2 (Recursive least squares method)** The recursive least squares algorithm is given by

$$\mathbf{L}(k) = \frac{\mathbf{P}(k-1)\boldsymbol{\varphi}(k)}{\frac{1}{\alpha_k} + \boldsymbol{\varphi}^T(k)\mathbf{P}(k-1)\boldsymbol{\varphi}(k)} \tag{2.97}$$

$$\widehat{\boldsymbol{\theta}}(k) = \widehat{\boldsymbol{\theta}}(k-1) + \mathbf{L}(k)\left[y(k) - \widehat{\boldsymbol{\theta}}^T(k-1)\boldsymbol{\varphi}(k)\right] \tag{2.98}$$

$$\mathbf{P}(k) = \mathbf{P}(k-1) - \mathbf{L}(k)\boldsymbol{\varphi}^T(k)\mathbf{P}(k-1) \tag{2.99}$$

where $k = k_0 + 1, k_0 + 2, k_0 + 3, ...$ The initial values $\widehat{\boldsymbol{\theta}}(k_0)$ and $\mathbf{P}(k_0)$ are obtained by using the LS on the first $k_0 > \dim\boldsymbol{\theta}$ samples

$$\mathbf{P}(k_0) = \left[\sum_{i=1}^{k_0}\alpha_i\boldsymbol{\varphi}(i)\boldsymbol{\varphi}^T(i)\right]^{-1} \tag{2.100}$$

$$\widehat{\boldsymbol{\theta}}(k_0) = \mathbf{P}(k_0)\sum_{i=1}^{k_0}\alpha_i\boldsymbol{\varphi}(i)y(i) \tag{2.101}$$

The RLS method is one of the most widely used recursive parameter estimation techniques, due to its robustness and easiness of implementation.

**Example 11 ($O_2$ dynamics: continued)** Let us consider the same problem as in Example 9 where the parameters of the following model were to be estimated:

$$\left[1 + aq^{-1}\right]\left[C_F(k) - \overline{C}_F\right] = bq^{-6}\left[Q_C(k) - \overline{Q}_C\right] + c \tag{2.102}$$

Using the recursive LS-method with the initial values $k_0 = 7$:

$$\widehat{\boldsymbol{\theta}}(7) = \begin{bmatrix} 0 \\ 0 \\ 0 \end{bmatrix} ; \mathbf{P}(7) = \begin{bmatrix} 10^9 & 0 & 0 \\ 0 & 10^9 & 0 \\ 0 & 0 & 10^9 \end{bmatrix} \tag{2.103}$$

and substituting for $k = 7, 8, ..., 91$

$$y(k) \leftarrow \left[C_F(k) - \overline{C}_F\right] ; \boldsymbol{\varphi}(k) \leftarrow \begin{bmatrix} C_F(k-1) - \overline{C}_F \\ Q_C(k-6) - \overline{Q}_C \\ 1 \end{bmatrix} \tag{2.104}$$

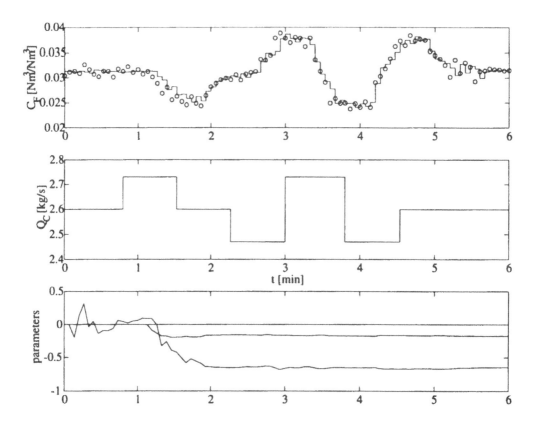

Figure 2.3: On-line prediction by the estimated model. Upper plot shows the predicted (solid line) and measured (circles) flue gas oxygen content; the middle plot shows the model input, fuel feed. The evolution of the values of the estimated parameters is shown at the bottom of the figure.

we have the following parameters at $k = 91$:

$$\widehat{\theta}(91) = \begin{bmatrix} 0.646 \\ -0.0172 \\ -0.0000 \end{bmatrix} \tag{2.105}$$

which are the same (up to two digits) as in Example 9. Fig. 2.3 illustrates the evolution of the parameters $a$, $b$ and $c$, as well as the on-line prediction by the model.

**Remark 2 (Factorization)** The covariance matrix must remain positive definite. However, even if the initial matrix $P(0)$ satisfies the second order condition of optimality (least squares optimization problem), the positive definiteness of $P(k)$ can be lost, owing the numerical round-off errors in a long term behavior (adaptive context, *etc.*). In order to maintain numerical accuracy it is more advisable to update the estimator in a factorized form which guarantees that $\mathbf{P}(k)$ remains positive definite and that the round-off errors, unavoidable in computer applications, do not affect the solution significantly. One of the most popular methods is the UD factorization which is based on the decomposition of $\mathbf{P}(k)$

$$\mathbf{P}(k) = \mathbf{U}(k)\mathbf{D}(k)\mathbf{U}^T(k)$$

where the factors $\mathbf{U}(k)$ and $\mathbf{D}(k)$ are, respectively, a unitary upper triangular matrix and a diagonal matrix.

### 2.3.3  *A posteriori* prediction error

In the previous developments, the RLS was derived using the *a priori* prediction error

$$\epsilon(k|k-1) = y(k) - \widehat{\boldsymbol{\theta}}^T(k-1)\boldsymbol{\varphi}(k) \tag{2.106}$$

In some cases, the *a posteriori* version may be preferred[51]

$$\epsilon(k|k) = y(k) - \widehat{\boldsymbol{\theta}}^T(k)\boldsymbol{\varphi}(k) \tag{2.107}$$

The connection between these can be obtained using (2.106) and (2.107)

$$
\begin{aligned}
\epsilon(k|k) &= y(k) - \widehat{\boldsymbol{\theta}}^T(k-1)\boldsymbol{\varphi}(k) \\
&\quad - \left[\widehat{\boldsymbol{\theta}}^T(k) - \widehat{\boldsymbol{\theta}}^T(k-1)\right]\boldsymbol{\varphi}(k) \\
&= \epsilon(k|k-1) - \boldsymbol{\varphi}^T(k)\left[\widehat{\boldsymbol{\theta}}(k) - \widehat{\boldsymbol{\theta}}(k-1)\right]
\end{aligned}
$$

$$\tag{2.108}$$
$$\tag{2.109}$$

From (2.98) we derive

$$\left[\widehat{\boldsymbol{\theta}}(k) - \widehat{\boldsymbol{\theta}}(k-1)\right] = \mathbf{L}(k)\epsilon(k|k-1) \tag{2.110}$$

Substituting (2.97) into this equation leads to

$$\left[\widehat{\boldsymbol{\theta}}(k) - \widehat{\boldsymbol{\theta}}(k-1)\right] = \frac{\mathbf{P}(k-1)\boldsymbol{\varphi}(k)}{\frac{1}{\alpha_k} + \boldsymbol{\varphi}^T(k)\mathbf{P}(k-1)\boldsymbol{\varphi}(k)}\epsilon(k|k-1) \tag{2.111}$$

Thus, substituting (2.111) into (2.109) gives

$$
\begin{aligned}
\epsilon(k|k) &= \epsilon(k|k-1) - \frac{\boldsymbol{\varphi}^T(k)\mathbf{P}(k-1)\boldsymbol{\varphi}(k)}{\frac{1}{\alpha_k} + \boldsymbol{\varphi}^T(k)\mathbf{P}(k-1)\boldsymbol{\varphi}(k)}\epsilon(k|k-1) \tag{2.112}\\
&= \frac{\epsilon(k|k-1)}{1 + \alpha_k\boldsymbol{\varphi}^T(k)\mathbf{P}(k-1)\boldsymbol{\varphi}(k)} \tag{2.113}
\end{aligned}
$$

which is the relation between *a priori* and *a posteriori* prediction errors. The modified RLS algorithm is then given by (2.97),

$$\epsilon\left(k|k-1\right) = y\left(k\right) - \widehat{\boldsymbol{\theta}}^T\left(k-1\right)\boldsymbol{\varphi}\left(k\right) \tag{2.114}$$

$$\epsilon\left(k|k\right) = \frac{\epsilon\left(k|k-1\right)}{1 + \alpha_k\boldsymbol{\varphi}^T\left(k\right)\mathbf{P}\left(k-1\right)\boldsymbol{\varphi}\left(k\right)} \tag{2.115}$$

$$\widehat{\boldsymbol{\theta}}\left(k\right) = \widehat{\boldsymbol{\theta}}\left(k-1\right) + \mathbf{L}\left(k\right)\epsilon\left(k|k\right) \tag{2.116}$$

and (2.99). It can be observed that $\epsilon\left(k|k\right)$ can tend to zero if $\boldsymbol{\varphi}\left(k\right)$ becomes unbounded, even if $\epsilon\left(k|k-1\right)$ doesn't.

## 2.4   RLS with exponential forgetting

The criterion (2.16) gives an estimate based on the average behavior of the system, as expressed by the samples used in the identification. This resulted in the Algorithms 1 and 2. However, if we believe that the system is time-varying, we need an estimate that is representative of the current properties of the system. This can be accomplished by putting more weight on newer samples, *i.e.* by forgetting old information. These types of algorithms are referred to as adaptive algorithms.

In the time-varying case, it is necessary to infer the model at the same time as the data is collected. The model is then updated at each time instant when some new data becomes available. The need to cope with time-varying processes is not the only motivation for adaptive algorithms. Adaptive identification may need to be considered, *e.g.*, for processes that are non-linear to the extent that one set of model parameters may not adequately describe the process over its operating region [85].

In order to obtain an estimate that is representative for the current properties of the system at sample instant $k$, consider a criterion where older measurements are discounted ([55], pp. 56-59):

$$J_k\left(\boldsymbol{\theta}\right) = \frac{1}{k}\sum_{i=1}^{k}\beta\left(k,i\right)\left[y\left(i\right) - \boldsymbol{\theta}^T\boldsymbol{\varphi}\left(i\right)\right]^2 \tag{2.117}$$

where $\beta\left(k,i\right)$ is increasing in $i$ for a given $k$. The criterion is still quadratic in $\boldsymbol{\theta}$ and the minimizing off-line estimate is given by

$$\widehat{\boldsymbol{\theta}}\left(k\right) = \left[\sum_{i=1}^{k}\beta\left(k,i\right)\boldsymbol{\varphi}\left(i\right)\boldsymbol{\varphi}^T\left(i\right)\right]^{-1}\sum_{i=1}^{k}\beta\left(k,i\right)\boldsymbol{\varphi}\left(i\right)y\left(i\right) \tag{2.118}$$

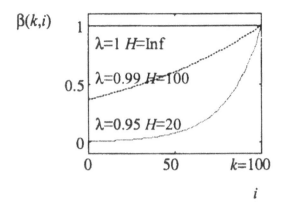

Figure 2.4: The effect of $\lambda$ ($\alpha_i = 1$ for all $i$).

Consider the following structure for $\beta(k, i)$:

$$\beta(k, i) = \lambda(k)\beta(k - 1, i) \tag{2.119}$$

where $1 \leq i \leq k - 1$ and $\lambda(k)$ is a scalar. This can also be written as

$$\beta(k, i) = \left[\prod_{j=i+1}^{k} \lambda(j)\right]\alpha_i \tag{2.120}$$

where

$$\beta(i, i) = \alpha_i \tag{2.121}$$

If $\lambda(i)$ is a constant $\lambda$, we get

$$\beta(k, i) = \lambda^{k-i}\alpha_i \tag{2.122}$$

which gives an exponential forgetting profile in the criterion (2.117). In such a case, the coefficient $\lambda$ is referred to as the *forgetting factor*. Figure 2.4 illustrates the weighting obtained using a constant $\lambda$.

The effect of $\lambda$ can be illustrated by computing the *equivalent memory horizon* $H = \frac{1}{1-\lambda}$ ($\alpha_i = 1$ for all $i$). A common choice of $\lambda$ is $0.95 - 0.99$. When $\lambda$ is close to 1, the time constant for the exponential decay is approximately $H$. Thus choosing $\lambda$ in the range $0.95 - 0.99$ corresponds, roughly, to remembering the $20 - 100$ most recent data.

## 2.4.1   Derivation

We are now ready to derive a recursive form for the previous equations. Let us introduce the following notation (see (2.77))

$$\mathbf{R}(k) = \sum_{i=1}^{k} \beta(k,i) \, \boldsymbol{\varphi}(i) \, \boldsymbol{\varphi}^T(i) \qquad (2.123)$$

Separating the old and the new information

$$\mathbf{R}(k) = \sum_{i=1}^{k-1} \beta(k,i) \, \boldsymbol{\varphi}(i) \, \boldsymbol{\varphi}^T(i) + \beta(k,k) \, \boldsymbol{\varphi}(k) \, \boldsymbol{\varphi}^T(k) \qquad (2.124)$$

and substituting (2.119) and (2.122) into this equation leads to

$$\mathbf{R}(k) = \sum_{i=1}^{k-1} \lambda\beta(k-1,i) \, \boldsymbol{\varphi}(i) \, \boldsymbol{\varphi}^T(i) + \alpha_k \boldsymbol{\varphi}(k) \, \boldsymbol{\varphi}^T(k) \qquad (2.125)$$

Using (2.123) for $\mathbf{R}(k-1)$, we have a recursive formula for $\mathbf{R}(k)$

$$\mathbf{R}(k) = \lambda\mathbf{R}(k-1) + \alpha_k \boldsymbol{\varphi}(k) \, \boldsymbol{\varphi}^T(k) \qquad (2.126)$$

In a similar way to the RLS, we can write a recursive formula for the parameter update

$$\widehat{\boldsymbol{\theta}}(k) = \widehat{\boldsymbol{\theta}}(k-1) + \mathbf{R}^{-1}(k) \, \alpha_k \boldsymbol{\varphi}(k) \left[ y(k) - \widehat{\boldsymbol{\theta}}^T(k-1) \, \boldsymbol{\varphi}(k) \right] \qquad (2.127)$$

This is exactly the same as (2.84). Again, we can denote $\mathbf{P}(k) = \mathbf{R}^{-1}(k)$ and use the matrix inversion lemma (Lemma 1) to avoid matrix inversion in (2.127) (select $\mathbf{A} \leftarrow \lambda\mathbf{P}^{-1}(k-1)$ and $\mathbf{B} \leftarrow \boldsymbol{\varphi}(k)$; $\mathbf{C} \leftarrow \alpha_k$ ; $\mathbf{D} \leftarrow \boldsymbol{\varphi}^T(k)$).

## 2.4.2   Algorithm

Now the recursive least squares algorithm with exponential forgetting can be given.

**Algorithm 3 (RLS with exponential forgetting)** The recursive least squares algorithm with exponential forgetting is given by

$$\mathbf{L}(k) = \frac{\mathbf{P}(k-1) \, \boldsymbol{\varphi}(k)}{\frac{\lambda}{\alpha_k} + \boldsymbol{\varphi}^T(k) \, \mathbf{P}(k-1) \, \boldsymbol{\varphi}(k)} \qquad (2.128)$$

$$\widehat{\boldsymbol{\theta}}\left(k\right)=\widehat{\boldsymbol{\theta}}\left(k-1\right)+\mathbf{L}\left(k\right)\left[y\left(k\right)-\widehat{\boldsymbol{\theta}}^{T}\left(k-1\right)\boldsymbol{\varphi}\left(k\right)\right] \qquad (2.129)$$

$$\mathbf{P}\left(k\right)=\frac{1}{\lambda}\left[\mathbf{P}\left(k-1\right)-\mathbf{L}\left(k\right)\boldsymbol{\varphi}^{T}\left(k\right)\mathbf{P}\left(k-1\right)\right] \qquad (2.130)$$

where $0 < \lambda \leq 1$, and $\lambda = 1$ gives the RLS algorithm with no forgetting.

The effect of the forgetting factor $\lambda$ is that the $\mathbf{P}\left(k\right)$ and hence the gain $\mathbf{L}\left(k\right)$ are kept larger. With $\lambda < 1$, the $\mathbf{P}\left(k\right)$ will not tend to zero and the algorithm will always be alert to changes in $\boldsymbol{\theta}$.

**Example 12 ($O_2$ dynamics:   continued)** Let us illustrate the performance of the RLS with exponential forgetting. Consider the identification problem in an FBC plant in Example 9, and let an unmeasured 20% decrease in the char feed occur (*e.g.*, due to an increase in the fuel moisture). Fig 2.5 illustrates the prediction and the on-line estimated parameters when using a forgetting factor $\lambda = 0.97$.

The change occurs at $t = 8$ min. The algorithm is able to follow the changes in the process.

There exists a large number of other forgetting schemes. Many (if not most) of them are inspired by the robustness of the Kalman filter, discussed in the next section.

## 2.5   Kalman filter

In the Bayesian approach to the parameter estimation problem, the parameter itself is thought of as a random variable. Based on the observations of other random variables that are correlated with the parameter, we may infer information about its value. The Kalman filter is developed in such a framework. The unobservable state vector is assumed to be correlated with the output of a system. So, based on the observations of the output, the value of the state vector can be estimated. In what follows, the Kalman filter is first introduced for state estimation. This is followed by an application to the parameter estimation problem.

Assume that a stationary stochastic vector signal $\{\mathbf{x}\left(k\right)\}$ can be described by the following Markov process

$$\mathbf{x}\left(k+1\right)=\mathbf{A}\left(k\right)\mathbf{x}\left(k\right)+\mathbf{v}\left(k\right) \qquad (2.131)$$

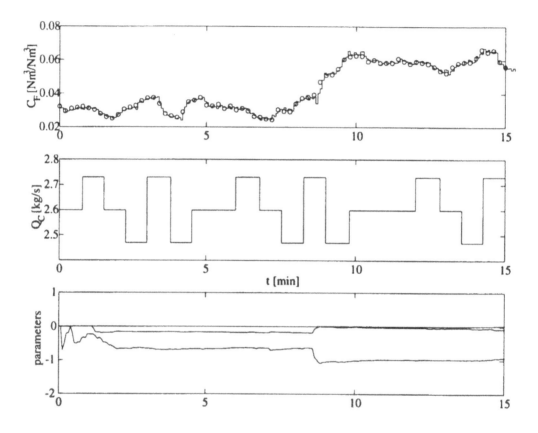

Figure 2.5: On-line prediction by the estimated model. Upper plot shows the predicted (solid line) and measured (circles) flue gas oxygen content (for clarity, only every third measurement is shown). The middle plot shows the model input, fuel feed. The evolution of the values of the estimated parameters is shown at the bottom of the figure.

with measurement equation

$$\mathbf{y}(k) = \mathbf{C}(k)\mathbf{x}(k) + \mathbf{e}(k) \qquad (2.132)$$

where $\mathbf{x}(k)$ is an $S \times 1$ dimensional column state vector, $\mathbf{v}(k)$ is a $S \times 1$ dimensional column vector containing the system noise; and $\mathbf{y}(k)$ and $\mathbf{e}(k)$ are $O \times 1$ dimensional column vectors of measurable outputs and the output noise. $\mathbf{A}(k)$ is an $S \times S$ dimensional system state transition matrix describing the internal dynamics of the system (Markov process). $\mathbf{C}(k)$ is the $O \times S$ output matrix, describing the relation between states and the measurable outputs. In state estimation, a stationary system is often assumed, $\mathbf{A}(k) = \mathbf{A}$, $\mathbf{C}(k) = \mathbf{C}$.

The objective is to estimate the state vector $\mathbf{x}(k)$ based on measurements of the outputs $\mathbf{y}(k)$, contaminated by noise $\mathbf{e}(k)$. The system model at sample instant $k$ is assumed to be known:

$$\mathbf{A}(k), \mathbf{C}(k) \qquad (2.133)$$

and the processes $\{\mathbf{v}(k)\}$ and $\{\mathbf{e}(k)\}$ are zero mean, independent Gaussian processes with known mean values and covariances:

$$E\{\mathbf{v}(k)\} = \mathbf{0}; E\{\mathbf{v}(k)\mathbf{v}^T(j)\} = \mathbf{V}(k)\delta_{kj} \qquad (2.134)$$

$$E\{\mathbf{e}(k)\} = \mathbf{0}; E\{\mathbf{e}(k)\mathbf{e}^T(j)\} = \mathbf{Y}(k)\delta_{kj} \qquad (2.135)$$

$$E\{\mathbf{e}(k)\mathbf{v}^T(j)\} = \mathbf{0} \qquad (2.136)$$

where $\delta_{ij}$ is the Kronecker delta function[7]. $\mathbf{v}(k)$ and $\mathbf{e}(k)$ have covariances $\mathbf{V}(k)$ and $\mathbf{Y}(k)$, respectively, which are non-negative and symmetric. It is assumed that $\{\mathbf{y}(k)\}$ is available to measurement, but $\{\mathbf{x}(k)\}$ is not. It is desirable to predict $\{\mathbf{x}(k)\}$ from the measurements of $\{\mathbf{y}(k)\}$.

The Kalman filter can be derived in a number of ways. In what follows, the mean square error approach for the Kalman predictor is considered [41]. We then proceed by giving the algorithm for the Kalman filter (the proof for the filter case is omitted as it is lengthy).

---

[7]Kronecker delta function is given by

$$\delta_{ij} = \{ \begin{array}{ll} 1 & \text{if } i = j \\ 0 & \text{otherwise} \end{array}$$

## 2.5.1   Derivation

Let us introduce the following *predictor* for the state $\mathbf{x}$ at instant $k + 1$

$$\widehat{\mathbf{x}}(k+1) = \mathbf{A}(k)\widehat{\mathbf{x}}(k) + \mathbf{K}(k)[\mathbf{y}(k) - \mathbf{C}(k)\widehat{\mathbf{x}}(k)] \qquad (2.137)$$

which consists of two terms: a *prediction* based on the system model and the previous estimate, and a *correction* term from the difference between the measured output and the output predicted using the system model. The gain matrix $\mathbf{K}(k)$ needs to be chosen.

Let us consider the following cost function to be minimized

$$J(k+1) = E\left\{\widetilde{\mathbf{x}}(k+1)\widetilde{\mathbf{x}}^T(k+1)\right\} \qquad (2.138)$$

where $\widetilde{\mathbf{x}}$ is the prediction error

$$\widetilde{\mathbf{x}}(k+1) = \widehat{\mathbf{x}}(k+1) - \mathbf{x}(k+1) \qquad (2.139)$$

The optimal solution is given by

$$\mathbf{K}(k) = \mathbf{A}(k)\mathbf{P}(k)\mathbf{C}^T(k)\left[\mathbf{Y}(k) + \mathbf{C}(k)\mathbf{P}(k)\mathbf{C}^T(k)\right]^{-1} \qquad (2.140)$$

where

$$\mathbf{P}(k+1) = \mathbf{A}(k)\mathbf{P}(k)\mathbf{A}^T(k) + \mathbf{V}(k) - \mathbf{K}(k)\mathbf{C}(k)\mathbf{P}(k)\mathbf{A}^T(k) \qquad (2.141)$$

**Proof.** Substituting (2.137) into (2.139) we have that

$$\begin{aligned}
\widetilde{\mathbf{x}}&(k+1) \\
&= \mathbf{A}(k)\widehat{\mathbf{x}}(k) + \mathbf{K}(k)[\mathbf{y}(k) - \mathbf{C}(k)\widehat{\mathbf{x}}(k)] - \mathbf{x}(k+1) \qquad (2.142) \\
&= [\mathbf{A}(k) - \mathbf{K}(k)\mathbf{C}(k)]\widehat{\mathbf{x}}(k) + \mathbf{K}(k)\mathbf{y}(k) - \mathbf{x}(k+1) \qquad (2.143)
\end{aligned}$$

and substituting (2.131)–(2.132) we have

$$\begin{aligned}
\widetilde{\mathbf{x}}(k+1) &= [\mathbf{A}(k) - \mathbf{K}(k)\mathbf{C}(k)]\widehat{\mathbf{x}}(k) + \mathbf{K}(k)\mathbf{C}(k)\mathbf{x}(k) \qquad (2.144) \\
&\quad + \mathbf{K}(k)\mathbf{e}(k) - \mathbf{A}(k)\mathbf{x}(k) - \mathbf{v}(k)
\end{aligned}$$

Reorganizing and using (2.139), we have the following prediction error dynamics

$$\widetilde{\mathbf{x}}(k+1) = [\mathbf{A}(k) - \mathbf{K}(k)\mathbf{C}(k)]\widetilde{\mathbf{x}}(k) + \mathbf{K}(k)\mathbf{e}(k) - \mathbf{v}(k) \qquad (2.145)$$

The cost function (2.138) can now be expressed as

$$
\begin{aligned}
J(k+1) \\
= \; & E\{[[\mathbf{A}(k) - \mathbf{K}(k)\,\mathbf{C}(k)]\,\tilde{\mathbf{x}}(k) + \mathbf{K}(k)\,\mathbf{e}(k) - \mathbf{v}(k)] \quad (2.146) \\
& [[\mathbf{A}(k) - \mathbf{K}(k)\,\mathbf{C}(k)]\,\tilde{\mathbf{x}}(k) + \mathbf{K}(k)\,\mathbf{e}(k) - \mathbf{v}(k)]^T\} \\
= \; & [\mathbf{A}(k) - \mathbf{K}(k)\,\mathbf{C}(k)]\,E\{\tilde{\mathbf{x}}(k)\,\tilde{\mathbf{x}}^T(k)\} \quad (2.147) \\
& \times [\mathbf{A}(k) - \mathbf{K}(k)\,\mathbf{C}(k)]^T \\
& + \mathbf{V}(k) + \mathbf{K}(k)\,\mathbf{Y}(k)\,\mathbf{K}^T(k)
\end{aligned}
$$

since $\mathbf{e}(k)$, $\mathbf{v}(k)$, and $\tilde{\mathbf{x}}(k)$ are statistically independent[8] and $\mathbf{K}(k)$ and $[\mathbf{A}(k) - \mathbf{K}(k)\,\mathbf{C}(k)]$ are known.

Let us use the following notation

$$
\begin{aligned}
\mathbf{P}(k) &= E\{\tilde{\mathbf{x}}(k)\,\tilde{\mathbf{x}}^T(k)\} \quad &(2.148) \\
\mathbf{Q}(k) &= \mathbf{Y}(k) + \mathbf{C}(k)\,\mathbf{P}(k)\,\mathbf{C}^T(k) \quad &(2.149)
\end{aligned}
$$

where $\mathbf{P}(k)$ is the covariance matrix of the estimation error. Rewrite (2.147) as

$$
\begin{aligned}
\mathbf{P}(k+1) = \; & \mathbf{A}(k)\,\mathbf{P}(k)\,\mathbf{A}^T(k) - \mathbf{K}(k)\,\mathbf{C}(k)\,\mathbf{P}(k)\,\mathbf{A}^T(k) \quad (2.150) \\
& - \mathbf{A}(k)\,\mathbf{P}(k)\,\mathbf{C}^T(k)\,\mathbf{K}^T(k) \\
& + \mathbf{V}(k) + \mathbf{K}(\mathbf{k})\,\mathbf{Q}(k)\,\mathbf{K}^T(k)
\end{aligned}
$$

By completing squares of terms containing $\mathbf{K}(k)$ we find

$$
\begin{aligned}
\mathbf{P}(k+1) = \; & \mathbf{A}(k)\,\mathbf{P}(k)\,\mathbf{A}^T(k) + \mathbf{V}(k) \quad (2.151) \\
& - \mathbf{A}(k)\,\mathbf{P}(k)\,\mathbf{C}^T(k)\,\mathbf{Q}^{-1}(k)\,\mathbf{C}(k)\,\mathbf{P}(k)\,\mathbf{A}^T(k) \\
& + \left[\mathbf{K}(k) - \mathbf{A}(k)\,\mathbf{P}(k)\,\mathbf{C}^T(k)\,\mathbf{Q}^{-1}(k)\right]\mathbf{Q}(k) \\
& \times \left[\mathbf{K}(k) - \mathbf{A}(k)\,\mathbf{P}(k)\,\mathbf{C}^T(k)\,\mathbf{Q}^{-1}(k)\right]^T
\end{aligned}
$$

Now only the last term of the sum depends on $\mathbf{K}(k)$, and minimization of $J$ can be done by choosing $\mathbf{K}(k)$ such that the last term disappears:

$$
\mathbf{K}(k) = \mathbf{A}(k)\,\mathbf{P}(k)\,\mathbf{C}^T(k)\left[\mathbf{Y}(k) + \mathbf{C}(k)\,\mathbf{P}(k)\,\mathbf{C}^T(k)\right]^{-1} \quad (2.152)
$$

---

[8] By assumption, $\mathbf{v}(k)$ and $\mathbf{e}(k)$ are statistically independent. $\tilde{\mathbf{x}}(k)$ is given by $\tilde{\mathbf{x}}(k) = \hat{\mathbf{x}}(k) - \mathbf{x}(k)$. The prediction $\hat{\mathbf{x}}(k)$ depends on the past measurement $\mathbf{y}(k-1)$, and is hence dependent on $\mathbf{e}(k-1)$. The state $\mathbf{x}(k)$ is dependent on noise $\mathbf{v}(k-1)$ disturbing the state. Thus the prediction error $\tilde{\mathbf{x}}(k)$ depends on $\mathbf{e}(k-1)$ and $\mathbf{v}(k-1)$, but not on $\mathbf{e}(k)$ or $\mathbf{v}(k)$. Thus, $\mathbf{v}(k)$, $\mathbf{e}(k)$, and $\tilde{\mathbf{x}}(k)$ are statistically independent.

Since the last term disappears, we have

$$\mathbf{P}(k+1) = \mathbf{A}(k)\mathbf{P}(k)\mathbf{A}^T(k) + \mathbf{V}(k) - \mathbf{K}(k)\mathbf{C}(k)\mathbf{P}(k)\mathbf{A}^T(k) \quad (2.153)$$

∎

Collecting the results, we have the following algorithm for an optimal estimate (in the mean square error sense) of the next state $\mathbf{x}(k+1)$, based on information up to $k$:

$$
\begin{aligned}
\mathbf{K}(k) &= \mathbf{A}(k)\mathbf{P}(k)\mathbf{C}^T(k)\left[\mathbf{Y}(k) + \mathbf{C}(k)\mathbf{P}(k)\mathbf{C}^T(k)\right]^{-1} &(2.154)\\
\widehat{\mathbf{x}}(k+1) &= \mathbf{A}(k)\widehat{\mathbf{x}}(k) - \mathbf{K}(k)\left[\mathbf{C}(k)\widehat{\mathbf{x}}(k) - \mathbf{y}(k)\right] &(2.155)\\
\mathbf{P}(k+1) &= \mathbf{A}(k)\mathbf{P}(k)\mathbf{A}^T(k) + \mathbf{V}(k) &(2.156)\\
&\quad -\mathbf{K}(k)\mathbf{C}(k)\mathbf{P}(k)\mathbf{A}^T(k)
\end{aligned}
$$

If the disturbances $\{\mathbf{e}(k)\}$ and $\{\mathbf{v}(k)\}$ as well as the initial state $\mathbf{x}(0)$ are Gaussian (with mean values $\mathbf{0}$, $\mathbf{0}$, and $\mathbf{x}_0$ and covariances $\mathbf{V}(k)$, $\mathbf{Y}(k)$ and $\mathbf{P}(0)$, respectively), the estimate $\widehat{\mathbf{x}}(k+1)$ is the mean of the conditional distribution of $\mathbf{x}(k+1)$, $\widehat{\mathbf{x}}(k+1) = E\{\mathbf{x}(k+1)|\mathbf{y}(0),\mathbf{y}(1),\cdots,\mathbf{y}(k)\}$. $\mathbf{P}(k+1)$ is the covariance of the conditional distribution of $\mathbf{x}(k+1)$.

### 2.5.2   Algorithm

Let us denote the estimate (2.155) based on information up to time $k$ by $\widehat{\mathbf{x}}(k+1|k)$. A Kalman *filter* can also be derived for estimating the state $\mathbf{x}(k+1)$, assuming now that the measurement $\mathbf{y}(k+1)$ has become available, *i.e.*

$$\widehat{\mathbf{x}}(k+1|k+1) = E\{\mathbf{x}(k+1)|\mathbf{y}(0),\mathbf{y}(1),\cdots,\mathbf{y}(k+1)\} \quad (2.157)$$

Consider now a filter of the form

$$\widehat{\mathbf{x}}(k+1|k+1) = \widehat{\mathbf{x}}(k+1|k) + \mathbf{K}(k+1)\left[\mathbf{y}(k+1) - \mathbf{C}(k)\widehat{\mathbf{x}}(k+1|k)\right]$$
$$(2.158)$$

The following algorithm can be derived. (Note that an extended state space model is used with an additional deterministic input $\mathbf{u}(k)$ and a noise transition matrix $\mathbf{G}(k)$.)

**Algorithm 4 (Kalman filter)** Estimate the state vectors $\mathbf{x}(k)$ of a system described by the following equations

$$\mathbf{x}(k+1) = \mathbf{A}(k)\mathbf{x}(k) + \mathbf{B}(k)\mathbf{u}(k) + \mathbf{G}(k)\mathbf{v}(k) \quad (2.159)$$

$$y(k) = C(k)x(k) + e(k) \qquad (2.160)$$

where $x(k)$ is an $S \times 1$ state vector: $u(k)$ and $v(k)$ are $I \times 1$ vectors containing the system inputs and Gaussian noise; $y(k)$ and $e(k)$ are $O \times 1$ vectors of measurable outputs and the output Gaussian noise, respectively. $A(k)$ is an $S \times S$ system state transition matrix; $B(k)$ and $G(k)$ are $S \times I$ and $S \times S$ system input and noise transition matrices; $C(k)$ is the $O \times S$ output matrix. The following are known for $k = k_0, k_0 + 1, k_0 + 2, ..., j \leq k$:

$$A(k), B(k), C(k), G(k) \qquad (2.161)$$

$$E\{v(k)\} = 0; E\{v(k)v^T(j)\} = V(k)\delta_{kj} \qquad (2.162)$$

$$E\{e(k)\} = 0; E\{e(k)e^T(j)\} = Y(k)\delta_{kj} \qquad (2.163)$$

$$E\{e(k)v^T(j)\} = 0 \qquad (2.164)$$

$$x(k_0) = x_{k_0}; \operatorname{cov}\{x(k_0)\} = P_{k_0} \qquad (2.165)$$

1. Set $k = k_0$. Initialize $\hat{x}(k_0|k_0) = x_{k_0}$ and $P(k_0|k_0) = P_{k_0}$.

2. **Time update**:

   Compute the state estimate at $k + 1$, given data up to $k$:

   $$\hat{x}(k+1|k) = A(k)\hat{x}(k|k) + B(k)u(k) \qquad (2.166)$$

   and update the covariance matrix of the error in $\hat{x}(k+1|k)$:

   $$P(k+1|k) = A(k)P(k|k)A^T(k) + G(k)V(k)G^T(k) \qquad (2.167)$$

3. **Measurement update**:

   Observe the new measurement $y(k+1)$, at time $t = kT$.

   Compute the Kalman filter gain matrix:

   $$\begin{aligned} K(k+1) = & \ P(k+1|k)C^T(k+1) \qquad (2.168) \\ & \times \left[Y(k+1) + C(k+1)P(k+1|k)C^T(k+1)\right]^{-1} \end{aligned}$$

   Correct the state estimate at $k + 1$, given data up to $k + 1$:

   $$\begin{aligned} \hat{x}(k+1|k+1) = & \ \hat{x}(k+1|k) \qquad (2.169) \\ & + K(k+1)[y(k+1) - C(k+1)\hat{x}(k+1|k)] \end{aligned}$$

and update the new error covariance matrix[9]:

$$
\begin{aligned}
\mathbf{P}(k+1|k+1) &= [\mathbf{I} - \mathbf{K}(k+1)\mathbf{C}(k+1)]\mathbf{P}(k+1|k) \\
&\quad \times [\mathbf{I} - \mathbf{K}(k+1)\mathbf{C}(k+1)]^T \qquad (2.170) \\
&\quad + \mathbf{K}(k+1)\mathbf{Y}(k+1)\mathbf{K}^T(k+1)
\end{aligned}
$$

4. Increase sample index $k = k+1$ and return to step 2.

### 2.5.3   Kalman filter in parameter estimation

Suppose that the data is generated according to

$$
y(k) = \boldsymbol{\varphi}^T(k)\boldsymbol{\theta} + e(k) \qquad (2.171)
$$

where $e(k)$ is a sequence of independent Gaussian variables with zero mean and variance $\sigma^2(k)$. Suppose also that the prior distribution of $\boldsymbol{\theta}$ is Gaussian with mean $\boldsymbol{\theta}_0$ and covariance $\mathbf{P}_0$. The model, (2.171), can be seen as a linear state-space model:

$$
\begin{aligned}
\boldsymbol{\theta}(k+1) &= \boldsymbol{\theta}(k) & (2.172) \\
y(k) &= \boldsymbol{\varphi}^T(k)\boldsymbol{\theta}(k) + e(k) & (2.173)
\end{aligned}
$$

Comparing with (2.159)-(2.165) shows that these equations are identical when making the following substitutions:

$$
\begin{aligned}
\mathbf{A}(k) &\leftarrow \mathbf{I}; \mathbf{x}(k) \leftarrow \boldsymbol{\theta}(k) & (2.174) \\
\mathbf{B}(k) &\leftarrow 0; \mathbf{u}(k) \leftarrow 0 & (2.175) \\
\mathbf{G}(k) &\leftarrow 0; \mathbf{v}(k) \leftarrow 0; \mathbf{V}(k) \leftarrow 0 & (2.176)
\end{aligned}
$$

$$
\begin{aligned}
\mathbf{C}(k) &\leftarrow \boldsymbol{\varphi}^T(k); \mathbf{y}(k) \leftarrow y(k); & (2.177) \\
\mathbf{e}(k) &\leftarrow e(k); \mathbf{Y}(k) \leftarrow \sigma^2(k) & (2.178)
\end{aligned}
$$

$$
\widehat{\mathbf{x}}(0|0) \leftarrow \boldsymbol{\theta}_0; \mathbf{P}(0|0) \leftarrow \mathbf{P}_0 \qquad (2.179)
$$

---

[9]This is a numerically better form of

$$
\mathbf{P}(k+1|k+1) = \mathbf{P}(k+1|k) - \mathbf{K}(k+1)\mathbf{C}(k+1)\mathbf{P}(k+1|k)
$$

(see [39], p. 270).

The Kalman filter algorithm, (2.159)-(2.170), is now given by (note that $\mathbf{P}(k) \leftarrow \mathbf{P}(k+1|k) = \mathbf{P}(k|k); \widehat{\boldsymbol{\theta}}(k); \widehat{\boldsymbol{\theta}} \leftarrow \widehat{\boldsymbol{\theta}}(k+1|k) = \widehat{\boldsymbol{\theta}}(k|k))$:

$$\mathbf{K}(k+1) = \frac{\mathbf{P}(k)\boldsymbol{\varphi}(k+1)}{\sigma^2(k) + \boldsymbol{\varphi}^T(k+1)\mathbf{P}(k)\boldsymbol{\varphi}(k+1)} \tag{2.180}$$

$$\widehat{\boldsymbol{\theta}}(k+1) = \widehat{\boldsymbol{\theta}}(k) + \mathbf{K}(k+1)\left(y(k+1) - \widehat{\boldsymbol{\theta}}^T(k)\boldsymbol{\varphi}(k+1)\right) \tag{2.181}$$

$$\mathbf{P}(k+1) = \mathbf{P}(k) - \mathbf{K}(k+1)\boldsymbol{\varphi}^T(k+1)\mathbf{P}(k) \tag{2.182}$$

Comparing with the RLS (Algorithm 2) shows that the Kalman filter holds the RLS as its special case. Note, that now the initial conditions of the RLS have a clear interpretation: $\widehat{\boldsymbol{\theta}}(0)$ is the *prior mean* and $\mathbf{P}(0)$ is the *prior covariance* of the parameters $\boldsymbol{\theta}$. Furthermore, the *posterior distribution of $\theta$ at sample instant $k$ is also Gaussian with mean $\widehat{\boldsymbol{\theta}}(k)$ and covariance* $\mathbf{P}(k)$ (see [55], pp. 33-36). The Kalman filter approach also shows that the optimal weighting factor $\alpha_k$ in the least squares criterion is the inverse of the variance of the noise term, $\alpha_k = 1/\sigma^2(k)$, at least when the noise is white and Gaussian.

If the dynamics of the system are changing with time, *i.e.* the model parameters are time-varying, we can assume that the parameter vector varies according to

$$\boldsymbol{\theta}(k+1) = \boldsymbol{\theta}(k) + \mathbf{v}(k) \tag{2.183}$$

Now $\mathbf{V} \neq 0$ and the covariance update becomes (see (2.167)):

$$\mathbf{P}(k+1) = \mathbf{P}(k) - \mathbf{K}(k+1)\boldsymbol{\varphi}^T(k+1)\mathbf{P}(k) + \mathbf{V} \tag{2.184}$$

This prevents the covariance matrix from tending to zero. In fact $\mathbf{P}(k) \rightarrow \mathbf{V}$ when the number of iterations increases, and the algorithm remains alert to changes in model parameters. For example, in [23] the addition (regularization) of a constant scaled identity matrix at each sample interval is suggested, $\mathbf{V} \leftarrow \alpha \mathbf{I}$. The bounded information algorithm [70] ensures both lower and upper bounds $a_{\min}$ and $a_{\max}$ for $\mathbf{P}(k)$

$$\mathbf{P}(k|k) = \frac{a_{\max} - a_{\min}}{a_{\max}}\mathbf{P}(k|k-1) + a_{\min}\mathbf{I} \tag{2.185}$$

An advantage of the Kalman filter approach, compared to the least squares algorithm with exponential forgetting, is that the nature of the parameter changes can be easily incorporated, and interpreted as the covariance matrix $\mathbf{V}$.

# Chapter 3

# Linear Dynamic Systems

In this chapter, our attention is focused on the discrete-time black-box modeling of linear dynamic systems. This type of model is commonly used in process identification, and is essential in digital process control. From the point of view of control, the simplicity of black-box models has established them as a fundamental means for obtaining input-output representations of processes.

The transfer function approach provides a basic tool for representing dynamic systems. Stochastic disturbance models provide a tool for characterizing (unmeasured) disturbances, present in all real systems.

## 3.1 Transfer function

Let us first consider two commonly used transfer function[1] representations of process dynamics.

### 3.1.1 Finite impulse response

A finite impulse response (FIR) system is given by

$$y(k) = B(q^{-1}) u(k - d) \tag{3.1}$$

where

---

[1] In order to avoid unnecessary complexity in notation and terminology, the backward shift operator, $q^{-1}$, notation will be used, $x(k-i) = q^{-i}x(k)$. Strictly speaking, a division of two polynomials in $q^{-1}$ is not meaningful (whereas the division of two functions in $z^{-1}$ is). However, the reader should consider this transfer operation as a symbolic notation (or as an equivalent $z$ transform). With this loose terminology, we allow ourselves to use the term 'transfer function' for descriptions that use polynomials in the backward shift operator.

- $\{y(k)\}$ is a sequence of system outputs, and

- $\{u(k)\}$ is a sequence of system inputs,

sampled from the process at instants $k = 1, 2, ...$ which are usually equidistant time intervals:

$$kT = t \tag{3.2}$$

where $t$ is the time and $T$ is the sampling interval (*e.g.*, in seconds). The process is characterized by

$$B\left(q^{-1}\right) = b_0 + b_1 q^{-1} + ... + b_{n_B} q^{-n_B} \tag{3.3}$$

which is a polynomial in the backward shift operator $q^{-1}$

$$q^{-1} x(k) = x(k-1) \tag{3.4}$$

$d$ is the time delay (in sampling instants) between process input and output. The system behavior is determined by its coefficients or parameters $b_n$, $n = 0, 1, 2, ..., n_B$, $b_n \in \Re$.

FIR structures are among the simplest used for describing dynamic processes. They involve:

- no complex calculations, and

- no assumption on the process order is required.

The parameters can be obtained directly from the elements of the impulse response of the system. The choice of $n_B$ and $d$ is less critical, if chosen large enough and small enough, respectively.

The disadvantages of FIR are that:

- unstable processes can not be modelled,

- a large number of parameters need to be estimated (especially for processes containing slow modes, *i.e.* slow dynamics).

## Residence time

Process engineers are often confronted with the calculation of residence time in continuous flow systems (reactors, columns, *etc.*) [62]. The residence time is the time needed for the fluid to travel from one end of the process to the other. The residence time is a convenient time base for normalization (usually, the states variables are made dimensionless and scaled to take the value of unity at their target value). The residence time is also directly related to the efficiency and productivity of a given chemical process.

Tracer tests (isotopic, *etc.*) are commonly used in chemical engineering for determining the residence time. An amount of tracer is fed into the process as quickly as possible (impulse input). The output is then measured and interpreted as the process impulse response. For linear systems, the residence time is directly calculated from the impulse response or from the parameters of their transfer function [97].

A linear system can be defined by its continuous-time impulse response $g(t)$. Its output equation is given by:

$$y(t) = \int_{\tau=0}^{\infty} g(t-\tau)u(\tau)d\tau \qquad (3.5)$$

where $y(t)$ and $u(t)$ represent respectively the output and the input. The residence time [97] is given by:

$$\tau_{res} = \frac{\int_{\tau=0}^{\infty} t\, g(t)dt}{\int_{\tau=0}^{\infty} g(t)dt} \qquad (3.6)$$

In continuous flow system, the residence time can be interpreted as the expected time it takes for a molecule to pass trough the flow system.

The residence time can also be connected to the input-output signals without using a phenomenological model description of the considered process. Based on the concept of impulse response function, the residence time can be calculated as follows:

- Continuous-time systems:

$$\tau_{res} = -\frac{TF'(0)}{TF(0)} \qquad (3.7)$$

where $TF(s)$ is the Laplace transform of the impulse response $g(t)$ and $TF'(\cdot)$ is the derivative of $TF(\cdot)$ with respect to $s$.

- Discrete-time systems:

$$\tau_{res} = \frac{\sum\limits_{k=0}^{\infty} k\, b_k}{\sum\limits_{k=0}^{\infty} b_k} \tag{3.8}$$

$B(z)$ represents the discrete impulse response function defined as

$$B(z) = \sum_{k=0}^{\infty} b_k z^{-k} \tag{3.9}$$

It is easy to verify that in the discrete case the residence time is

$$\tau_{res} = -\frac{B'(1)}{B(1)} \tag{3.10}$$

where $B'(\cdot)$ is the derivative of $B(\cdot)$ with respect to $z$.

The results concerning the calculation of the residence time for linear systems can also be extended to multidimensional continuous flow in non linear systems [65].

## 3.1.2   Transfer function

A more general structure is the transfer function (TF) structure. It holds the FIR structure as its special case.

**Definition 3 (Transfer function)** Transfer function is given by

$$y(k) = \frac{B(q^{-1})}{A(q^{-1})} u(k - d) \tag{3.11}$$

where $A$ is a monic polynomial of degree $n_A$

$$A(q^{-1}) = 1 + a_1 q^{-1} + \ldots + a_{n_A} q^{-n_A} \tag{3.12}$$

and $B$ is a polynomial of degree $n_B$

$$B(q^{-1}) = b_0 + b_1 q^{-1} + \ldots + b_{n_B} q^{-n_B} \tag{3.13}$$

where $a_n \in \Re$, $n = 1, 2, \ldots, n_A$ and $b_n \in \Re$, $n = 0, 1, \ldots, n_B$.

The main advantages of the TF model are that:

- a minimal number of parameters is required,

- both stable and unstable processes can be described.

Disadvantages include that

- an assumption of process order, $n_A$, is needed (in addition to $n_B$ and $d$),

- the prediction is more complex to compute.

*Poles* and *zeros* give a convenient way of characterizing the behavior of systems described using transfer functions. Note, that switching into z-transform gives

$$\frac{Y(z^{-1})}{U(z^{-1})} = z^{-d}\frac{B(z^{-1})}{A(z^{-1})} \tag{3.14}$$

$$= z^{-d}\frac{b_0 + b_1 z^{-1} + ... + b_{n_B} z^{-n_B}}{1 + a_1 z^{-1} + ... + a_{n_A} z^{-n_A}} \tag{3.15}$$

Multiplying the numerator and the denominator by $z^{n_B + n_A + d}$ gives

$$\frac{Y(z)}{U(z)} = \frac{z^{n_A}\left(b_0 z^{n_B} + b_1 z^{n_B-1} + ... + b_{n_B}\right)}{z^{n_B+d}\left(z^{n_A} + a_1 z^{n_A-1} + ... + a_{n_A}\right)} = \frac{B(z)}{A(z)} \tag{3.16}$$

The roots of the polynomials give the poles (roots of $A(z) = 0$) and the zeros (roots of $B(z) = 0$) of the system.

**Definition 4 (Poles and zeros)** For a transfer function (Definition 3) the $n_B$ zeros of the system are obtained from the roots of

$$B(z) = b_0 z^{n_B} + b_1 z^{n_B-1} + ... + b_{n_B} = 0 \tag{3.17}$$

The $n_A$ poles of the system are obtained from the roots of

$$A(z) = z^{n_A} + a_1 z^{n_A-1} + ... + a_{n_A} = 0 \tag{3.18}$$

$A(z)$ can be represented as

$$A(z) = \prod_{n=1}^{n_R}(z - p_n) \cdot \prod_{n=1}^{n_C}\left(z^2 + \alpha_n z + \beta_n\right) \tag{3.19}$$

where $p_n$ are the $n_R$ real poles and $z^2 + \alpha_n z + \beta_n$ contain the $n_C$ complex pairs of poles of the system. In a similar way, $B(z)$ can be represented as

$$B(z) = b_0 \prod_{n=1}^{n_R}(z - r_n) \cdot \prod_{n=1}^{n_C}\left(z^2 + \alpha_n z + \beta_n\right) \tag{3.20}$$

where $r_n$ are the $n_R$ real zeros and $z^2 + \alpha_n z + \beta_n$ contain the $n_C$ complex pairs of zeros of the system.

The *steady-state gain* is obtained when $z \to 1$

$$\lim_{z \to 1} \frac{Y(z)}{U(z)} \tag{3.21}$$

From (3.16) it is simple to derive the following result.

**Algorithm 5 (Steady-state gain)** The steady-state gain of a system described using a transfer function (Definition 3) is given by

$$K_{ss} = \frac{\sum_{n=0}^{n_B} b_n}{1 + \sum_{n=1}^{n_A} a_n} \tag{3.22}$$

where $K_{ss} \in \Re$ denotes the steady-state gain of the system.

**Example 13 (Pole and steady-state gain)** Consider the following first-order system:

$$y(k) = ay(k-1) + u(k-1) \tag{3.23}$$

The system can be written as

$$\frac{Y(z^{-1})}{U(z^{-1})} = \frac{B(z^{-1})}{A(z^{-1})} = \frac{z^{-1}}{1 - az^{-1}} \tag{3.24}$$

and

$$\frac{Y(z)}{U(z)} = \frac{B(z)}{A(z)} = \frac{1}{z - a} \tag{3.25}$$

This system has one pole in $z = a$.

The steady-state gain is given by

$$K_{ss} = \frac{1}{1 - a} \tag{3.26}$$

In general, a system is stable[2] if all its poles are located inside the unit circle. If at least one pole is on or outside the unit circle, the system is not stable.

**Example 14 (Stability)** Consider the system in Example 13 with initial condition $y(0) = y_0$ and control input $u(k) \equiv 0$.

The future values of the system for $k = 1, 2, \ldots$ are given by

$$y(k) = a^k y_0 \tag{3.27}$$

If $|a| < 1$, then $y(k)$ tends to zero and the system is stable.

---

[2]BIBO stability: A system is BIBO stable, if for every bounded input, we have a bounded output.

## 3.2 Deterministic disturbances

In general, a real process is always subjected to disturbances. The effects of the system environment and approximation errors are modelled as disturbance processes. Models of disturbance processes should capture the essential characteristics of the disturbances. In control, the disturbances that affect the control performance without making the resulting controller implementation uneconomical, are of interest.

Consider a TF structure with a disturbance:

$$y\left(k\right) = \frac{B\left(q^{-1}\right)}{A\left(q^{-1}\right)} u\left(k-d\right) + \xi\left(k\right) \tag{3.28}$$

where $\xi\left(k\right)$ represents the totality of all disturbances at the output of the process. It is the sum of both deterministic and stochastic disturbances.

In some cases, deterministic disturbances are exactly predictable. Assume that the disturbances are described by the following model

$$C_\xi\left(q^{-1}\right) \xi\left(k\right) = 0 \tag{3.29}$$

Typical exactly predictable deterministic disturbances include

- a constant

$$C_\xi\left(q^{-1}\right) = 1 - q^{-1} \tag{3.30}$$

- a sinusoid

$$C_\xi\left(q^{-1}\right) = 1 - 2q^{-1}\cos\left(\omega T_s\right) + q^{-2} \tag{3.31}$$

**Example 15 (Constant deterministic disturbance)** A constant disturbance gives

$$\xi\left(k\right) = \xi\left(k-1\right) \tag{3.32}$$

Thus, the effect of a disturbance at sampling instant $k-1$ remains also at instant $k$.

## 3.3 Stochastic disturbances

The most serious difficulty in applying identification and control techniques to industrial processes is the lack of good models. The effect of the environment of the process to be modeled, and approximation errors, are modeled as disturbance processes. These disturbances are classified into two categories: measured (*e.g.*, ambient temperature) and unmeasured (*e.g.*, particle size distributions, or composition of raw materials).

Usually, random disturbances are assumed to be stationary. Let us recall the definition of stationary processes.

**Definition 5 (Stationary process)** A process $\{x(k), k \in T\}$ is said to be stationary if, for any $\{k_1, k_2, ..., k_N\}$, *i.e.* any finite subset of $T$, the joint distribution $(x(k_1 + \tau), x(k_2 + \tau), ..., x(k_N + \tau))$ of $x(k + \tau)$ does not depend upon $\tau$.

The modeling of unmeasured perturbation is based on a sequence $\{e(k)\}$ of independent random variables with zero mean, $E\{e(k)\} = 0$, and variance $\sigma^2 = 1$.

These assumptions are not restrictive. In fact, a random sequence $\{b(k)\}$ such that $E\{b(k)\} = m$ and $E\{b(k)^2\} = \sigma^2$ can be expressed as a function of $e(k)$ as follows

$$b(k) = \sigma e(k) + m \tag{3.33}$$

**Remark 3 (Gaussian stationary processes)** The usual argument given in favor of Gaussian stationary processes hinges upon the *central limit theorem*. Roughly, a large number of small independent fluctuations, when averaged, give a Gaussian random variable. Notice also that linear operations upon Gaussian process leave it Gaussian. Physically independent sources (linear systems or linear regime) of small disturbances produce Gaussian processes.

**Example 16 (Fluidized bed)** Consider a bubbling fluidized bed [20]. Theoretically it is possible to understand and predict the mechanism and coalescence for two or three isolated bubbles in a deterministic manner. However, we are unable to extend the deterministic model to accurately predict the behavior of a large swarm of bubbles, since we do not have exact and complete knowledge about the initial conditions (start-up of a fluidized bed) and external forces acting on the system (particle size distributions, *etc.*). Such a process appears to us to be stochastic, and we speak of the random coalescence and movement of the bubbles, which leads to pressure and density fluctuations.

### 3.3.1 Offset in noise

The following model

$$\xi(k) = C\left(q^{-1}\right) e(k) \tag{3.34}$$

where $C$ is a polynomial in the backward shift operator $q^{-1}$, can be used to describe the noise affecting the plant under consideration. The model consists of a zero mean random noise sequence, $\{e(k)\}$, colored by the polynomial $C$. The offset is not modeled by (3.34).

To take the offset into account, the following solution has been proposed

$$\xi(k) = C\left(q^{-1}\right) e(k) + d \tag{3.35}$$

where $d$ is a constant depending on the plant operating point. However, it has been shown that even if $d$ is a constant, or slowly varying, there are inherent problems in estimating its value (appearance of 1 in the regressor, which is not a persistently exciting signal). Thus, the parameter $d$ is inherently different from the other parameters of the model.

A better solution, which does not involve the estimation of the offset, is to assume that the perturbation process has stationary increments, *i.e.*

$$\xi(k) = \frac{C\left(q^{-1}\right)}{\Delta\left(q^{-1}\right)} e(k) \tag{3.36}$$

where $\Delta$ is the difference operator

$$\Delta\left(q^{-1}\right) = 1 - q^{-1} \tag{3.37}$$

This disturbance model is more realistic. It can be interpreted as random step disturbances occurring at random intervals (*e.g.*, sudden change of load or variation in the quality of feed flow).

The model described in (3.36) corresponds to the inherent inclusion of an integrator in the closed loop system. In general, the perturbation is described by $\frac{C}{D}$ where $D$ is a polynomial in $q^{-1}$. The choice of $D = \Delta D^{*}$ allows the incorporation of an explicit integral action into the design, where $D^{*}$ is a polynomial in $q^{-1}$. In particular, the choice $D = \Delta A$ is common. Various system structures with stochastic disturbances will be considered in the following.

### 3.3.2 Box–Jenkins

The representation of process dynamics is usually achieved with a disturbed linear discrete-time model. Practically all linear black-box SISO model structures can be seen as special cases of the Box–Jenkins (BJ) model structure.

**Definition 6 (Box–Jenkins)** Box–Jenkins (BJ) structure is given by

$$y(k) = \frac{B(q^{-1})}{A(q^{-1})} u(k-d) + \frac{C(q^{-1})}{D(q^{-1})} e(k) \qquad (3.38)$$

where $\{y(k)\}$ is a sequence of process outputs, $\{u(k)\}$ is a sequence of process inputs, and $\{e(k)\}$ is a discrete white noise sequence (zero mean with finite variance $\sigma^2$) sampled from the process at instants $k = 1, 2, ...$;

$$
\begin{aligned}
A(q^{-1}) &= 1 + a_1 q^{-1} + ... + a_{n_A} q^{-n_A} & (3.39) \\
B(q^{-1}) &= b_0 + b_1 q^{-1} + ... + b_{n_B} q^{-n_B} & (3.40) \\
C(q^{-1}) &= 1 + c_1 q^{-1} + ... + c_{n_C} q^{-n_C} & (3.41) \\
D(q^{-1}) &= 1 + d_1 q^{-1} + ... + d_{n_D} q^{-n_D} & (3.42)
\end{aligned}
$$

are polynomials in the backward shift operator $q^{-1}$, $q^{-1}x(k) = x(k-1)$.

Basically, this type of black-box system is used for four main purposes:

1. characterizing (understanding) the input-output behavior of a process,

2. predicting future responses of a plant,

3. developing control systems and tuning controllers, and

4. filtering and smoothing of signals.

Items 1-2 are related to process modeling (monitoring, fault detection, *etc.*) and items 2-3 to process control (controller design, especially model-based control). The fourth topic concerns signal processing (handling of measurements in process engineering).

$d$ is the time delay (in sampling instants) between the process input and the output:

- In process modeling $d \geq 1$ assures causality: process output can not change before (or exactly at the same time) when a change in process input occurs.

- $d \leq 0$ is used in filtering (smoothing) signals. $d = 0$ can be used in on-line filtering to remove measurement noise; $d < 0$ can be applied only in off-line filtering (to compute the filtered signal, future values of the signal are required).

In what follows, interest is focused on process modeling, $d \geq 1$, where $d$ is not a design parameter, but depends on the time delay observed in the process to be modeled.

Assume that the current sample instant is $k$, and that the following information is available:

- current and past process inputs $u(k), u(k-1), ..., u(k-n_B)$,

- process outputs $y(k), y(k-1), ..., y(k-n_A)$.

Let us denote by $\widehat{y}(k+1)$ the prediction of $y(k+1)$ obtained using the model. Let us assume further that

- the predictions $\widehat{y}(k), \widehat{y}(k-1), ..., \widehat{y}(k - \max(n_A, n_C))$

are available as well.

In practice, an exact mathematical description of the dynamic response of an industrial process may be either impossible or impractical. The use of linear models involves a loss of information (approximation errors, neglected dynamics). When selecting a structure for a stochastic process model, an assumption on the effect of noise is made. In the following, some commonly used transfer function models (input-output model of the process) with stochastic noise models (effect of unmeasured noise to the process output) are discussed.

### 3.3.3 Autoregressive exogenous

A variety of real-world processes can be described well by the autoregressive (AR) model. The AR process can be viewed as a result of passing of the white noise through a linear all-pole filter. In the acronym ARX, the X denotes the presence of an exogenous variable.

**Definition 7 (ARX structure)** ARX (autoregressive exogenous) structure is obtained by setting $C = 1$, $D = A$ in the general structure (Definition 6):

$$y(k) = \frac{B(q^{-1})}{A(q^{-1})} u(k-d) + \frac{1}{A(q^{-1})} e(k) \qquad (3.43)$$

Let us rewrite the ARX system for $k+1$ and multiply by $A$:

$$A(q^{-1}) y(k+1) = B(q^{-1}) u(k-d+1) + e(k+1) \qquad (3.44)$$

For the system output at $k+1$ we get

$$y(k+1) = B\left(q^{-1}\right) u(k-d+1) - A_1\left(q^{-1}\right) y(k) + e(k+1) \qquad (3.45)$$

where $A = 1 + q^{-1}A_1$. Noticing that the first two terms on the right side can be calculated exactly from the available data up to time $k$, and the noise term $e(k+1)$ will act on the process in the future, we have that

$$\widehat{y}(k+1) = B\left(q^{-1}\right) u(k-d+1) - A_1\left(q^{-1}\right) y(k) \qquad (3.46)$$

which minimizes the expected squared prediction error[3].

**Algorithm 6 (ARX predictor)** Predictor for an ARX system (Definition 7) is given by:

$$\widehat{y}(k+1) = B\left(q^{-1}\right) u(k-d+1) - A_1\left(q^{-1}\right) y(k) \qquad (3.48)$$

where $A_1\left(q^{-1}\right) = a_1 + ... + a_{n_A} q^{-(n_A-1)}$. The prediction is a function of the process measurements.

---

[3]The objective is to find a linear predictor depending on the information available up to and including $k$ which minimizes the expectation of the squared prediction error, *i.e.*

$$\widehat{y}(k+1) = \arg\min_{\widehat{y}} E\left\{[y(k+1) - \widehat{y}]^2\right\} \qquad (3.47)$$

where $E\{\cdot\}$ represents the conditional expectation (on the available data). Introducing (3.45) in (3.47), we have

$$
\begin{aligned}
&E\left\{[y(k+1) - \widehat{y}]^2\right\} \\
=\ &E\left\{[B\left(q^{-1}\right)u(k-d+1) - A_1\left(q^{-1}\right)y(k) + e(k+1) - \widehat{y}]^2\right\} \\
=\ &E\left\{[(B\left(q^{-1}\right)u(k-d+1) - A_1\left(q^{-1}\right)y(k) - \widehat{y}) + e(k+1)]^2\right\} \\
=\ &E\left\{[B\left(q^{-1}\right)u(k-d+1) - A_1\left(q^{-1}\right)y(k) - \widehat{y}]^2\right\} \\
&+2E\left\{(B\left(q^{-1}\right)u(k-d+1) - A_1\left(q^{-1}\right)y(k) - \widehat{y})e(k+1)\right\} \\
&+E\left\{e^2(k+1)\right\}
\end{aligned}
$$

Since $e(k+1)$ is independent with respect to $u(k-d), u(k-d-1), ...$ and $y(k), y(k-1), ...$ , and a linear combination of these variables generating $\widehat{y}$, the second term will be zero. The third term does not depend on the choice of $\widehat{y}$ and the criterion will be minimized if the first term becomes null. This leads to (3.46).

The prediction given by the ARX structure can further be written out as scalar computations:

$$
\begin{aligned}
\widehat{y}(k+1) =\ & b_0 u(k-d+1) \\
& + b_1 u(k-d) + \ldots \\
& + b_{n_B} u(k-d+1-n_B) \\
& - a_1 y(k) \\
& - a_2 y(k-1) - \ldots \\
& - a_{n_A} y(k-n_A+1)
\end{aligned} \tag{3.49}
$$

Note, that the predictor can be written as a linear regression

$$
\widehat{y}(k+1) = \widehat{\boldsymbol{\theta}}^T \boldsymbol{\varphi}(k+1) \tag{3.50}
$$

where $\widehat{\boldsymbol{\theta}}^T = \left[\widehat{b}_0, ..., \widehat{b}_{n_B}, \widehat{a}_1, ...\widehat{a}_{n_A}\right]$ and $\boldsymbol{\varphi}(k+1) = [u(k-d+1), ..., y(k),$ $...]^T$, and the LS method can be used for estimating the parameters.

In general, the predictor can be written as

$$
\widehat{y}(k+1) = f(u(k-d+1), ..., y(k), ...) \tag{3.51}
$$

where f is a linear function of the process inputs and past process outputs. If f is a non-linear function, these models are referred to as NARX models. The prediction is a function of the process inputs and the past (real, measured) process outputs. This avoids the model to drift far from the true process in the case of modeling errors.

### 3.3.4 Output error

**Definition 8 (OE structure)** Output error (OE) structure is obtained by setting $C = D = 1$ in the general system (Definition 6):

$$
y(k) = \frac{B(q^{-1})}{A(q^{-1})} u(k-d) + e(k) \tag{3.52}
$$

In the OE system, the process output is disturbed by white noise only.

Let us calculate the output of such a system at the future sampling instant $k+1$ (one-step-ahead prediction); assume that $A$ and $B$ are known. We can rewrite the OE structure for $k+1$

$$
y(k+1) = \frac{B(q^{-1})}{A(q^{-1})} u(k-d+1) + e(k+1) \tag{3.53}
$$

The first term on the right side is a deterministic process, with $u(k - d + 1)$, $u(k - d - 2),...$ available. The second term is unknown (not available at instant $k$) but $\{e(k)\}$ is assumed to have zero mean. Thus, we get

$$\widehat{y}(k + 1) = \frac{B(q^{-1})}{A(q^{-1})}u(k - d + 1) \qquad (3.54)$$

where the hat in $\widehat{y}$ indicates that a prediction of $y$ is considered. It is easy to show that this predictor minimizes the expected squared prediction error[4].

**Algorithm 7 (OE 1-step-ahead predictor)** Predictor for an OE system (Definition 8) is given by

$$\widehat{y}(k + 1) = \frac{B(q^{-1})}{A(q^{-1})}u(k - d + 1) \qquad (3.55)$$

The predictor operates 'in parallel' with the process. Only a sequence of system inputs is required and the measured process outputs $y(k)$ are not needed.

The predictor can be written in a more explicit way as

$$\widehat{y}(k + 1) = B(q^{-1})u(k - d + 1) - A_1(q^{-1})\widehat{y}(k) \qquad (3.56)$$

where $A_1(q^{-1}) = a_1 + ... + a_{n_A}q^{-(n_A - 1)}$ (containing the model coefficients corresponding to past predictions), *i.e.* given by

$$A(q^{-1}) = 1 + q^{-1}A_1(q^{-1}) \qquad (3.57)$$

---

[4]Let us minimize the expected squared prediction error,

$$\widehat{y}(k + 1) = \arg\min_{\widehat{y}} E\left\{[y(k + 1) - \widehat{y}]^2\right\}$$

Substituting (3.52) to the above, we get

$$
\begin{aligned}
&E\left\{[y(k + 1) - \widehat{y}]^2\right\} \\
= \quad &E\left\{\left[\left(\frac{B(q^{-1})}{A(q^{-1})}u(k + 1 - d) - \widehat{y}\right)\right]^2\right\} \\
&+ 2E\left\{\left(\frac{B(q^{-1})}{A(q^{-1})}u(k + 1 - d) - \widehat{y}\right)e(k + 1)\right\} \\
&+ E\left\{e^2(k + 1)\right\}
\end{aligned}
$$

The second term will be zero. The third term does not depend on the choice of $\widehat{y}$. The criterion will be minimized if the first term becomes null. This leads to (3.54) [51].

The prediction given by the OE-structure can further be written out as scalar computations:

$$
\begin{aligned}
\widehat{y}(k+1) \;=\; & b_0 u\,(k-d+1) \\
& +b_1 u\,(k-d)+\dots \\
& +b_{n_B} u\,(k-d+1-n_B) \\
& -a_1 \widehat{y}\,(k) \\
& -a_2 \widehat{y}\,(k-1)-\dots \\
& -a_{n_A} \widehat{y}\,(k-n_A+1)
\end{aligned}
\tag{3.58}
$$

Note, that the prediction has the form

$$
\widehat{y}(k+1) = \mathrm{f}\left(u\,(k-d+1),\dots,\widehat{y}\,(k),\dots\right)
\tag{3.59}
$$

where f is a linear function (superposition) of the process inputs and past predictions. Nonlinear models are referred to as NOE models (non-linear output error). The prediction is a function of the past predicted outputs. Notice that the output measurement noise does not affect the prediction. Notice also that we can write the predictor as $\widehat{y}(k+1) = \widehat{\boldsymbol{\theta}}^T \boldsymbol{\varphi}(k)$; however, the $\widehat{y}$s in the regression vector are functions of the parameters $\widehat{\boldsymbol{\theta}}$ (see Section 3.3.8 how this affects the parameter estimation).

### 3.3.5 Other structures

A third important system structure is the ARMAX (autoregressive moving average exogenous) structure.

**Definition 9 (ARMAX structure)** The ARMAX structure is obtained by setting $D = A$ in the general structure (Definition 6):

$$
A\left(q^{-1}\right) y\,(k) = B\left(q^{-1}\right) u\,(k-d) + C\left(q^{-1}\right) e\,(k)
\tag{3.60}
$$

Let us again rewrite the system for $k+1$

$$
A\left(q^{-1}\right) y\,(k+1) = B\left(q^{-1}\right) u\,(k-d+1) + C\left(q^{-1}\right) e\,(k+1)
\tag{3.61}
$$

Defining $C_1$ $(C = 1 + q^{-1}C_1)$ and $A_1$ ($C$ and $A$ are monic), we can write

$$
\begin{aligned}
y\,(k+1) \;=\; & -A_1\left(q^{-1}\right) y\,(k) + B\left(q^{-1}\right) u\,(k-d+1) \\
& +e\,(k+1) + C_1\left(q^{-1}\right) e\,(k)
\end{aligned}
\tag{3.62}
$$

Taking into account that the random variable $e(k+1)$ will act on the process (system) in the future, we obtain an expression of the ARMAX predictor

$$\widehat{y}(k+1) = -A_1\left(q^{-1}\right)y(k) + B\left(q^{-1}\right)u(k-d+1) + C_1\left(q^{-1}\right)e(k) \quad (3.63)$$

which is the prediction minimizing the expected squared error[5].

In view of (3.62), it follows that the prediction error is equal to

$$e(k) = y(k) - \widehat{y}(k) \qquad (3.64)$$

The past noise terms can be calculated from data. Alternatively, we can obtain the expression from (3.60) for computing past noise terms

$$e(k) = \frac{A\left(q^{-1}\right)}{C\left(q^{-1}\right)}\left[y(k) - \frac{B\left(q^{-1}\right)}{A\left(q^{-1}\right)}u(k-d)\right] \qquad (3.65)$$

**Algorithm 8 (ARMAX predictor)** Predictor for an ARMAX structure (Definition 9) is given by:

$$\widehat{y}(k+1) = -A_1\left(q^{-1}\right)y(k) + B\left(q^{-1}\right)u(k-d+1) + C_1\left(q^{-1}\right)e(k) \quad (3.66)$$

where $e(k) = y(k) - \widehat{y}(k)$. It is a function of three terms: the output measurements, system inputs, and known errors.

Substituting (3.64) to (3.63) and reorganizing, we can see that the prediction can be written as

$$\begin{aligned}
\widehat{y}(k+1) &= \left(C_1\left(q^{-1}\right) - A_1\left(q^{-1}\right)\right)y(k) \\
&\quad + B\left(q^{-1}\right)u(k-d+1) \\
&\quad - C_1\left(q^{-1}\right)\widehat{y}(k)
\end{aligned} \qquad (3.67)$$

---

[5]Let us minimize the expected squared prediction error. Let the ARMAX system be given by (3.62). We have that

$$\begin{aligned}
J &= E\left\{[y(k+1) - \widehat{y}]^2\right\} \\
&= E\{[-A_1\left(q^{-1}\right)y(k) + B\left(q^{-1}\right)u(k-d+1) \\
&\quad + C_1\left(q^{-1}\right)e(k) + e(k+1) - \widehat{y}]^2\}
\end{aligned}$$

Reorganizing gives

$$\begin{aligned}
J &= E\left\{[-A_1\left(q^{-1}\right)y(k) + B\left(q^{-1}\right)u(k-d+1) + C_1\left(q^{-1}\right)e(k) - \widehat{y}]^2\right\} \\
&\quad + E\left\{[e(k+1)]^2\right\} \\
&\quad + 2E\{e(k+1) \times \\
&\quad [-A_1\left(q^{-1}\right)y(k) + B\left(q^{-1}\right)u(k-d+1) + C_1\left(q^{-1}\right)e(k) - \widehat{y}]\}
\end{aligned}$$

where the last term vanishes since $e(k+1)$ is independent of all previous observations. The minimum of $J$ is obtained at (3.63).

which has the form

$$\widehat{y}(k+1) = f(u(k-d+1),...,y(k),...,\widehat{y}(k),...) \qquad (3.68)$$

where f is a linear function of the process inputs, past process outputs, and past predictions; non-linear models are referred to as NARMAX models.

Another important form is obtained by rewriting the noise term

$$\xi(k+1) = \frac{C(q^{-1})}{A(q^{-1})}e(k+1) \qquad (3.69)$$

Using definitions for $C_1$, $A_1$

$$\xi(k+1) + A_1(k)\xi(k) = e(k+1) + C_1 e(k) \qquad (3.70)$$

and $\xi(k) = \frac{C}{A}e(k)$, from (3.69), we have

$$\xi(k+1) = e(k+1) + \left[C_1(q^{-1}) - A_1(q^{-1})\frac{C(q^{-1})}{A(q^{-1})}\right]e(k) \qquad (3.71)$$

From (3.65) we get an expression for the past noise terms. Substituting (3.65) to (3.71) and reorganizing we have for the noise term

$$\begin{aligned}\xi(k+1) = \ & e(k+1) \qquad (3.72)\\ & +\frac{C_1(q^{-1}) - A_1(q^{-1})}{C(q^{-1})}\left[y(k) - \frac{B(q^{-1})}{A(q^{-1})}u(k-d)\right]\end{aligned}$$

Substituting (3.72) for the noise term we obtain another expression for the ARMAX predictor.

**Algorithm 9 (ARMAX predictor: continued)** Predictor for an ARMAX structure (Definition 9) is given by:

$$\begin{aligned}\widehat{y}(k+1) = \ & \frac{B(q^{-1})}{A(q^{-1})}u(k+1-d) \qquad (3.73)\\ & +\frac{C_1(q^{-1}) - A_1(q^{-1})}{C(q^{-1})}\left[y(k) - \frac{B(q^{-1})}{A(q^{-1})}u(k-d)\right]\end{aligned}$$

Thus the ARMAX predictor can be seen as consisting of an OE predictor and a correction term.

**Example 17 (ARMA)** Let us consider the following stochastic process [2]

$$y(k) + ay(k-1) = e(k) + ce(k-1) \qquad (3.74)$$

where $\{e(k)\}$ is a sequence of equally distributed normal random variables with zero mean.

The process can be written as

$$y(k) = \frac{1 + cq^{-1}}{1 + aq^{-1}} e(k) \tag{3.75}$$

Consider the situation at sampling instant $k$ when $y(k), y(k-1), ...$ are observed and we want to determine $y(k+1)$. (3.75) gives

$$\begin{aligned} y(k+1) &= \frac{1 + cq^{-1}}{1 + aq^{-1}} e(k+1) \tag{3.76} \\ &= e(k+1) + \frac{c-a}{1 + aq^{-1}} e(k) \tag{3.77} \end{aligned}$$

The term $e(k+1)$ is independent of all observations. The last term is a linear combination of $e(k), e(k-1), ...$ to be computed from the available data:

$$e(k) = \frac{1 + aq^{-1}}{1 + cq^{-1}} y(k) \tag{3.78}$$

Eliminating $e(k)$ from (3.77), we obtain

$$y(k+1) = e(k+1) + \frac{c-a}{1 + cq^{-1}} y(k) \tag{3.79}$$

The problem now is to find the prediction $\hat{y}(k+1)$ of $y(k+1)$, based on the available data at instant $k$, such that the criterion

$$J = E\left\{\varepsilon^2(k+1)\right\} \tag{3.80}$$

is minimized, where $\varepsilon(k+1)$ is the prediction error

$$\varepsilon(k+1) = y(k+1) - \hat{y}(k+1) \tag{3.81}$$

Equations (3.79)-(3.81) lead to

$$\begin{aligned} J &= E\left\{e^2(k+1)\right\} \\ &+ E\left\{\left[\frac{c-a}{1 + cq^{-1}} y(k) - \hat{y}\right]^2\right\} \tag{3.82} \\ &+ 2E\left\{e(k+1)\left[\frac{c-a}{1 + cq^{-1}} y(k) - \hat{y}\right]\right\} \end{aligned}$$

As $e(k+1)$ is independent of the observations available at instant $k$, it follows that the last term vanishes. Hence, we can write

$$J = E\left\{\varepsilon^2(k+1)\right\} \geq E\left\{e^2(k+1)\right\} \tag{3.83}$$

where the equality is obtained for

$$\hat{y} = \hat{y}(k+1) = \frac{c-a}{1+cq^{-1}}y(k) \tag{3.84}$$

The prediction error is given by $\varepsilon(k+1) = e(k+1)$.

**Example 18 (ARMA: continued)** Let us obtain the same result using Algorithm 9. From the system given by (3.75) we get

$$
\begin{align}
C\left(q^{-1}\right) &= 1+cq^{-1} \tag{3.85} \\
B\left(q^{-1}\right) &= 0 \tag{3.86} \\
A\left(q^{-1}\right) &= 1+aq^{-1} \tag{3.87}
\end{align}
$$

Using Algorithm 8 we get

$$\hat{y}(k+1) = \frac{C_1\left(q^{-1}\right) - A_1\left(q^{-1}\right)}{C\left(q^{-1}\right)}y(k)$$

Substituting $C$, $C_1 = c$ and $A_1 = a$ gives

$$\hat{y}(k+1) = \frac{c-a}{1+cq^{-1}}y(k) \tag{3.88}$$

**Definition 10 (ARIMAX structure)** ARIMAX (autoregressive integral moving average exogenous) structure is obtained by setting $C = 1$, $D = \Delta A$ in the general structure (Definition 6):

$$y(k) = \frac{B\left(q^{-1}\right)}{A\left(q^{-1}\right)}u(k-d) + \frac{C\left(q^{-1}\right)}{\Delta A\left(q^{-1}\right)}e(k) \tag{3.89}$$

where $\Delta = 1 - q^{-1}$.

Multiplying (3.89) by $\Delta A$, reorganizing, and shifting to $k+1$ gives

$$\Delta A\left(q^{-1}\right)y(k+1) = B\left(q^{-1}\right)\Delta u(k-d+1) + C\left(q^{-1}\right)e(k+1) \tag{3.90}$$

The ARIMAX system can be seen as an ARMAX process, where $A\left(q^{-1}\right) \longleftarrow \Delta A\left(q^{-1}\right)$ and $u(k) \longleftarrow \Delta u(k)$. Then, using Algorithm 8, we have the predictor.

**Algorithm 10 (ARIMAX predictor)** The predictor for an ARIMAX system is given by

$$
\begin{aligned}
\widehat{y}\left(k+1\right) \;=\; & -\left[\Delta A\right]_1\left(q^{-1}\right)y\left(k\right)+B\left(q^{-1}\right)\Delta u\left(k-d+1\right) \quad (3.91)\\
& +C_1\left(q^{-1}\right)e\left(k\right)
\end{aligned}
$$

where $e\left(k\right)=y\left(k\right)-\widehat{y}\left(k\right)$ and $\left[\Delta A\right]=1+q^{-1}\left[\Delta A\right]_1$. In the ARIMAX process, the noise (filtered by $C$) is integrated to the process output, which makes it possible to model disturbances of random walk type.

The ARIMAX model (also referred to as the CARIMA model) is used in the Generalized Predictive Control (GPC). Due to the integral term present in the noise model, an additional integral-of-error term is not needed in the controller.

### 3.3.6   Diophantine equation

Prediction is intimately related to the separation procedure of available and unavailable data. This separation procedure is performed by Diophantine equation which will be presented next.

The Diophantine equation

$$
\frac{X\left(q^{-1}\right)}{Y\left(q^{-1}\right)}=E_i\left(q^{-1}\right)+q^{-i}\frac{F_i\left(q^{-1}\right)}{Y\left(q^{-1}\right)} \tag{3.92}
$$

is used for separating a transfer operator into future and known parts (available and unavailable information). The solution to this equation will be needed in the next sections. Equation (3.92) can be solved in a recursive fashion, so that polynomials $E_{i+1}$ and $F_{i+1}$ are obtained given the values of $E_i$ and $F_i$. In the following, this recursive solution will be derived.

Let us assume that $Y$ is monic. Hence, the polynomials are given by

$$
\begin{aligned}
Y\left(q^{-1}\right) &= 1+y_1q^{-1}+...+y_{n_Y}q^{-n_Y} & (3.93)\\
X\left(q^{-1}\right) &= x_0+x_1q^{-1}+...+x_{n_Y}q^{-n_X} & (3.94)\\
E_i\left(q^{-1}\right) &= e_{i,0}+e_{i,1}q^{-1}+...+e_{i,n_{E_i}}q^{-n_{E_i}} & (3.95)\\
F_i\left(q^{-1}\right) &= f_{i,0}+f_{i,1}q^{-1}+...+f_{i,n_{F_i}}q^{-n_{F_i}} & (3.96)
\end{aligned}
$$

Consider two Diophantine equations

$$
\begin{aligned}
X\left(q^{-1}\right) &= Y\left(q^{-1}\right)E_{i+1}\left(q^{-1}\right)+q^{-(i+1)}F_{i+1}\left(q^{-1}\right) & (3.97)\\
X\left(q^{-1}\right) &= Y\left(q^{-1}\right)E_i\left(q^{-1}\right)+q^{-i}F_i\left(q^{-1}\right) & (3.98)
\end{aligned}
$$

Subtracting (3.97) from (3.98) yields

$$0 = Y\left(q^{-1}\right)\left[E_{i+1}\left(q^{-1}\right) - E_i\left(q^{-1}\right)\right] \tag{3.99}$$
$$+q^{-i}\left[q^{-1}F_{i+1}\left(q^{-1}\right) - F_i\left(q^{-1}\right)\right]$$

The polynomial $E_{i+1} - E_i$ can be split into two parts (by simply taking out one element)

$$E_{i+1}\left(q^{-1}\right) - E_i\left(q^{-1}\right) = \tilde{E}\left(q^{-1}\right) + e_{i+1,i}q^{-i} \tag{3.100}$$

Substituting (3.100) into (3.99) gives

$$0 = Y\left(q^{-1}\right)\left[\tilde{E}\left(q^{-1}\right) + e_{i+1,i}q^{-i}\right] \tag{3.101}$$
$$+q^{-i}\left[q^{-1}F_{i+1}\left(q^{-1}\right) - F_i\left(q^{-1}\right)\right]$$
$$= Y\left(q^{-1}\right)\tilde{E}\left(q^{-1}\right) \tag{3.102}$$
$$+q^{-i}\left[q^{-1}F_{i+1}\left(q^{-1}\right) - F_i\left(q^{-1}\right) + Y\left(q^{-1}\right)e_{i+1,i}\right]$$

Hence, it follows that

$$\tilde{E}\left(q^{-1}\right) = 0 \tag{3.103}$$

and

$$q^{-1}F_{i+1}\left(q^{-1}\right) - F_i\left(q^{-1}\right) + Y\left(q^{-1}\right)e_{i+1,i} = 0 \tag{3.104}$$

In order to derive the coefficients of the polynomial $q^{-1}F_{i+1}$, let us rewrite this equation into the following form:

$$q^{-1}\left[f_{i+1,0} + f_{i+1,1}q^{-1} + \dots + f_{i+1,n_{F_{i+1}}}q^{-n_{F_{i+1}}}\right] \tag{3.105}$$
$$-\left[f_{i,0} + f_{i,1}q^{-1} + \dots + f_{i,n_{F_i}}q^{-n_{F_i}}\right]$$
$$+\left[1 + y_1q^{-1} + \dots + y_{n_Y}q^{-n_Y}\right]e_{i+1,i}$$
$$= 0$$

Finally, we obtain

$$\begin{aligned} e_{i+1,i} &= f_{i,0} \\ f_{i+1,0} &= f_{i,1} - y_1e_{i+1,i} \\ f_{i+1,1} &= f_{i,2} - y_2e_{i+1,i} \\ &\vdots \end{aligned} \tag{3.106}$$

$$f_{i+1,j} = f_{i,j+1} - y_{j+1}e_{i+1,i} \tag{3.107}$$

$$\vdots$$

where (3.107) is for $j = 1, 2, ...$ Thus, a recursive formula for computing $F_{i+1}$ was obtained. Using (3.100) and (3.103), we also obtain a recursive formula for $E_{i+1}$

$$E_{i+1}\left(q^{-1}\right) = E_i\left(q^{-1}\right) + e_{i+1,i}q^{-i} \tag{3.108}$$

Now all that is needed are the initial values $E_1$ and $F_1$ for the recursive formula. Setting $i = 1$ in (3.92) gives

$$\frac{X\left(q^{-1}\right)}{Y\left(q^{-1}\right)} = E_1\left(q^{-1}\right) + q^{-1}\frac{F_1\left(q^{-1}\right)}{Y\left(q^{-1}\right)} \tag{3.109}$$

$$X\left(q^{-1}\right) = E_1\left(q^{-1}\right)Y\left(q^{-1}\right) + q^{-1}F_1\left(q^{-1}\right) \tag{3.110}$$

Since $Y$ is monic, we get

$$E_1\left(q^{-1}\right) = x_0 \tag{3.111}$$

and substituting (3.111) into (3.110) gives

$$F_1\left(q^{-1}\right) = q\left[X\left(q^{-1}\right) - x_0 Y\left(q^{-1}\right)\right] \tag{3.112}$$

The Diophantine equation (3.92) for (3.93)-(3.96) can thus be solved starting from the initial values $E_1$ and $F_1$ given by (3.111) and (3.112). The solutions $E_i$ and $F_i$, $i = 2, 3, ...$ , are then obtained recursively using (3.106), (3.107), and (3.108) using $i = 1, 2, 3, ...$

**Algorithm 11 (Solution of the Diophantine equation)** The solution of the Diophantine equation

$$\frac{X\left(q^{-1}\right)}{Y\left(q^{-1}\right)} = E_i\left(q^{-1}\right) + q^{-i}\frac{F_i\left(q^{-1}\right)}{Y\left(q^{-1}\right)} \tag{3.113}$$

where

$$Y\left(q^{-1}\right) = 1 + y_1 q^{-1} + ... + y_{n_Y} q^{-n_Y} \tag{3.114}$$
$$X\left(q^{-1}\right) = x_0 + x_1 q^{-1} + ... + x_{n_X} q^{-n_X} \tag{3.115}$$
$$E_i\left(q^{-1}\right) = e_{i,0} + e_{i,1} q^{-1} + ... + e_{i,n_{E_i}} q^{-n_{E_i}} \tag{3.116}$$
$$F_i\left(q^{-1}\right) = f_{i,0} + f_{i,1} q^{-1} + ... + f_{i,n_{F_i}} q^{-n_{F_i}} \tag{3.117}$$

$n_Y > 0$, can be computed recursively using

$$E_1\left(q^{-1}\right) = x_0 \tag{3.118}$$
$$F_1\left(q^{-1}\right) = q\left[X\left(q^{-1}\right) - x_0 Y\left(q^{-1}\right)\right] \tag{3.119}$$

and for $i = 1, 2, \ldots$ and $j = 0, 1, \ldots, \max(n_X - i, n_Y - 1) - 1$

$$
\begin{align}
e_{i+1,i} &= f_{i,0} \tag{3.120} \\
f_{i+1,j} &= f_{i,j+1} - y_{j+1} e_{i+1,i} \tag{3.121} \\
E_{i+1}\left(q^{-1}\right) &= E_i\left(q^{-1}\right) + e_{i+1,i} q^{-i} \tag{3.122}
\end{align}
$$

The degrees of the polynomials are given by

$$
\begin{align}
n_{E_i} &= i - 1 \tag{3.123} \\
n_{F_i} &= \max(n_X - i, n_Y - 1) \tag{3.124}
\end{align}
$$

## 3.3.7 *i*-step-ahead predictions

Let us consider a Box–Jenkins structure (Definition 6)

$$
y(k) = \frac{B\left(q^{-1}\right)}{A\left(q^{-1}\right)} u(k - d) + \frac{C\left(q^{-1}\right)}{D\left(q^{-1}\right)} e(k) \tag{3.125}
$$

where the disturbance is given by

$$
\xi(k) = \frac{C\left(q^{-1}\right)}{D\left(q^{-1}\right)} e(k) \tag{3.126}
$$

and let us calculate a 'one-step' algorithm for obtaining $i$-step ahead predictions (see [88]). Thus, we wish to have a prediction $\widehat{y}(k + i)$ for the plant output $y(k + i)$, provided with information up to instant $k$: $y(k), y(k-1), \ldots$; $u(k), u(k-1), \ldots$ and $\widehat{y}(k), \widehat{y}(k-1), \ldots$. Observe that the future output values

$$
y(k + i) = \frac{B\left(q^{-1}\right)}{A\left(q^{-1}\right)} u(k + i - d) + \frac{C\left(q^{-1}\right)}{D\left(q^{-1}\right)} e(k + i) \tag{3.127}
$$

can only be predicted with uncertainty since the future noise terms $e(k + 1)$, $e(k + 2), \ldots, e(k + i)$ are unknown. The minimization of such an uncertainty is the objective of the predictor design problem. This is a crucial issue in the predictive control, to be discussed in later chapters.

### Separation of disturbance

Let us start by separating unknown terms (future) and known terms by introducing the Diophantine equation for the disturbance process

$$
\frac{C\left(q^{-1}\right)}{D\left(q^{-1}\right)} = E_i\left(q^{-1}\right) + q^{-i} \frac{F_i\left(q^{-1}\right)}{D\left(q^{-1}\right)} \tag{3.128}
$$

where

$$
\deg E_i \left( q^{-1} \right) = \left\{ \begin{array}{ll} i - 1 & \text{if } n_D > 0 \\ \min \left( i - 1, n_C \right) & \text{otherwise} \end{array} \right. \tag{3.129}
$$

$$
\deg F_i \left( q^{-1} \right) = \max \left( n_C - i, n_D - 1 \right) \tag{3.130}
$$

The disturbance at $k + i$ can be decomposed into unknown (future) and known (current and past) parts

$$
\xi \left( k + i \right) = E_i \left( q^{-1} \right) e \left( k + i \right) + \frac{F_i \left( q^{-1} \right)}{D \left( q^{-1} \right)} e \left( k \right) \tag{3.131}
$$

The polynomials $E_i$ and $F_i$ are usually solved recursively (see 3.3.6). Assume that the solutions $E_i$ and $F_i$ are available. The second term on the right side can be computed by multiplying (3.125) by $\frac{F_i}{C}$

$$
\frac{F_i \left( q^{-1} \right)}{C \left( q^{-1} \right)} y \left( k \right) = \frac{F_i \left( q^{-1} \right)}{C \left( q^{-1} \right)} \frac{B \left( q^{-1} \right)}{A \left( q^{-1} \right)} u \left( k - d \right) + \frac{F_i \left( q^{-1} \right)}{C \left( q^{-1} \right)} \frac{C \left( q^{-1} \right)}{D \left( q^{-1} \right)} e \left( k \right) \tag{3.132}
$$

and rearranging

$$
\frac{F_i \left( q^{-1} \right)}{D \left( q^{-1} \right)} e \left( k \right) = \frac{F_i \left( q^{-1} \right)}{C \left( q^{-1} \right)} \left[ y \left( k \right) - \frac{B \left( q^{-1} \right)}{A \left( q^{-1} \right)} u \left( k - d \right) \right] \tag{3.133}
$$

The process output $i$-steps ahead then becomes

$$
\begin{aligned}
y \left( k + i \right) &= q^{-d} \frac{B \left( q^{-1} \right)}{A \left( q^{-1} \right)} u \left( k + i \right) \\
&\quad + \frac{F_i \left( q^{-1} \right)}{C \left( q^{-1} \right)} \left[ y \left( k \right) - \frac{B \left( q^{-1} \right)}{A \left( q^{-1} \right)} u \left( k - d \right) \right] \\
&\quad + E_i \left( q^{-1} \right) e \left( k + i \right)
\end{aligned} \tag{3.134}
$$

The third term depends on future noise terms $e \left( k + i \right)$, which are unknown. However, $\{ e \left( k \right) \}$ was assumed to have zero mean, and we can take the conditional expectation of $y \left( k + i \right)$, given all data up to $k$ and the future process inputs. The best $i$-step ahead predictor (in the sense that the variance of the prediction error is minimal) then becomes

$$
\begin{aligned}
\hat{y} \left( k + i \right) &= q^{-d} \frac{B \left( q^{-1} \right)}{A \left( q^{-1} \right)} u \left( k + i \right) \\
&\quad + \frac{F_i \left( q^{-1} \right)}{C \left( q^{-1} \right)} \left[ y \left( k \right) - \frac{B \left( q^{-1} \right)}{A \left( q^{-1} \right)} u \left( k - d \right) \right]
\end{aligned} \tag{3.135}
$$

Notice, that (3.135) represents the $i$-step-ahead prediction as a function of system inputs and prediction errors.

The prediction error for the $i$'th predictor is given by

$$\varepsilon\left(k+i\right) = y\left(k+i\right) - \widehat{y}\left(k+i\right) = E_i\left(q^{-1}\right) e\left(k+i\right) \qquad (3.136)$$

which consists of future noise only (white noise with zero mean and variance $\sigma^2$). The variance is given by

$$E\left[\varepsilon\left(k+i\right)^2\right] = E\left[\left(E_i\left(q^{-1}\right) e\left(k+i\right)\right)^2\right] = \sigma^2 \sum_{j=0}^{n_{E_i}} e_{i,j} \qquad (3.137)$$

where $e_{i,j}$ is the $j$'th element of $E_i$. Thus, the variance of the prediction error is minimal.

Let us continue a bit further and write (3.135) strictly as a function of system inputs and past outputs. Multiplying both sides of the Diophantine (3.128) with $BD/AC$ we obtain

$$\frac{B\left(q^{-1}\right)}{A\left(q^{-1}\right)} = \frac{B\left(q^{-1}\right) D\left(q^{-1}\right) E_i\left(q^{-1}\right)}{A\left(q^{-1}\right) C\left(q^{-1}\right)} + q^{-i}\frac{B\left(q^{-1}\right) F_i\left(q^{-1}\right)}{A\left(q^{-1}\right) C\left(q^{-1}\right)} \qquad (3.138)$$

which with (3.135) yields:

$$\begin{aligned}
\widehat{y}\left(k+i\right) &= q^{-d}\left[\frac{B\left(q^{-1}\right) D\left(q^{-1}\right) E_i\left(q^{-1}\right)}{A\left(q^{-1}\right) C\left(q^{-1}\right)} + q^{-i}\frac{B\left(q^{-1}\right) F_i\left(q^{-1}\right)}{A\left(q^{-1}\right) C\left(q^{-1}\right)}\right] u\left(k+i\right) \\
&\quad + \frac{F_i\left(q^{-1}\right)}{C\left(q^{-1}\right)}\left[y\left(k\right) - \frac{B\left(q^{-1}\right)}{A\left(q^{-1}\right)}u\left(k-d\right)\right]
\end{aligned} \qquad (3.139)$$

Simple algebraic calculations lead to the following $i$-step-ahead predictor.

**Algorithm 12 ($i$-step-ahead BJ predictor)** The $i$-step-ahead predictor for a Box–Jenkins system (Definition 6) is given by

$$\widehat{y}\left(k+i\right) = q^{-d}\frac{B\left(q^{-1}\right) D\left(q^{-1}\right) E_i\left(q^{-1}\right)}{A\left(q^{-1}\right) C\left(q^{-1}\right)}u\left(k+i\right) + \frac{F_i\left(q^{-1}\right)}{C\left(q^{-1}\right)}y\left(k\right) \qquad (3.140)$$

where $E_i$ and $F_i$ are obtained from the Diophantine equation

$$\frac{C\left(q^{-1}\right)}{D\left(q^{-1}\right)} = E_i\left(q^{-1}\right) + q^{-i}\frac{F_i\left(q^{-1}\right)}{D\left(q^{-1}\right)} \qquad (3.141)$$

**Example 19 (1-step-ahead OE predictor)** Let us derive the one-step-ahead predictor for an OE system (Definition 8).

The Diophantine equation becomes

$$1 = E_1 \left( q^{-1} \right) + q^{-1} F_1 \left( q^{-1} \right) \tag{3.142}$$

for which the solution is

$$E_1 \left( q^{-1} \right) = 1 \tag{3.143}$$
$$F_1 \left( q^{-1} \right) = 0 \tag{3.144}$$

The predictor becomes

$$\widehat{y} \left( k + 1 \right) = \frac{B \left( q^{-1} \right)}{A \left( q^{-1} \right)} u \left( k - d + 1 \right) \tag{3.145}$$

If $A$ is a factor of $D$, numerical problems may occur (notice that in the ARX and ARMAX structures $D = A$, in the ARIMAX $D = \Delta A$.) To avoid these problems, let us rewrite the algorithm for this particular case.

**Algorithm 13 ($i$-step-ahead BJ predictor: continued)** If $A$ is a factor of $D$, denote

$$D \left( q^{-1} \right) = D_1 \left( q^{-1} \right) A \left( q^{-1} \right) \tag{3.146}$$

The $i$-step-ahead predictor for a Box–Jenkins system (Definition 6) is then given by

$$\widehat{y} \left( k + i \right) = q^{-d} \frac{B \left( q^{-1} \right) D_1 \left( q^{-1} \right) E_i \left( q^{-1} \right)}{C \left( q^{-1} \right)} u \left( k + i \right) + \frac{F_i \left( q^{-1} \right)}{C \left( q^{-1} \right)} y \left( k \right) \tag{3.147}$$

where $E_i$ and $F_i$ are obtained from the Diophantine equation

$$\frac{C \left( q^{-1} \right)}{D \left( q^{-1} \right)} = E_i \left( q^{-1} \right) + q^{-i} \frac{F_i \left( q^{-1} \right)}{D \left( q^{-1} \right)} \tag{3.148}$$

**Example 20 (1-step-ahead ARX predictor)** Let us derive a one-step-ahead predictor for an ARX system (Definition 7).

Since $A = D$, $D_1 = 1$. The Diophantine equation becomes

$$\frac{1}{A \left( q^{-1} \right)} = E_1 \left( q^{-1} \right) + q^{-1} \frac{F_1 \left( q^{-1} \right)}{A \left( q^{-1} \right)} \tag{3.149}$$

The solution for the Diophantine is given by

$$E_1 \left( q^{-1} \right) = 1 \tag{3.150}$$
$$F_1 \left( q^{-1} \right) = q \left[ 1 - A \left( q^{-1} \right) \right] = -A_1 \left( q^{-1} \right) \tag{3.151}$$

The predictor becomes

$$\widehat{y} \left( k + 1 \right) = B \left( q^{-1} \right) u \left( k - d + i \right) - A_1 \left( q^{-1} \right) y \left( k \right) \tag{3.152}$$

**Separation of inputs**

In control, the future process inputs are of interest (they are to be determined by the controller). The future and known signals in (3.140) can be further separated into future and known parts using a Diophantine equation:

$$\frac{B\left(q^{-1}\right)D\left(q^{-1}\right)E_i\left(q^{-1}\right)}{A\left(q^{-1}\right)C\left(q^{-1}\right)} = G_i\left(q^{-1}\right) + q^{-i+d}\frac{H_i\left(q^{-1}\right)}{A\left(q^{-1}\right)C\left(q^{-1}\right)} \tag{3.153}$$

which gives the algorithm for the $i$-step ahead prediction.

**Algorithm 14 ($i$-step-ahead BJ predictor: continued)** Using a BJ model with separated available and unavailable information, the $i$-step ahead prediction is given by

$$\hat{y}\left(k+i\right) = G_i\left(q^{-1}\right)u\left(k-d+i\right) \tag{3.154}$$
$$+\frac{H_i\left(q^{-1}\right)}{A\left(q^{-1}\right)C\left(q^{-1}\right)}u\left(k\right)$$
$$+\frac{F_i\left(q^{-1}\right)}{C\left(q^{-1}\right)}y\left(k\right)$$

where $E_i$ and $F_i$ are obtained from the Diophantine equation

$$\frac{C\left(q^{-1}\right)}{D\left(q^{-1}\right)} = E_i\left(q^{-1}\right) + q^{-i}\frac{F_i\left(q^{-1}\right)}{D\left(q^{-1}\right)} \tag{3.155}$$

and $H_i$ and $G_i$ are obtained from the Diophantine equation

$$\frac{B\left(q^{-1}\right)D\left(q^{-1}\right)E_i\left(q^{-1}\right)}{A\left(q^{-1}\right)C\left(q^{-1}\right)} = G_i\left(q^{-1}\right) + q^{-i+d}\frac{H_i\left(q^{-1}\right)}{A\left(q^{-1}\right)C\left(q^{-1}\right)} \tag{3.156}$$

Finally, let us give the corresponding algorithm for the case of having $A$ as a factor of $D$.

**Algorithm 15 ($i$-step-ahead BJ predictor: continued)** Consider a BJ model with separated available and unavailable information and where $A$ is a factor of $D$. Denote

$$D\left(q^{-1}\right) = D_1\left(q^{-1}\right)A\left(q^{-1}\right) \tag{3.157}$$

The $i$-step-ahead predictor for a Box–Jenkins system (Definition 6) is then given by

$$\hat{y}\left(k+i\right) = G_i\left(q^{-1}\right)u\left(k-d+i\right) \tag{3.158}$$
$$+\frac{H_i\left(q^{-1}\right)}{C\left(q^{-1}\right)}u\left(k\right)$$
$$+\frac{F_i\left(q^{-1}\right)}{C\left(q^{-1}\right)}y\left(k\right)$$

where $E_i$ and $F_i$ are obtained from the Diophantine equation

$$\frac{C\left(q^{-1}\right)}{D\left(q^{-1}\right)} = E_i\left(q^{-1}\right) + q^{-i}\frac{F_i\left(q^{-1}\right)}{D\left(q^{-1}\right)} \tag{3.159}$$

and $H_i$ and $G_i$ are obtained from the Diophantine equation

$$\frac{B\left(q^{-1}\right)D_1\left(q^{-1}\right)E_i\left(q^{-1}\right)}{C\left(q^{-1}\right)} = G_i\left(q^{-1}\right) + q^{-i+d}\frac{H_i\left(q^{-1}\right)}{C\left(q^{-1}\right)} \tag{3.160}$$

## 3.3.8   Remarks

Let us conclude this chapter by making a few remarks concerning the practical use of the stochastic time-series models.

### Incremental estimation

In practice, differencing of data is often preferred, *i.e.* working with signals $\Delta y(k)$ and $\Delta u(k)$, where $\Delta = 1 - q^{-1}$. However, differencing data with high frequency noise components degrades the signal-to-noise ratio. It is possible (simple solution) to overcome this with appropriate signal filtering.

### Gradients

The estimation of the parameters in the polynomials $A$ and $B$ of the process model is usually based on gradient-based techniques. For the ARX structure, the predictor is given by

$$\widehat{y}(k+1) = B\left(q^{-1}\right)u(k-d+1) - A_1\left(q^{-1}\right)y(k) \tag{3.161}$$

Since the inputs are independent of the predictor parameters, LS, RLS, *etc.* (see Chapter 2) can be used.

In the OE structure (as well as ARMAX, *etc.*), the predictor output depends on the past predictions and the regression vector is thus a function of the parameters themselves. In order to estimate the parameters, alternative methods must be used. Following chapters will present the prediction error methods (non-linear LS methods), for which the gradients of the predictor output with respect to the parameters are required.

The OE predictor is given by

$$\widehat{y}(k+1) = B\left(q^{-1}\right)u(k-d+1) - A_1\left(q^{-1}\right)\widehat{y}(k) \tag{3.162}$$

The gradients with respect to the parameters in the feed forward part $B$ are given by

$$\frac{\partial \widehat{y}(k)}{\partial b_n} = u(k - d - n) - \sum_{m=1}^{n_A} a_m \frac{\partial \widehat{y}(k - m)}{\partial b_n} \qquad (3.163)$$

where $n = 0, 1, ..., n_B$; and with respect to parameters in the feedback part $A$

$$\frac{\partial \widehat{y}(k)}{\partial a_n} = -\widehat{y}(k - n) - \sum_{m=1}^{n_A} a_m \frac{\partial \widehat{y}(k - m)}{\partial a_n} \qquad (3.164)$$

where $n = 1, 2, ..., n_A$.[6]

---

[6]The gradients with respect to the parameters in the feedforward part $B$ are given by

$$\frac{\partial \widehat{y}(k)}{\partial a_n} = \frac{\partial}{\partial a_n} B\left(q^{-1}\right) u(k - d) - \frac{\partial}{\partial a_n} A_1\left(q^{-1}\right) \widehat{y}(k - 1)$$

The first term on the right hand side does not depend on $a_n$. For the second term, since $A_1 = a_1 + a_2 q^{-1} + ... + a_{n_A} q^{-(n_A - 1)}$, we can write

$$\begin{aligned}
\frac{\partial \widehat{y}(k)}{\partial a_n} &= -\frac{\partial}{\partial a_n} a_1 \widehat{y}(k - 1) - ... - \frac{\partial}{\partial a_n} a_n \widehat{y}(k - n) - ... - \frac{\partial}{\partial a_n} a_{n_A} \widehat{y}(k - n_A) \\
&= -a_1 \frac{\partial \widehat{y}(k - 1)}{\partial a_n} - ... - \left[ a_n \frac{\partial \widehat{y}(k - n)}{\partial a_n} + \widehat{y}(k - n) \right] - ... - a_{n_A} \frac{\partial \widehat{y}(k - n_A)}{\partial a_n}
\end{aligned}$$

which can be written as (3.164). Similarly, the gradient with respect to parameters in $B$ are given by

$$\frac{\partial \widehat{y}(k)}{\partial b_n} = \frac{\partial}{\partial b_n} B\left(q^{-1}\right) u(k - d) - \frac{\partial}{\partial b_n} A_1\left(q^{-1}\right) \widehat{y}(k - 1)$$

The first term on the right hand side gives

$$\begin{aligned}
\frac{\partial}{\partial b_n} B\left(q^{-1}\right) u(k - d) &= \frac{\partial}{\partial b_n} b_0 u(k - d) + ... + \frac{\partial}{\partial b_n} b_n u(k - d - n) + ... \\
&\quad + \frac{\partial}{\partial b_n} b_{n_B} u(k - d - n_B) \\
&= u(k - d - n)
\end{aligned}$$

and the second term gives

$$-\frac{\partial}{\partial b_n} A_1\left(q^{-1}\right) \widehat{y}(k - 1) = -\sum_{m=1}^{n_A} a_m \frac{\partial \widehat{y}(k - m)}{\partial b_n}$$

Combining these, we have (3.163).

Assuming that the parameters change slowly during the estimation[7], the past gradients can be stored and the computations performed in a recursive fashion. Let us collect the results in this more convenient form.

**Algorithm 16 (Gradients of the OE predictor)** The derivatives of the output of the OE predictor with respect to the parameters in $A$ and $B$ are given by

$$\frac{\partial \widehat{y}}{\partial b_n}(k) \; \leftarrow \; \Psi_n^{\mathrm{b}}(k) \, ; \frac{\partial \widehat{y}}{\partial a_n}(k) \leftarrow \Psi_n^{\mathrm{a}}(k) \tag{3.165}$$

$$\Psi_n^{\mathrm{b}}(k) \;\; = \;\; u\left(k - d - n\right) - \sum_{m=1}^{n_A} a_m \Psi_n^{\mathrm{b}}\left(k - m\right) \tag{3.166}$$

$$\Psi_n^{\mathrm{a}}(k) \;\; = \;\; -\widehat{y}\left(k - n\right) - \sum_{m=1}^{n_A} a_m \Psi_n^{\mathrm{a}}\left(k - m\right) \tag{3.167}$$

where $n = 0, 1, ..., n_B$ and $n = 1, 2, ..., n_A$, respectively.

Notice that the system needs to be stable, since otherwise the gradients will grow (unbounded).

**Estimation of noise polynomials**

In the system model

$$y(k) = \frac{B(q^{-1})}{A(q^{-1})} u(k - d) + \frac{C(q^{-1})}{D(q^{-1})} e(k) \tag{3.168}$$

only the process dynamics, $B$ and $A$, are usually identified. $D$ is a design parameter, the selection of which results in the OE structure, ARX structure, *etc.* Estimating $C$ is generally difficult in practice, because of the nature of $C$ and the fact that $e(k)$ is never available and must be approximated by *a priori* or *a posteriori* prediction errors, thus reducing the convergence rate of the parameters.

For estimating $C$, a simple solution is to filter the data (using the prior information about the process noise) with a low pass filter, $F$, thus removing high frequency components of the signals. It is then possible to use a fixed estimate of $C$ (often denoted by $T$), representing prior knowledge about the process noise. One interpretation of $T$ is that of a fixed observer. In the estimation, *e.g.*, the RLS can be used.

---

[7]The assumptions are that $\frac{\partial \widehat{y}}{\partial b_n}(k - m)|_{\theta=\theta(k)} \approx \frac{\partial \widehat{y}}{\partial b_n}(k - m)|_{\theta=\theta(k-m)}$ and $\frac{\partial \widehat{y}}{\partial a_n}(k - m)|_{\theta=\theta(k)} \approx \frac{\partial \widehat{y}}{\partial a_n}(k - m)|_{\theta=\theta(k-m)}$, where $\theta$ contains the time-varying components of the model.

# Chapter 4

# Non-linear Systems

Identification can be justified by the reduced time and effort required in building the models, and the flexibility of parameterized experimental models in real-world modeling problems. For simple input-output relations, linear models are a relatively robust alternative. Linear models are simple and efficient also when extending to the identification of adaptive and/or dynamic models, and readily available control design methods can be found from the literature. However, most industrial processes are non-linear.

If the non-linear characteristics of the process are known, a seemingly non-linear identification problem may often be converted to a linear identification problem. Using the available *a priori* knowledge of the non-linearities, the model input–output data can be pre-processed, or the model re-parameterized. This is in fact what is often done in gray-box modeling. As the processes become more complex, a sufficiently accurate non-linear input–output behavior is more difficult to obtain using linear descriptions. If more detailed models are required, then the engineer needs to turn to methods of identification of non-linear systems.

Many types of model structures have been considered for the identification of non-linear systems. Traditionally, model structures with constrained non-linearities have been considered (see, *e.g.*, [78]). Lately, a number of new structures have been proposed (see, *e.g.*, [86]) and shown to be useful in applications. Particular interest has been focused on fields such as neural computation [29][27] and fuzzy systems [47] [73]. These fields, among many other topics, are a part of the field of artificial intelligence.

In this chapter, a brief introduction to some basic topics in the identification of non-linear systems is given. The target of this chapter is to provide the reader with a basic understanding and overview of some common parameterized (black-box) structures used for approximating non-linear functions. In particular, the basis function networks are introduced. They provide a

general framework for most non-linear model structures, which should help
the reader in understanding and clustering the multitude of different specific
paradigms, structures and methods available. The power series, one-hidden-
layer sigmoid neural networks and 0–order Sugeno fuzzy models are considered in detail, including linearization of the mappings and the computation
of gradients.

# 4.1    Basis function networks

In this section, the basis function networks [86] are introduced. They provide
a general framework for most non-linear model structures.

## 4.1.1    Generalized basis function network

Most non-linear model structures can be presented as decomposed into two
parts:

- a mapping $\varphi$ from the input space to regressors; and

- a mapping f from regressors to model output.

The selection of regressors $\varphi$ is mainly based on utilizing physical insight
to the problem. Obviously, all the necessary input signals should be included. Some transformation (pre-processing, filtering) of the raw measurements could also be used in order to facilitate the estimation of the parameters. In dynamic time-series modeling, the 'orders' of the system (number of
past inputs, outputs and predictions) need to be chosen. Such semi-physical
regressors are formed in view of what is known about the system. In the
remaining sections, we will be interested in the mapping f.

The non-linear mapping f can be viewed as function expansions [86]. In
a generalized basis function network [31], the mapping f is formed by

$$\widehat{y}(k) = f(\varphi(k), \cdot) = \sum_{h=1}^{H} h_h(\varphi(k), \cdot) g_h(\varphi(k), \cdot) \qquad (4.1)$$

where $g_h$ are the *basis functions* and $h_h$ are *weighting functions*, $h = 1, 2, ..., H$.
$\widehat{y}$ denotes the model output[1]. The dot indicates that there may be some parameters associated with these functions. The output of each basis function

---

[1]Usually the models are to be used as predictors. We will use this notation throughout
the remaining chapters.

is multiplied by the weighting function and these values are summed to form the function output. With constants as weighting functions, the structure is referred to as a *standard basis function network*. The $k$'s in (4.1) refer to the fact that these models will be used for sampled systems. The mapping f, however, is not dependent on the sampling, just as the operations of multiplication and summing in linear systems are not dependent on the sampling. In the remainder of this chapter, simplified notation will be used

$$\widehat{y} = f(\varphi, \cdot) = \sum_{h=1}^{H} h_h(\varphi, \cdot) g_h(\varphi, \cdot) \tag{4.2}$$

An important interpretation of the basis function network is that of *local models* [31]: In (4.2), ...

> ... each function $h_h$ can be viewed as a local model, validity of which is defined by the activation value of $g_h$. Hence $g_h$'s partition the input space into operating regions on each of which a local model is defined. The network smoothly joins these local models together through interpolation to form an overall global model f.

## 4.1.2 Basis functions

Usually the basis functions are obtained by parameterizing a *single-variable mother basis function*, $\kappa$, and repeating it a large number of times in the expansion. Single-variable basis functions can be classified into local and global basis functions. *Local basis functions* have a gradient with a bounded support (at least in a practical sense), whereas *global basis functions* have an infinitely spreading gradient. This means, roughly, that with local basis functions there are large areas in the input space where a change in the input variable causes no change in the function output; a change in the input of a global basis function always causes a change in the function output. Different kinds of single-variable basis functions are illustrated in Fig. 4.1.

In the *multi-variable* case, the basis functions can be classified into three main groups [86]: tensor products, radial constructions and ridge constructions.

- The *tensor product* construction is the product of single-variable functions

$$g_h(\varphi) = \prod_{i=1}^{I} \kappa(\varphi_i, \cdot) \tag{4.3}$$

where the subscript $i$ indexes the elements of the regression vector.

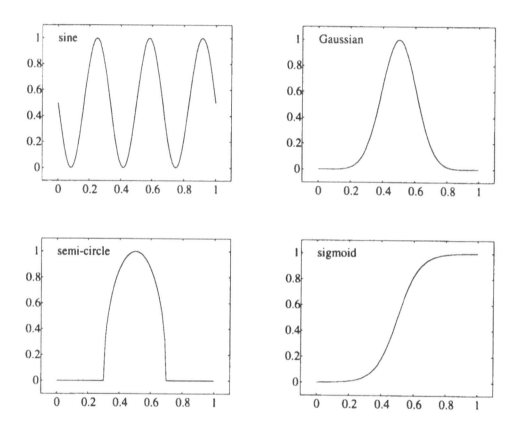

Figure 4.1: Examples of single-variable basis functions, $\kappa$. A global basis function (sine) has a gradient with an infinite support. Local basis functions (semi-circle) have a bounded support, at least in a practical sense (Gaussian and sigmoid functions).

- *Radial construction* is based on taking some norm on the space of the regression vector and passing the result through a single-variable function

$$g_h\left(\varphi\right) = \kappa\left(\left\|\varphi - \beta_h\right\|_{\gamma_h}\right) \tag{4.4}$$

- In *ridge constructions*, a linear combination of the regression vector is passed through a single-variable function

$$g_h\left(\varphi\right) = \kappa\left(\beta_h^T\varphi + \gamma_h\right) \tag{4.5}$$

The parameters $\gamma_h$ and $\beta_h$ are typically related to the scale and position of $g_h$.

### 4.1.3 Function approximation

The powerful function approximation capabilities of some basis function networks are a major reason for their popularity in the identification of nonlinear systems.

Let us call by a *universal approximator* something that can uniformly approximate continuous functions to any degree of accuracy on compact sets [12]. Proofs of universal approximation for basis function networks have been published by several authors. Hornik [30] showed that the multi-layer feedforward networks with one hidden layer using arbitrary squashing functions (*e.g.*, sigmoid neural networks) are capable of approximating any measurable function from finite dimensional space to another. This can be done to any desired degree of accuracy, provided that sufficiently many basis functions are available. The function approximation capability can be explained in the following intuitive way [29]:

> Any reasonable function f{$x$} can be represented by a linear combination of localized bumps that are each non-zero only in a small region of the domain {$x$}. Such bumps can be constructed with local basis functions and the associated weighting functions.

Not surprisingly, universal function approximation capability can be proved for many types of networks. All the proofs are existence proofs, showing that approximations are possible: There exists a set of basis functions with a set of parameters that produces a mapping with given accuracy.

Unfortunately, less can be said about how to find this mapping: How to find the correct parameters from data, or what is a (smallest) sufficient number of basis functions for a particular problem. A typical framework is to approximate an unknown function F

$$y = F(\varphi) + e \qquad (4.6)$$

based on sampled data $\varphi(k)$, $y(k)$, $k = 1, 2, ..., K$, where the observed outputs are corrupted by zero mean noise $e(k)$ with finite variance. Notice, that in a standard basis function network

$$\widehat{y} = f(\varphi) = \sum_{h=1}^{H} \alpha_h g_h(\varphi, \beta_h, \gamma_h) \qquad (4.7)$$

the parameters $\alpha_h$ appear linearly. If only $\alpha_h$ are of interest, these can be estimated from data, *e.g.*, using the least squares (the regressor containing the evaluated basis functions). If there are parameters in the basis functions to be estimated $(\beta_h, \gamma_h)$ they typically appear non-linearly. In some cases, these types of parameters are commonly estimated using iterative gradient-based methods (see Chapter 6).

The structure selection problem (roughly, the selection of $H$) can also be guided by data (see, *e.g.*, [18][26]). The main obstacle in structure selection is the fundamental trade-off between bias (due to insufficient model structure) and variance (due to noise in a finite data set), the *bias-variance dilemma*. With increased network size the bias decreases but the variance increases, and *vice versa*. In practice the performance of data driven structure selection (smoothing) algorithms can be computationally expensive and sometimes questionable, however, and it is more common to experiment with several fixed network sizes $H$. The 'optimal' network size is then found as the smallest network which gives sufficient accuracy both on the data and on independent test data (roughly, cross-validation). The bias-variance dilemma can also be tackled in parameter estimation by posing constraints on the functional form of the mapping (see Chapter 6).

## 4.2   Non-linear black-box structures

Non-linear system identification can be difficult because a non-linear system can be non-linear in so many different ways. Traditionally only model structures with constrained non-linearities have had success in practice. Lately, a number of new model structures have been proposed and shown to be useful in applications (see, *e.g.*, [34]). Most interest has been focused on artificial neural networks (such as sigmoid neural networks and radial basis function networks), and fuzzy systems.

To start with, recall the structure of the generalized basis function network (4.2)

$$\hat{y} = \text{f}(\boldsymbol{\varphi}, \cdot) = \sum_{h=1}^{H} \text{h}_h(\boldsymbol{\varphi}, \cdot)\, \text{g}_h(\boldsymbol{\varphi}, \cdot) \qquad (4.8)$$

The overall mapping is obtained by taking a weighted sum of the activation of the $H$ basis functions. In what follows, some commonly used structures are presented and shown to fit to the above generalized basis function network scheme.

## 4.2.1 Power series

When global basis functions are used, each weighting function $\text{h}_h$ has an effect on the model outcome at every operating region. Typical examples include the linear and multi-linear models, special cases of power series, or polynomial developments. In power series, the powers of the regressor generate the basis functions; in multi-linear systems only first order terms of each regressor component are used. The static mapping can be seen as a special case of the identification of non-linear dynamic systems using Volterra series (see Chapter 5). Other common structures include the Fourier series, for example. These belong to the class of *series estimators*, an extension of linear regression where the components of the regression vector represent the basis functions. A convenient feature of these structures is that all the parameters appear linearly, and can be estimated, *e.g.*, using the least squares method.

**Linear regression**

A *linear regression* model uses global basis functions $\boldsymbol{\varphi}$:

$$\hat{y} = \widehat{\boldsymbol{\theta}}^T \boldsymbol{\varphi} \qquad (4.9)$$

where $y$ is the model output, $\boldsymbol{\varphi} = \left[\varphi_1, \varphi_2, ..., \varphi_I, \varphi_{I+1} \equiv 1\right]^T$ are the $I$ inputs to the model with bias, and $\widehat{\boldsymbol{\theta}} = \left[\widehat{\theta}_1, \widehat{\theta}_2, ..., \widehat{\theta}_I, \widehat{\theta}_{I+1}\right]^T$ are the corresponding parameters. A linear model can be presented in the framework of the generalized basis function network by assigning

$$
\begin{aligned}
\widehat{y} &\equiv \widehat{y} & H &= I+1 \\
g_1\left(\varphi,\cdot\right) &\leftarrow \varphi_1 & h_1\left(\varphi,\cdot\right) &\leftarrow \widehat{\theta}_1 \\
&\;\vdots & &\;\vdots \\
g_h\left(\varphi,\cdot\right) &\leftarrow \varphi_i & h_h\left(\varphi,\cdot\right) &\leftarrow \widehat{\theta}_i \\
&\;\vdots & &\;\vdots \\
g_{H-1}\left(\varphi,\cdot\right) &\leftarrow \varphi_I & h_{H-1}\left(\varphi,\cdot\right) &\leftarrow \widehat{\theta}_I \\
g_H\left(\varphi,\cdot\right) &\equiv 1 & h_H\left(\varphi,\cdot\right) &\leftarrow \widehat{\theta}_{I+1}
\end{aligned}
\tag{4.10}
$$

Quite obviously, only linear functions can be mapped using the above model structure.

Alternatively, we can also consider using the observed data points as basis functions. Assume that a linear model is based on $K$ available data points $\left(\varphi\left(k\right),y\left(k\right)\right)$, $k = 1,2,...,K$. Let a linear model be given by (4.9) where $\widehat{\theta} = \left[\Phi^T\Phi\right]^{-1}\Phi^T\mathbf{y}$ (see Section 2.2.3). Then

$$
\widehat{y} = \varphi^T\mathbf{Z}\mathbf{y}
\tag{4.11}
$$

where $\mathbf{Z} = \left[\Phi^T\Phi\right]^{-1}\Phi^T$. Denote the $k^{\text{th}}$ column of $\mathbf{Z}$ by $\mathbf{Z}_k$. The presentation in the framework of the generalized basis function network is obtained by assigning

$$
\begin{aligned}
\widehat{y} &\equiv \widehat{y} & H &= K \\
g_1\left(\varphi,\cdot\right) &\leftarrow \varphi^T\mathbf{Z}_1 & h_1\left(\varphi,\cdot\right) &\leftarrow y\left(1\right) \\
&\;\vdots & &\;\vdots \\
g_h\left(\varphi,\cdot\right) &\leftarrow \varphi^T\mathbf{Z}_k & h_h\left(\varphi,\cdot\right) &\leftarrow y\left(k\right) \\
&\;\vdots & &\;\vdots \\
g_H\left(\varphi,\cdot\right) &\leftarrow \varphi^T\mathbf{Z}_K & h_H\left(\varphi,\cdot\right) &\leftarrow y\left(K\right)
\end{aligned}
\tag{4.12}
$$

This type of formulation is important in smoothing ([26] [25]). The smoothed values for each observed data point are given by

$$
\widehat{\mathbf{y}} = \Phi\left[\Phi^T\Phi\right]^{-1}\Phi^T\mathbf{y}
\tag{4.13}
$$

where $\mathbf{S} = \Phi\left[\Phi^T\Phi\right]^{-1}\Phi^T$ (a $K \times K$ *smoother matrix*), and its rows are referred to as *equivalent kernels* of a linear smoother.

## Multi-linear systems

In many practical cases, *multi-linear developments* are sufficient. A function $g\left(\varphi\right)$, $\varphi = \left[\varphi_1,...,\varphi_i,...,\varphi_I\right]^T$, is multi-linear if it is linear in each component

$\varphi_i$, when all other components $\varphi_j$, $j \neq i$ , are fixed. A general form is given by

$$f(\varphi) = \widehat{\theta}_0 + \sum_{i_1=1}^{I} \widehat{\theta}_{i_1}\varphi_{i_1} + \sum_{\{i_1,i_2\}=1;i_1<i_2}^{I} \widehat{\theta}_{i_1,i_2}\varphi_{i_1}\varphi_{i_2} + \cdots \qquad (4.14)$$

$$+ \sum_{\{i_1=1,\cdots,i_{I-1}\}=1;i_1<i_2<\cdots<i_{I-1}}^{I} \widehat{\theta}_{i_1,i_2,\cdots,i_{I-1}}\varphi_{i_1}\varphi_{i_2}\cdots\varphi_{i_{I-1}}$$

$$+\widehat{\theta}_{1,2,\cdots,I}\varphi_1\varphi_2\cdots\varphi_I \qquad (4.15)$$

where $\sum_{\{a,b,\cdots,c\}=1;a<b<\cdots<c}^{D}$ denotes multiple summations $\sum_{a=1}^{D} \sum_{b=1}^{D} \cdots \sum_{c=1}^{D}$ under the conditions $a < b < \cdots < c$ .

**Example 21 (Multilinear system)** For a two-input system $(I = 2)$, a multi-linear development is given by

$$\widehat{y} = \widehat{\theta}_0 + \widehat{\theta}_1\varphi_1 + \widehat{\theta}_2\varphi_2 + \widehat{\theta}_{1,2}\varphi_1\varphi_2 \qquad (4.16)$$

**Example 22 (Fluidized bed combustion)** In an FBC (see Appendix B), the steady-state relation between the fuel feed rate, $Q_C$, and the combustion power is given by

$$P = H_C (1 - V) Q_C + H_V V Q_C \qquad (4.17)$$

Let us now consider $Q_C$ and $V$ (fraction of volatiles in fuel) as non-constant inputs to the system. Using the following input transformations

$$\varphi_1 \leftarrow Q_C; \varphi_2 \leftarrow V \qquad (4.18)$$

the equation can be written as

$$P = H_C\varphi_1 + (H_V - H_C)\varphi_1\varphi_2 \qquad (4.19)$$

which is a multi-linear mapping from $\varphi$ to $P$.

**Example 23 (Multilinear system: continued)** For a three-input system $(I = 3)$ a multi-linear development is given by

$$\widehat{y} = \widehat{\theta}_0 + \widehat{\theta}_1\varphi_1 + \widehat{\theta}_2\varphi_2 + \widehat{\theta}_3\varphi_3 + \qquad (4.20)$$
$$+\widehat{\theta}_{1,2}\varphi_1\varphi_2 + \widehat{\theta}_{1,3}\varphi_1\varphi_3 + \widehat{\theta}_{2,3}\varphi_2\varphi_3 +$$
$$+\widehat{\theta}_{1,2,3}\varphi_1\varphi_2\varphi_3$$

**Power series**

In *power series*, the basis functions are formed by taking tensor products of integer-valued powers $(j_i = 0, 1, 2, ...; i = 1, 2, ..., I)$ of the input variables

$$g_h(\varphi, \cdot) = \prod_{i=1}^{I} \varphi_i^{j_i} \qquad (4.21)$$

up to a given order $j$

$$j = \sum_{i=1}^{I} j_i \qquad (4.22)$$

The model is then produced by taking a weighted sum of the activations of the basis functions and the associated parameters.

**Example 24 (Power series)** In many practical situations, a second order development $(j = 2)$ is sufficient. For a two-input system $(I = 2)$, a polynomial development model would be as follows

$$\widehat{y} = \widehat{\theta}_0 + \widehat{\theta}_1 \varphi_1 + \widehat{\theta}_2 \varphi_2 + \widehat{\theta}_{1,1} \varphi_1^2 + \widehat{\theta}_{1,2} \varphi_1 \varphi_2 + \widehat{\theta}_{2,2} \varphi_2^2 \qquad (4.23)$$

The corresponding presentation in the framework of the generalized basis function network is obtained by substituting

$$
\begin{array}{ll}
\widehat{y} \equiv \widehat{y} & H = 6 \\
g_1(\varphi, \cdot) \equiv 1 & h_1(\varphi, \cdot) \leftarrow \widehat{\theta}_0 \\
g_2(\varphi, \cdot) \leftarrow \varphi_1 & h_2(\varphi, \cdot) \leftarrow \widehat{\theta}_1 \\
g_3(\varphi, \cdot) \leftarrow \varphi_2 & h_3(\varphi, \cdot) \leftarrow \widehat{\theta}_1 \\
g_4(\varphi, \cdot) \leftarrow \varphi_1^2 & h_4(\varphi, \cdot) \leftarrow \widehat{\theta}_{1,1} \\
g_5(\varphi, \cdot) \leftarrow \varphi_1 \varphi_2 & h_5(\varphi, \cdot) \leftarrow \widehat{\theta}_{1,2} \\
g_6(\varphi, \cdot) \leftarrow \varphi_2^2 & h_6(\varphi, \cdot) \leftarrow \widehat{\theta}_{2,2}
\end{array}
\qquad (4.24)
$$

In multi-linear and polynomial developments, the model output $\widehat{y}$ is linear with respect to the parameters $\widehat{\theta}$, yet non-linear with respect to inputs $\varphi_i$. From parameter estimation point of view, the methods could also be seen as a method of pre-processing the input data, and then applying simple linear regression. However, non-linear functions can be mapped using the above model constructions.

**Inverse power series**

The inverse of a function is commonly needed in applications of control, for example. In practice it is often simplest to choose the structure of the inverse model (of the power series expansion) and to estimate their coefficients. As a side note, however, we discuss in the following a less known approach for finding the inverse of a power series.

The Bürman–Lagrange series constitutes a generalization of Taylor series. These series appear when we expand an analytical function $f(\varphi)$ into a series of ascending powers of another analytical function $w(\varphi)$:

$$f(\varphi) = \sum_{h=0}^{\infty} \alpha_h w^h(\varphi) \qquad (4.25)$$

For $n \geq 1$, it follows:

$$\alpha_n = \frac{1}{n!} \lim_{\varphi \to a} \frac{d^{n-1}}{d\varphi^{n-1}} \left\{ f'(\varphi) \frac{(\varphi - a)^n}{w^n(\varphi)} \right\}, (n = 1, 2, ...) \qquad (4.26)$$

where

$$f'(\varphi) = \frac{df(\varphi)}{d\varphi}. \qquad (4.27)$$

**Example 25 (Function approximation)** Consider a function $f(x) = x^2$ to be approximated with $w(x) = x$ around $x = 0$. Then $f'(x) = 2x$ and using (4.26) we have for the coefficients:

$$\alpha_2 = \frac{1}{2} \lim_{x \to a} \frac{d}{dx} \left\{ 2x \frac{(x-a)^2}{x^2} \right\} = \lim_{x \to 0} \frac{d}{dx} x = 1 \qquad (4.28)$$

$$\alpha_h = 0 \text{ for } h = 0, h = 2, 3, 4, ... \qquad (4.29)$$

Let us apply this result for inverting a power series. Consider the following power series:

$$w(\varphi) = c_1 \varphi + c_2 \varphi^2 + ... + c_n \varphi^n + ... \qquad (4.30)$$

$c_1 \neq 0$, which is convergent in the neighborhood of the point $\varphi = 0$. Find the expansion of the function $\varphi(w)$ with respect to the ascending powers of $w$

$$\varphi(w) = \alpha_0 + \alpha_1 w + ... + \alpha_n w^n + ... \qquad (4.31)$$

This particular problem can be solved using the Bürman–Lagrange series.

$$\alpha_n = \frac{1}{n!} \lim_{\varphi \to 0} \frac{d^{n-1}}{d\varphi^{n-1}} \left(\frac{\varphi}{w}\right)^n , (n = 1, 2, ...) \qquad (4.32)$$

because in our case

$$f(\varphi) \equiv \varphi \text{ and } f'(\varphi) = 1. \qquad (4.33)$$

**Example 26 (Numerical example)** Consider the following power series model

$$y = w(x) = c_1 x + c_2 x^2 \qquad (4.34)$$

For an approximation of its inverse in the neighborhood of $x = 0$

$$x = v(y) = \alpha_0 + \alpha_1 y + \alpha_2 y^2 + \cdots \qquad (4.35)$$

the following coefficients are obtained using (4.32)

$$\alpha_0 = 0 \qquad (4.36)$$

$$\alpha_1 = \lim_{x \to 0} \left(\frac{x}{c_1 x + c_2 x^2}\right) = \frac{1}{c_1} \qquad (4.37)$$

$$\alpha_2 = \frac{1}{2} \lim_{x \to 0} \left(\left(\frac{x}{c_1 x + c_2 x^2}\right)^2\right) = -\frac{c_2}{c_1^3} \qquad (4.38)$$

$$\alpha_3 = 2\frac{c_2^2}{c_1^5}, \alpha_4 = -5\frac{c_2^3}{c_1^7}, \alpha_5 = 14\frac{c_2^4}{c_1^9}, \cdots \qquad (4.39)$$

Figure 4.2 illustrates the approximation with $c_1 = 1$ and $c_2 = 10$.

**Example 27 (Exponential function)** Consider the following series

$$w(x) = xe^{-a\varphi} = \sum_{n=0}^{\infty} \frac{(-a)^n}{n!} x^{n+1} \qquad (4.40)$$

Using (4.32) we derive

$$\alpha_n = \frac{1}{n!} \lim_{x \to 0} \frac{d^{n-1}}{dx^{n-1}} \exp(anx) = \frac{(-an)^{n-1}}{n!} \qquad (4.41)$$

which leads to:

$$\varphi = \sum_{n=1}^{\infty} \frac{(an)^{n-1}}{n!} w^n. \qquad (4.42)$$

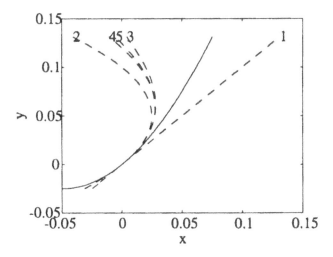

Figure 4.2: A power series (solid line) and its inverses (dotted lines) approximated around $x = 0$ using $n = 1, 2, 3, 4$ and 5 first terms of the series.

## 4.2.2 Sigmoid neural networks

Neural networks consist of multiple techniques related loosely to each other by the background of the algorithms: the neural circuitry in a living brain. There are three basic perspectives to *neural networks*. On can consider them as a form of artificial intelligence, as a means for enabling computers to perform intelligent tasks. On the other hand, neural networks can be seen from a biological point of view, as a way of modeling the neural circuitry observed in living creatures. The approach taken in engineering, the more practical view, considers neural networks from a purely technical perspective of data classification, filtering and identification of non-linear systems.

For a large part the research in neural computation overlaps with the fields of statistical analysis and optimization. In general, neural networks are modeling structures characterized by:

- A large number of simple interconnected elements (units, nodes, neurons); and

- A learning mechanism for adjusting the connections (weights, parameters) between the nodes, based on observed patterns of the system behavior.

There are several alternative ways to categorize neural network models and techniques. From the pragmatic point of view, we can categorize them roughly into two classes:

- Multi-layer perceptron networks, such as sigmoid neural networks (SNN), deal with function approximation. Perhaps the most important result brought by the neural research has been to show that any reasonable function can be approximated to any degree of accuracy; and to provide model structures that are also viable in practice.

- Self-organizing maps (SOM) consider the problems of clustering and quantization. Among the main new contributions is the introduction of an internal topology into the clustering process.

In what follows, we will focus on function approximation tasks. The SOM will be briefly discussed in connection of nearest neighbor methods (section 4.2.3).

*Sigmoid neural networks* are probably the most common neural network structure used for non-linear function approximation. They are also commonly referred to as multi-layer perceptrons (MLPs), backpropagation networks, or feed-forward artificial neural nets. These names come from the different properties of the standard sigmoid neural network.

Sigmoid neural networks use (practically) local basis functions. However, due to the use of the ridge construction (4.5), the interpretation as local models is not very useful as this type of structure estimates non-linear hypersurfaces, rather than local models for various operating regions. In practice, the fact that hypersurfaces are estimated provides advantages in interpolation. This type of structure is often referred to as semi-global. Other examples of similar structures include perceptrons or hinging hyperplanes, for example.

In sigmoid neural networks, the basis functions have a sigmoidal shape. The network units are organized as layers. A typical structure is that of a layered[2] feed-forward[3] sigmoid neural network. In the neural network terminology, the model inputs reside at the input layer. The input layer then feeds the hidden layer units. The network units, at the hidden layer, compute a linear combination of the input variables and pass this sum through a sigmoid function

$$g_h\left(\varphi, \beta_h, \gamma_h\right) = \frac{1}{1 + e^{-\beta_h^T \varphi - \gamma_h}} \qquad (4.43)$$

The outputs of the multiple units at the hidden layer are then fed to the output unit(s). The output unit computes a weighted sum of the activations

---

[2]In the simplest class of layered networks, every unit feeds signals only to the units located at the next layer (and receives signals only from the units located at the preceding layer). Hence there are no connections leading from a unit to units in preceeding layers, nor to other units in the same layer, nor to units more than one layer ahead.

[3]A network topology is feed-forward, if it does not contain any closed loops (feedback).

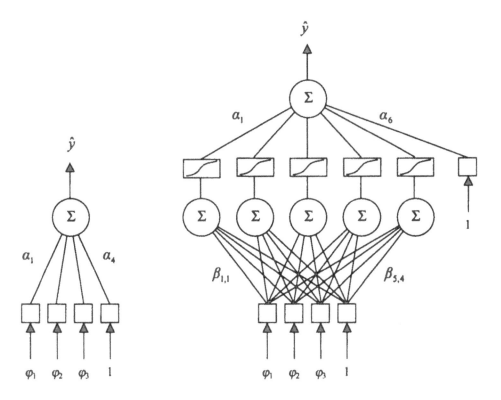

Figure 4.3: Two network constructions. Both models compute a function, $\hat{y} = f(\varphi)$. Left: linear model consisting of a single summing node. Right: standard sigmoid neural network with five hidden nodes (consisting of a summing element followed by a non-linear sigmoid element) at the hidden layer, and a single summing output node.

of the hidden layer units. The value of the weighted sum is then the output of the model.

Figure 4.3 illustrates a standard sigmoid neural network. Also structures with multiple hidden layers can be constructed, where the basic mappings are further convolved with each other by treating basis function outputs as new regressors.

## One-hidden-layer sigmoid neural net

For most practical purposes in process engineering, a single-hidden-layer network topology is sufficient. Let us consider a standard one-hidden-layer sigmoid neural net with $H$ hidden units (see Fig. 4.3) including bias parameters. The network computes a function f from an $I$ dimensional column vector $\varphi$

of model inputs:

$$\hat{y} = f(\varphi, \alpha, \beta) = \sum_{h=1}^{H} \alpha_h g_h(\varphi, \beta_h) + \alpha_{H+1} \qquad (4.44)$$

where

$$g_h(\varphi, \beta_h) = \cfrac{1}{1 + \exp\left(-\sum_{i=1}^{I} \beta_{h,i}\varphi_i - \beta_{h,I+1}\right)} \qquad (4.45)$$

The $H + 1 + H(I + 1)$ network parameters are contained in an $H + 1$ dimensional column vector $\alpha$:

$$\alpha^T = [\alpha_1, \alpha_2, ...\alpha_{H+1}] \qquad (4.46)$$

and a matrix $\beta$:

$$\beta = \begin{bmatrix} \beta_{1,1} & \cdots & \beta_{1,I+1} \\ \vdots & \beta_{h,i} & \vdots \\ \beta_{H,1} & \cdots & \beta_{H,I+1} \end{bmatrix} \qquad (4.47)$$

Note that, for convenience, the bias parameters $\gamma$ in (4.43) are now integrated into the structure of the matrix $\beta$. It is common to include bias constants in the linear summing.

Usually, the parameters in the sigmoid neural network are estimated using some gradient-based method. To do this, the *derivatives with respect to the parameters* need to be computed. Let us derive the required gradients.

For the parameters $\alpha$ at the output node we have

$$\frac{\partial}{\partial \alpha_h}f = \frac{\partial}{\partial \alpha_h}\sum_{h=1}^{H}\alpha_h g_h(\varphi, \beta_h) = g_h(\varphi, \beta_h) \qquad (4.48)$$

$$\frac{\partial}{\partial \alpha_{H+1}}f = 1 \qquad (4.49)$$

For the parameters $\beta$ at the hidden layer nodes, the derivative can be written as

$$\frac{\partial}{\partial \beta_{h,i}}f = \frac{\partial}{\partial \beta_{h,i}}\sum_{j=1}^{H}\alpha_j g_j(\varphi, \beta_j) = \alpha_h\frac{\partial}{\partial \beta_{h,i}}g_h(\varphi, \beta_h) \qquad (4.50)$$

The derivative $\frac{\partial}{\partial \beta_{h,i}} g_h (\varphi, \beta_h)$ of a sigmoid function $\kappa$ and ridge construction $\kappa (\phi)$ can be rewritten using the chain rule

$$\frac{\partial}{\partial \beta_{h,i}} g_h (\varphi, \beta_h) = \frac{\partial}{\partial \beta_{h,i}} \kappa (\phi (\varphi, \beta_h)) = \frac{\partial}{\partial \phi_h} \kappa (\phi_h) \frac{\partial}{\partial \beta_{h,i}} \phi_h (\varphi, \beta_h) \quad (4.51)$$

For $\kappa$ as the sigmoid function

$$\kappa (\phi_h) = \frac{1}{1 + \exp (-\phi_h)} \quad (4.52)$$

the derivative can be expressed in terms of the function output

$$\frac{\partial}{\partial \phi_h} \kappa (\phi_h) = \kappa (\phi_h) [1 - \kappa (\phi_h)] \quad (4.53)$$

For the linear sum $\phi_h$

$$\phi_h = \sum_{i=1}^{I} \beta_{h,i} \varphi_i + \beta_{h,I+1} \quad (4.54)$$

we have that

$$\frac{\partial}{\partial \beta_{h,i}} \phi_h = \varphi_i \quad (4.55)$$

$$\frac{\partial}{\partial \beta_{h,I+1}} \phi_h = 1 \quad (4.56)$$

Using (4.50), (4.51), (4.53) and (4.55) the derivative $\frac{\partial}{\partial \beta_{h,i}} f$ can be written as

$$\frac{\partial}{\partial \beta_{h,i}} f = \alpha_h g_h (\varphi, \beta_h) [1 - g_h (\varphi, \beta_h)] \varphi_i \quad (4.57)$$

where $g_h$ is given by (4.45). For the bias parameters we have, using (4.50), (4.51), (4.53) and (4.56)

$$\frac{\partial}{\partial \beta_{h,I+1}} f = \alpha_h g_h (\varphi, \beta_h) [1 - g_h (\varphi, \beta_h)] \quad (4.58)$$

In a similar way, it is possible to linearize the one-hidden layer sigmoid neural net in the neighborhood of its operating point $\tilde{\varphi}$ using the Taylor series approximation:

$$\tilde{f}(\varphi, \tilde{\alpha}) = \tilde{\alpha}_1 \varphi_1 + ... + \tilde{\alpha}_I \varphi_I + \tilde{\alpha}_{I+1} \quad (4.59)$$

where $\tilde{\alpha}_i = \frac{\partial}{\partial \varphi_i} \mathbf{f}(\tilde{\varphi}, \boldsymbol{\alpha}, \boldsymbol{\beta})$. To do this, the *derivatives with respect to the system inputs* need to be calculated:

$$\tilde{\alpha}_i = \frac{\partial}{\partial \varphi_i} \mathbf{f} = \frac{\partial}{\partial \varphi_i} \sum_{h=1}^{H} \alpha_h \frac{\partial}{\partial \varphi_i} g_h\left(\boldsymbol{\varphi}, \boldsymbol{\beta}_h\right) \tag{4.60}$$

Again, the chain rule can be applied. The sigmoid and its derivative have already been given in (4.52)–(4.53). For the linear sum, we have that

$$\frac{\partial}{\partial \varphi_i} \phi_h = \beta_{h,i} \tag{4.61}$$

Substituting these into (4.60) and evaluating at the point of linearization $\tilde{\varphi}$, the linearized parameters are obtained from

$$\tilde{\alpha}_i = \sum_{h=1}^{H} \alpha_h g_h\left(\tilde{\varphi}, \boldsymbol{\beta}_h\right)\left[1 -_h g_h\left(\tilde{\varphi}, \boldsymbol{\beta}_h\right)\right] \beta_{h,i} \tag{4.62}$$

For zero error at the operating point $\varphi$, the bias can be taken as

$$\tilde{\alpha}_{I+1} = \mathbf{f}\left(\tilde{\varphi}, \boldsymbol{\alpha}, \boldsymbol{\beta}\right) - \sum_{i=1}^{I} \tilde{\alpha}_i \tilde{\varphi}_i \tag{4.63}$$

Let us collect the results.

**Algorithm 17 (One-hidden-layer sigmoid neural network)** The output of a one-hidden-layer sigmoid neural network with $H$ hidden nodes is given by

$$\hat{y} = \mathbf{f}(\boldsymbol{\varphi}, \cdot) = \sum_{h=1}^{H} \alpha_h g_h\left(\boldsymbol{\varphi}, \boldsymbol{\beta}_h\right) + \alpha_{H+1} \tag{4.64}$$

where

$$g_h\left(\boldsymbol{\varphi}, \boldsymbol{\beta}_h\right) = \frac{1}{1 + \exp\left(-\sum_{i=1}^{I} \beta_{h,i} \varphi_i - \beta_{h,I+1}\right)} \tag{4.65}$$

and $\varphi$ is the $I$ dimensional input vector. The parameters of the network are contained in a $H + 1$ dimensional vector $\boldsymbol{\alpha}$ and $H \times (I + 1)$ dimensional matrix $\beta$. The gradients with respect to the parameters are given by

$$\frac{\partial}{\partial \alpha_h} f = g_h(\varphi, \beta_h) \tag{4.66}$$

$$\frac{\partial}{\partial \alpha_{H+1}} f = 1 \tag{4.67}$$

$$\frac{\partial}{\partial \beta_{h,i}} f = \alpha_h g_h(\varphi, \beta_h)[1 - g_h(\varphi, \beta_h)]\varphi_i \tag{4.68}$$

$$\frac{\partial}{\partial \beta_{h,I+1}} f = \alpha_h g_h(\varphi, \beta_h)[1 - g_h(\varphi, \beta_h)] \tag{4.69}$$

where $h = 1, 2, ..., H$ and $i = 1, 2, ..., I$. A linearized approximation in the neighborhood of an operating point $\widetilde{\varphi}$ is given by

$$\widetilde{f}(\varphi, \widetilde{\alpha}) = \widetilde{\alpha}_1(\widetilde{\varphi})\varphi_1 + ... + \widetilde{\alpha}_I(\widetilde{\varphi})\varphi_I + \widetilde{\alpha}_{I+1}(\widetilde{\varphi}) \tag{4.70}$$

where the linearized parameters are given by

$$\widetilde{\alpha}_i(\widetilde{\varphi}) = \sum_{h=1}^{H} \alpha_h g_h(\widetilde{\varphi}, \beta_h)[1 - g_h(\widetilde{\varphi}, \beta_h)]\beta_{h,i} \tag{4.71}$$

$$\widetilde{\alpha}_{I+1} = f(\widetilde{\varphi}, \alpha, \beta) - \sum_{i=1}^{I} \widetilde{\alpha}_i \widetilde{\varphi}_i \tag{4.72}$$

$i = 1, 2, ..., I$.

### 4.2.3 Nearest neighbor methods

There exists a large variety of paradigms using local basis functions. Probably the simplest paradigm is the *nearest neighbor method*. Let us consider a pool of data of $K$ input–output measurement pairs: $\Phi = (\varphi(1), y(1))$, $(\varphi(2), y(2)), ..., (\varphi(K), y(K))$. In order to obtain an estimate for the output given an input pattern $\varphi$, $\varphi$ is compared with all the input patterns $\varphi(1), ..., \varphi(K)$ in the data pool. A pattern in the pool is found that is closest (*e.g.*, in the Euclidean sense) to the given input (competition between the patterns). Denote this closest pattern by $\varphi(c)$, $c \in \{1, 2, ..., K\}$. The estimate of the output for the given $\varphi$ is then $\widehat{y} = y(c)$.

Let us choose $\kappa$ as the indicator function in (4.4):

$$g_k(\varphi) = \left\{ \begin{array}{ll} 1 & \text{if } k \in \Gamma(\varphi) \\ 0 & \text{otherwise} \end{array} \right. \tag{4.73}$$

where $\Gamma(\varphi) = \arg\min_{k=1,2,\ldots,K} \|\varphi - \varphi(k)\|_2$; and in (4.2):

$$\widehat{y} = \sum_{k=1}^{K} y(k) \, g_k(\varphi) \tag{4.74}$$

the standard basis function network has as many basis functions as there are data points, $H \leftarrow K$, the local models are given by the observed data, $h_h \leftarrow y(k)$. As can be easily seen, this type of approach belongs to the radial constructions [86].

In order to cope with noise and storage limits, an estimate can be computed based on *prototypes* representing average local behavior, instead of direct observations. In this case, $H < K$. The problem then is to find a set of suitable centers $\beta_h$ of the basis functions. There are several ways to look for centers. The $\beta_{h,i}$s can be spread on the domain of each $i$ using, *e.g.*, equidistant intervals, thus forming a grid of points. In the *Kohonen networks* (learning vector quantization, self-organizing map (SOM) [49], see also [32] [80]), the distribution of the basis function centers resembles the probability distribution of the data patterns $k = 1, 2, \ldots, K$.

All the above methods find a *single winner* among the basis functions; there is competition between the basis functions (non-overlapping partitions). One can also consider *multiple winners* at the same time (overlapping partitions). For example, in the *k-nearest neighbors* estimate, the $\widehat{y}$ is taken to be the average of those $\lambda$ observed $y(k)$'s that are associated with the $\lambda$ inputs $\varphi(k)$ closest to the given $\varphi$. The $k$-nearest neighbors method can be represented by using (4.2) with:

$$g_k(\varphi) = \begin{cases} \frac{1}{\lambda} & \text{if } k \in \Gamma(\varphi) \\ 0 & \text{otherwise} \end{cases} \tag{4.75}$$

where $\Gamma(\varphi)$ contains the indexes for the $\lambda$-nearest neighbors of $\varphi$. Notice that the nearest neighbors estimators have no parameters to be estimated, and that no *a priori* assumption on the shape of the function is made. Therefore this type of methods are typical examples of *non-parametric regression* methods [18][25][26]. The $\lambda$ is seen as a *smoothing parameter* concerning structure selection. Proper selection of $\lambda$ is important: notice that as $\lambda$ increases the bias increases and the variance decreases, and *vice versa*. The linear smoother is given by $\widehat{y} = \mathbf{S}y$, where $\mathbf{S}$ is, roughly, a $\lambda$-banded matrix (provided a suitable ordering). For $\lambda = 1$, the *smoother matrix* is given by the identity matrix, $\mathbf{S} = \mathbf{I}$). For $\lambda > 1$, the *equivalent kernel* is a rectangular one; note that many other types of kernels can be considered.

When each basis function is associated with a constant output at the model output space, piecewise constant functions can be constructed. With overlapping partitioning the resolution of the mapping is enhanced, since the model outcome is an average of the activated constant values. However, the result is still a piecewise constant function. Often, this is an undesirable property for a model. Consider applications of control, for example: Based on a piecewise constant type of model, the gain of the system is either zero or undefined! There are many ways to arrive at smoother models. It is common to use prototypes (instead of direct data points), multiple winners (instead of a single winner) and Gaussian kernels (instead of rectangular kernels), as in the radial basis function networks to be considered in the next subsection.

**Radial basis function networks**

In radial-basis function (RBF) networks, the input space is partitioned using radial basis functions. The center and width of each basis function is adjustable. Each basis function is associated with an adjustable weight and the output estimate is produced by computing a weighted sum at the output unit. Hence, the output of the network is a linear superposition of the activities of all the radial functions in the network.

The RBF network is often given in the normalized form:

$$f(\varphi, \beta, \gamma) = \sum_{h=1}^{H} \alpha_h \widetilde{g}_h (\varphi, \beta_h, \gamma_h) = \frac{\sum_{h=1}^{H} \alpha_h g_h (\varphi, \beta_h, \gamma_h)}{\sum_{h=1}^{H} g_h (\varphi, \beta_h, \gamma_h)} \tag{4.76}$$

The normalization makes the total sum of all basis functions $\widetilde{g}_h$ unity, in the whole operating region. Gaussian functions are typically used with RBF networks:

$$g_h (\varphi, \beta_h, \gamma_h) = \exp \left( \frac{-\sum (\varphi_i - \beta_{h,i})^2}{\gamma_h^2} \right) \tag{4.77}$$

The use of the RBF networks requires that a suitable partitioning of the input space can be obtained. In simple methods, the locations of the Gaussians are taken from arbitrary data points or from a uniform lattice in the input space. Alternatively, clustering techniques can be used to choose the centers, such as the SOM. In the orthogonal least-squares method, locations are selected from data points one by one, maximizing the increment of the explained variance of the desired output.

Fig. 4.4 illustrates function approximation with non-overlapping partitioning of the input space (left), and overlapping partition (right). The

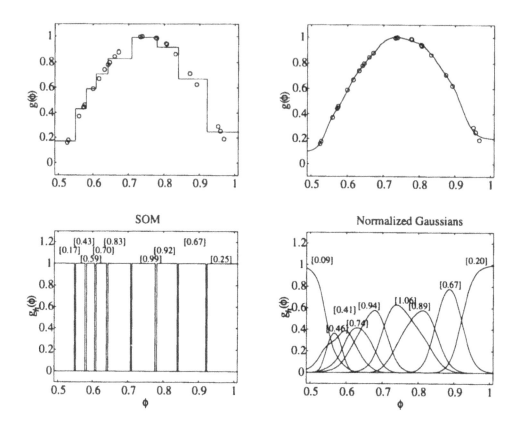

Figure 4.4: Examples of identification using local basis functions.

lower pictures illustrate the partitioning found by SOM (left), and normalized Gaussians (right) placed at the same centers as well as the associated values of the LS weighting constants. The upper figures depict the data patterns (dots) and the mapping obtained at the output of the structure (solid line).

## 4.2.4   Fuzzy inference systems

Fuzzy modeling [73][50] stems from advances in logic and cybernetics. Originally, fuzzy systems were developed in the 1960s as an outcome of fuzzy set theory. The fuzzy sets are a mathematical means to represent vague information. In applications to process modeling and control, the uncertainty handling aspects have, however, received less interest. Instead, the focus has been in extending the interpolation capabilities of rule-based expert systems. In what follows, we will focus on fuzzy rule-based systems. For more general fuzzy systems, please see remarks at the end of this section.

In *rule-based* inference systems, the universe is partitioned using concepts, modeled via sets. Reasoning is then based on expressions of logical relationships between the concepts: if-then rules. In expert systems, binary-valued logic is applied; fuzzy systems belong to the class of multi-valued logic systems.

*Expert systems* are rule-based systems. The knowledge about the process is represented using rules, such as

$$\text{if } premise \text{ then } consequent \tag{4.78}$$

Traditional expert systems use crisp rules based on two-state logic, where elements either belong or do not belong to a given class. The propositions (*premise* and *consequent*) represented by the rule can be either true or false, $\mu \in \{0, 1\}$.

In real life, the classes are ill-defined, overlapping, or fuzzy, and a pattern may belong to more than one class. Such nuances can be described with the help of fuzzy sets. In a fuzzy context, a pattern may be assigned a degree of membership value, which represents its degree of membership in a fuzzy set, $\mu \in [0, 1]$.

**Example 28 (Fuzzy and crisp sets)** For example, it might be difficult to classify the speed of a car as 'fast' or 'not fast' because human reasoning recognizes different shades of meaning between the two concepts [50]. Fig. 4.5 illustrates crisp and fuzzy concepts of 'fast'.

The domain knowledge is expressed as if-then rules, which relate the input fuzzy sets with the model outcome (if *speed is fast* then *move away*). Note, how the use of the adjective *'fast'* to characterize the *speed* of an approaching car is entirely sufficient to signal the necessity to move away; the precise velocity of the car at this moment is not important.

The if-then rule structure of fuzzy inference systems is convenient in that it is easily comprehensible as being close to one of the ways humans store knowledge. It also provides explanations for the model outcome since it is always possible to find out the exact rules that were fired, and to convert these into semantically meaningful sentences. In this sense, the fuzzy inference systems also provide insight and understanding of the considered process, and support 'what if' type of analysis.

**Fuzzy systems in process modeling**

From the process modeling point of view, the applications have shown two main contributions due to the use of fuzzy systems:

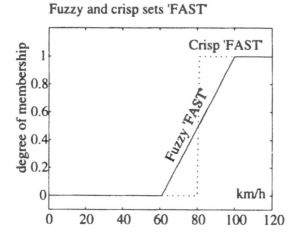

Figure 4.5: Examples of a crisp set and a fuzzy set describing the 'fast' speed of a car. A crisp set has sharp boundaries and its membership function asssumes binary values in {0,1}. A fuzzy set has vague boundaries and its membership function takes values in [0,1].

- A reduction of the complexity of systems, based on the use of fuzzy sets; and

- A transparent form of reasoning (similar to the conscious reasoning by humans).

Neural networks have been shown to be very efficient in their function approximation capabilities (see *e.g.* [27][29]), that is in mimicking the observed input-output behavior. Unfortunately, neural networks appear as black-box models to the developer and end-user. The disadvantage of black-box models is that, although they seem to provide the correct functional mappings, they do not easily give any additional explanation on what this mapping is composed of, or make it easier to understand the nature of the relation between the function inputs and outputs. This *lack of transparency* might lead to difficulties if human intervention or man–machine interaction is required or expected. This is often the case when models are utilized for optimization or monitoring purposes. Obviously, some transparency would also help the model developer to evaluate the validity of the model and to locate unsatisfactory behavior when further model development is needed. The need for transparency has motivated the use of fuzzy systems.

In process modeling, fairly simple fuzzy models have been applied. In general, fuzzy modeling can be an efficient way to quickly build a model or a controller for a process, when only rough information is available. Also non-linear systems can be considered without extra effort. In a 'standard' learning approach:

1. fuzzy sets and rules are stated by the experts (plant operators, engineers),

2. the system structure is established, and

3. the membership functions and/or output constants are fine-tuned using data.

This allows us to build a model of a system based on experimental human knowledge. Alternatively, one may start from a nominal model, in which case the motivation for using the fuzzy approach comes from the easiness of validation and the possibility to tune the system manually.

### Sugeno and Mamdani fuzzy models

Fuzzy models can also be seen as based on local basis functions [86]. In a fuzzy system, the input partitioning is given by the premises of rules. In *Mamdani fuzzy models*, both the premise and the consequent of a fuzzy rule are specified using fuzzy sets:

$$\text{if } \{(\varphi_1 \text{ is } A_{h,1}) \text{ and } ... \text{ and } (\varphi_I \text{ is } A_{h,I})\} \text{ then } (\widehat{y} \text{ is } B_h) \qquad (4.79)$$

where $A_{h,i}$ and $B_h$ are fuzzy sets specifying the $h$'th rule and $I$ is the input dimension. In order to get a crisp output from the fuzzy inference, defuzzification is needed to convert the inferred fuzzy output into a crisp singleton.

In *Sugeno fuzzy models*, the consequent of a rule is a parameterized function of the input variables. Hence the rules assume the form:

$$\text{if } \{(\varphi_1 \text{ is } A_{h,1}) \text{ and } ... \text{ and } (\varphi_I \text{ is } A_{h,I})\} \text{ then } (\widehat{y} = f_h(\varphi, \cdot)) \qquad (4.80)$$

Typically, the functions $f_h$ are constants (0-order Sugeno model) or linear polynomials. With Sugeno models, the consequences of multiple rules are combined by summing, weighting the rules with the normalized activation level of each rule. 0-order Sugeno models can be viewed as a special case of Mamdani models, in which each rule's consequent is specified by a constant (a singleton fuzzy set).

In order to compute the fuzzy rules, the operations on fuzzy sets (is, and) need to be specified, as well as the inference (if *premise* then *consequent*). Common choice is to implement the '$\varphi_i$ is $A_{h,i}$' by evaluating a triangular membership function (or Gaussian, or bell-shaped function), and the 'p and q' operation as a product (or minimum). The if-then inference (implication) is usually seen as a binary-valued relation, true for the sets contained in the rule, zero elsewhere.

### Fuzzy inference systems

Let us have a brief look at the logic background concerning fuzzy sets and reasoning in fuzzy systems. For more information, see *e.g.*, [40].

A *fuzzy set* $A$ of $X$ is expressed by its membership function $\mu_A$ from the universe of discourse to the unit interval

$$\mu_A : X \rightarrow [0, 1] \tag{4.81}$$

$\mu_A(\varphi)$ expresses the extent to which $\varphi$ fulfills the category specified by $\mu_A$, where $X$ is the universe of discourse (domain) of $\varphi$.

Fuzzy inference systems consist of five blocks

- fuzzification

- data base

- rule base

- decision logic

- defuzzification

The *fuzzification* block converts the system input $\varphi = \varphi_0 \in \Re$ into a fuzzy set $A'$ on $X$. Its membership function $\mu_{A'}(\varphi)$ is usually defined by the point fuzzification

$$\mu_{A'}(\varphi) = \{ \begin{array}{ll} 1 & \text{if } \varphi = \varphi_0 \\ 0 & \text{otherwise} \end{array} \tag{4.82}$$

Alternative fuzzifications can be used if information about the uncertainty of the measurement $\varphi = \varphi_0$ is available, or if the measurement itself is not crisp.

The *data base* contains information about the fuzzy sets $\mu_{A'_i}(\varphi_i)$'s (fuzzification), $\mu_{A_i}(\varphi_i)$'s and $\mu_B(y)$'s (rules), and the associated linguistic terms, $A_i$'s and $B$'s (rules). The *rule base* is a set of linguistic statements: rules. The rules assume the form

$$\text{if } (\varphi_1 \text{ is } A_1) \text{ and } ... \text{ and } (\varphi_I \text{ is } A_I) \text{ then } (\widehat{y} \text{ is } B) \tag{4.83}$$

where $A$ and $B$ are linguistic terms defined by fuzzy sets in the data base. This can be translated into a simpler form using fuzzy and[4]

$$\mu_A(\varphi) = \mathrm{T}\left(\mu_{A_1}(\varphi_1), ..., \mu_{A_I}(\varphi_I)\right) \qquad (4.84)$$

A rule can be seen as a fuzzy implication function I

$$\mathrm{I}\left(\mu_A(\varphi), \mu_B(\widehat{y})\right) \qquad (4.85)$$

which is often modeled using a t-norm.

The *decision logic* processes the input fuzzy signals using linguistic rules. Let us derive the *modus ponens* inference assuming a point fuzzification.

$$\begin{array}{l} \text{if } (\varphi \text{ is } A) \text{ then } (\widehat{y} \text{ is } B) \\ \underline{\varphi \text{ is } A'} \\ \Longrightarrow \widehat{y} \text{ is } B' \end{array} \qquad (4.86)$$

where

$$\mu_{B'}(\widehat{y}) = \sup_{\varphi}\left(\mathrm{T}\left\{\mu_{A'}(\varphi), \mathrm{I}\left[\mu_A(\varphi), \mu_B(\widehat{y})\right]\right\}\right) \qquad (4.87)$$

Since the input is a point $\varphi_0$ (we assume point fuzzification here), then

$$\mu_{A'}(\varphi) = \left\{ \begin{array}{ll} 1 & \text{if } \varphi = \varphi_0 \\ 0 & \text{otherwise} \end{array} \right. \qquad (4.88)$$

and the result can be expressed as

$$\begin{aligned} \mu_{B'}(\widehat{y}) &= \mathrm{T}\left\{1, \mathrm{I}\left[\mu_A(\varphi_0), \mu_B(\widehat{y})\right]\right\} & (4.89) \\ &= \mathrm{I}\left[\mu_A(\varphi_0), \mu_B(\widehat{y})\right] & (4.90) \end{aligned}$$

In general, the inference of the $h$'th rule, $h = 1, 2, ..., H$, for input $\varphi_0$ can be expressed as

$$\mu_{B'_h}(\widehat{y}) = \left\{ \begin{array}{ll} 0 & \text{if } \mu_{A_h}(\varphi_0) = 0 \\ \mathrm{I}\left[\mu_{A_h}(\varphi_0), \mu_{B_h}(\widehat{y})\right] & \text{otherwise} \end{array} \right. \qquad (4.91)$$

---

[4]Basic operations (intersection $\Rightarrow$ fuzzy and, union $\Rightarrow$ fuzzy or) on fuzzy sets can be defined using t-norms and s-norms. T-norms are monotonic non-decreasing: $a \leq b \Rightarrow \mathrm{T}(a, c) \leq \mathrm{T}(b, c)$; commutative: $\mathrm{T}(a, b) = \mathrm{T}(b, a)$; associative: $\mathrm{T}(a, \mathrm{T}(b, c)) = \mathrm{T}(\mathrm{T}(a, b), c)$; and have 1 as unit element $\mathrm{T}(1, a) = a$. Any t-norm is related to its dual s-norm (t-conorm) by the deMorgan law $\mathrm{S}(a, b) = 1 - \mathrm{T}(1 - a, 1 - b)$. The commonly used t-norms include product: $\mathrm{T}(a, b) = ab$ and minimum: $\mathrm{T}(a, b) = \min(a, b)$. The related s-norms are probabilistic sum: $\mathrm{S}(a, b) = a + b - ab$ and maximum: $\mathrm{S}(a, b) = \max(a, b)$.

which in the case of t-norm implications can be further simplified to

$$\mu_{B'_h}(\widehat{y}) = \mathrm{T}\left[\mu_{A_h}(\varphi_0), \mu_{B_h}(\widehat{y})\right] \tag{4.92}$$

The combination of all fuzzy inferences is made by means of an s-norm

$$\mu_{B'}(\widehat{y}) = \mathrm{S}\left[\left\{\mu_{B'_h}(\widehat{y})\right\}_{h=1,2,\dots,H}\right] \tag{4.93}$$

The *defuzzification* determines (converts) the fuzzy output of the decision logic into a crisp output value. A common choice is the center of area method

$$\widehat{y} = \frac{\int_{\widehat{y} \in Y} \widehat{y}\mu_{B'}(\widehat{y})\, d\widehat{y}}{\int_{\widehat{y} \in Y} \mu_{B'}(\widehat{y})\, d\widehat{y}} \tag{4.94}$$

In the case of 0-order Sugeno fuzzy models, if $(\varphi$ is $A_h)$ then $(\widehat{y} = \overline{y}_h)$, we can think of the output sets in (4.92) as given by singleton sets: $\mu_{B_h}(\widehat{y}) = 1$ if $\widehat{y} = \overline{y}_h$ , zero elsewhere. The defuzzification can then be replaced by a weighted average

$$\widehat{y} = \frac{\sum\limits_{h=1}^{H} \overline{y}_h \mu_{A_h}(\varphi_0)}{\sum\limits_{h=1}^{H} \mu_{A_h}(\varphi_0)} \tag{4.95}$$

### 0-order Sugeno fuzzy model

0-order Sugeno fuzzy models are common in many process engineering applications. They represent a simple case of the more general fuzzy inference systems. Very often, the following choices are made:

- system inputs are crisp,

- product is chosen for the fuzzy and, and

- weighted sum is chosen for the defuzzification.

With these choices we arrive to the following 0-order Sugeno fuzzy model.

**Definition 11 (0-order Sugeno fuzzy model)** A 0-order Sugeno model with $H$ rules is a function f from an $I$ dimensional column vector of model inputs $\varphi$

$$\widehat{y} = \mathrm{f}(\varphi, \alpha, \cdot) = \frac{\sum\limits_{h=1}^{H} \alpha_h g_h(\varphi, \cdot)}{\sum\limits_{h=1}^{H} g_h(\varphi, \cdot)} \tag{4.96}$$

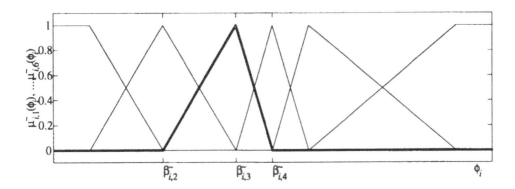

Figure 4.6: Add-one partition of the domain of the $i$'th input $\varphi_i$ ($i = 1, 2, ..., I$) using triangular fuzzy sets. The centers of the sets are given by $\widetilde{\beta}_{i,p}$ ($p = 1, 2, ..., P_i$). The bold line shows the membership function $\widetilde{\mu}_{i,3}(\varphi_i)$.

where

$$g_h(\varphi, \cdot) = \prod_{i=1}^{I} \mu_{h,i}(\varphi_i, \cdot) \qquad (4.97)$$

where $\mu_{h,i}(\varphi_i, \cdot) \in [0, 1]$ is the degree of membership of the $i$'th input $\varphi_i \in \Re$ in the premise of the $h$'th rule.

Let us change slightly the notation. Assume that an *add-one partitioning* is used, see Fig. 4.6, where the domain of each input $\varphi_i$ is partitioned separately such that

$$\sum_{p_i=1}^{P_i} \widetilde{\mu}_{i,p_i}(\varphi_i, \cdot) = 1 \text{ for all } \varphi_i \qquad (4.98)$$

where $P_i$ is the number of fuzzy sets used for partitioning the domain of the $i$'th input. Notice that it is usually simple to set the membership functions such that an add-one partitioning is obtained. The tilde emphasizes that difference in the notation (strong fuzzy partition). The following result can be derived:

**Theorem 1 (Add-one partition)** Assume that each input domain $i$ is partitioned such that

$$\sum_{p_i=1}^{P_i} \widetilde{\mu}_{i,p_i}(\varphi_i, \cdot) = 1 \qquad (4.99)$$

where $\widetilde{\mu}_{i,p_i} \in [0,1]$ are the $P_i$ membership functions used for partitioning the domain of the $i$'th input $\varphi_i \in \Re$, $i = 1, 2, ..., I$. In addition, suppose that the rule-base is complete and that the product t-norm is used. Then, the sum of basis functions in a 0-order Sugeno model is given by

$$\sum_{h=1}^{H} g_h(\varphi, \cdot) = \sum_{p_1=1}^{P_1} \cdots \sum_{p_I=1}^{P_I} \prod_{i=1}^{I} \widetilde{\mu}_{i,p_i}(\varphi_i, \cdot) \qquad (4.100)$$

where $h = 1, 2, ..., H = \prod_{i=1}^{I} P_i$ are the $H$ rules. The sum of basis functions is equal to one

$$\sum_{h=1}^{H} g_h(\varphi, \cdot) = 1 \qquad (4.101)$$

for all $\varphi_i$.

A typical add-one partition is obtained using triangular membership functions

$$\mu_{h,i}(\varphi_i, \cdot) = \widetilde{\mu}_{i,p_i}\left(\varphi_i, \widetilde{\boldsymbol{\beta}}_i\right) \qquad (4.102)$$

$$= \max\left(\min\left(\frac{\varphi_i - \widetilde{\beta}_{i,p_i-1}}{\widetilde{\beta}_{i,p_i} - \widetilde{\beta}_{i,p_i-1}}, \frac{\widetilde{\beta}_{i,p_i+1} - \varphi_i}{\widetilde{\beta}_{i,p_i+1} - \widetilde{\beta}_{i,p_i}}\right), 0\right) \qquad (4.103)$$

where $\widetilde{\beta}_{i,p_i-1} < \widetilde{\beta}_{i,p_i}$. Observe, how any crisp input $\varphi_i$ can have non-zero degrees of membership only in at most two fuzzy sets $\widetilde{\mu}_{i,p_i}$. Hence, we have the following simpler result.

**Algorithm 18 (0-order add-1 Sugeno fuzzy model)** Assume crisp system inputs, an add-one partition, product t-norm, and weighted average defuzzification. Then, a 0-order Sugeno model is given by

$$\widehat{y} = \sum_{h=1}^{H} \alpha_h \prod_{i=1}^{I} \mu_{h,i}(\varphi_i, \cdot) \qquad (4.104)$$

where $\mu_{h,i}(\varphi_i, \cdot)$ is the membership function associated with the $h$'th rule and the $i$'th input. Equivalently, we can write

$$\widehat{y} = \sum_{p_1=1}^{P_1} \sum_{p_2=1}^{P_2} \cdots \sum_{p_I=1}^{P_I} \widetilde{\alpha}_{p_1,p_2,\cdots,p_I} \widetilde{\mu}_{1,p_1}(\varphi_1, \cdot) \widetilde{\mu}_{2,p_2}(\varphi_2, \cdot) \cdots \widetilde{\mu}_{I,p_I}(\varphi_I, \cdot) \qquad (4.105)$$

where $\widetilde{\mu}_{i,p_i}(\varphi_i, \cdot)$ is the membership function associated with the $p_i$'th set partitioning the $i$'th input $(p_i = 1, 2, \cdots, P_i)$.

Notice, that at each point in the input space which is a center of some triangular fuzzy set, *i.e.* $\varphi_i = \tilde{\beta}_{i,p_i} \forall i$, the output of the system is given by $\tilde{\alpha}_{p_1,p_2,\cdots,p_I}$.

**Example 29 (Fuzzy PI controller)** A PI controller is given by

$$u(k) = K_{\mathrm{P}}e(k) + K_{\mathrm{I}}\sum e(k) \tag{4.106}$$

and can be rewritten in an incremental form

$$\Delta u(k) = K_{\mathrm{P}}\Delta e(k) + K_{\mathrm{I}}e(k) \tag{4.107}$$

where the control applied to the plant is given by $u(k) = u(k-1) + \Delta u(k)$.

Let us develop a fuzzy PI-type controller. Clearly, the system has two inputs: the error $e(k)$ and the change-of-error $\Delta e(k)$, $I = 2$. For simplicity, let us choose $P_1 = 5$ with linguistic labels negative big (NB), negative small (NS), zero (Z), positive small (PS), positive large (PL) defined by the centers of triangular add-1 fuzzy sets $\boldsymbol{\beta}_1 = [\beta_{1,1},\cdots,\beta_{1,5}]$, and $P_2 = 3$ (negative (N), zero (Z), positive (P) set by $\boldsymbol{\beta}_2 = [\beta_{2,1},\beta_{2,2},\beta_{2,3}]$).

This can be written as a Sugeno model

$$\Delta u(k) = \sum_{p_1=1}^{5}\sum_{p_2=1}^{3}\tilde{\alpha}_{p_1,p_2}\tilde{\mu}_{1,p_1}\left(e(k),\boldsymbol{\beta}_1\right)\tilde{\mu}_{2,p_2}\left(\Delta e(k),\boldsymbol{\beta}_2\right) \tag{4.108}$$

where $\tilde{\mu}_{1,1}$, $\tilde{\mu}_{1,2}$, ... represent the degrees of membership for the propositions (fuzzy predicates)

$$\begin{array}{r}\textit{error} \text{ is } \textit{negative big} \\ \textit{error} \text{ is } \textit{negative small}\end{array} \tag{4.109}$$

$$\vdots$$

*etc.* Similarly, the products $\tilde{\mu}_{1,1}\tilde{\mu}_{2,1}, \tilde{\mu}_{1,1}\tilde{\mu}_{2,2},\cdots$ can be interpreted as the truth values of the propositions

$$\begin{array}{r}\textit{error} \text{ is } \textit{negative big} \text{ and } \textit{change-of-error} \text{ is } \textit{negative} \\ \textit{error} \text{ is } \textit{negative big} \text{ and } \textit{change-of-error} \text{ is } \textit{zero}\end{array} \tag{4.110}$$

$$\vdots$$

*etc.* The entire rule base can be collected in a table format, showing the values of $\Delta u(k)$

$$
\begin{array}{c|ccc}
e(k) \backslash \Delta e(k) & N & Z & P \\
\hline
\text{NB} & \widetilde{\alpha}_{1,1} & \widetilde{\alpha}_{1,2} & \widetilde{\alpha}_{1,3} \\
\text{NS} & \widetilde{\alpha}_{2,1} & \widetilde{\alpha}_{2,2} & \widetilde{\alpha}_{2,3} \\
\text{Z} & \widetilde{\alpha}_{3,1} & \widetilde{\alpha}_{3,2} & \widetilde{\alpha}_{3,3} \\
\text{PS} & \widetilde{\alpha}_{4,1} & \widetilde{\alpha}_{4,2} & \widetilde{\alpha}_{4,3} \\
\text{PB} & \widetilde{\alpha}_{5,1} & \widetilde{\alpha}_{5,2} & \widetilde{\alpha}_{5,3}
\end{array}
\tag{4.111}
$$

Often, linguistic labels are also assigned for the output singletons $\widetilde{\alpha}_{p_1,p_2}$, in order to further enhance the transparency of the controller.

Next, let us consider the derivatives of a 0-order add-1 Sugeno model. The model is given by

$$
\mathbf{f}(\boldsymbol{\varphi}, \cdot) = \sum_{p_1=1}^{P_1} \sum_{p_2=1}^{P_2} \cdots \sum_{p_I=1}^{P_I} \widetilde{\alpha}_{p_1,p_2,\cdots,p_I} \prod_{i=1}^{I} \widetilde{\mu}_{i,p_i}(\varphi_i, \cdot)
\tag{4.112}
$$

The derivatives with respect to the parameters $\widetilde{\alpha}_{p_1,p_2,\cdots,p_I}$ are simple to calculate. It follows

$$
\frac{\partial \mathbf{f}(\boldsymbol{\varphi}, \cdot)}{\partial \widetilde{\alpha}_{p_1,p_2,\cdots,p_I}} = \prod_{i=1}^{I} \widetilde{\mu}_{i,p_i}(\varphi_i, \cdot)
\tag{4.113}
$$

If only $\widetilde{\alpha}$s are of interest, notice that these parameters appear linearly and, e.g., least squares can be used for their estimation. Also the gradients with respect to parameters $\widetilde{\beta}$ can be calculated. However, the tuning of fuzzy sets using data is more complicated due to various reasons (especially the transparency of the model may easily be lost). Therefore, this is omitted here.

In order to get a linearized approximation of the Sugeno model in the neighborhood of its operating point $\widetilde{\boldsymbol{\varphi}}$

$$
\widetilde{\mathbf{f}}\left(\boldsymbol{\varphi}, \widetilde{\widetilde{\boldsymbol{\alpha}}}\right) = \widetilde{\widetilde{\alpha}}_1 \varphi_1 + \ldots + \widetilde{\widetilde{\alpha}}_I \varphi_I + \widetilde{\widetilde{\alpha}}_{I+1}
\tag{4.114}
$$

the derivatives with respect to the inputs need to be computed:

$$
\widetilde{\widetilde{\alpha}}_i = \frac{\partial}{\partial \varphi_i} \mathbf{f}(\boldsymbol{\varphi}, \cdot) = \frac{\partial}{\partial \varphi_i} \sum_{p_1=1}^{P_1} \sum_{p_2=1}^{P_2} \cdots \sum_{p_I=1}^{P_I} \widetilde{\alpha}_{p_1,p_2,\cdots,p_I} \prod_{i=1}^{I} \widetilde{\mu}_{i,p_i}(\varphi_i, \cdot)
\tag{4.115}
$$

Separating terms not depending on $\varphi_i$ and moving the derivation operator inside the summation gives

$$\frac{\partial}{\partial \varphi_i} f(\varphi, \cdot) = \sum_{p_1=1}^{P_1} \sum_{p_2=1}^{P_2} \cdots \sum_{p_I=1}^{P_I} \widetilde{\alpha}_{p_1,p_2,\cdots,p_I} \prod_{j=1;j\neq i}^{I} \widetilde{\mu}_{j,p_j}(\varphi_j, \cdot) \frac{\partial}{\partial \varphi_i} \widetilde{\mu}_{i,p_i}(\varphi_i, \cdot)$$

(4.116)

where the derivative of the triangular membership function, (4.103), is given by

$$\frac{\partial}{\partial \varphi_i} \widetilde{\mu}_{i,p_i}\left(\varphi_i, \widetilde{\beta}_i\right) = \begin{cases} 1/\left(\widetilde{\beta}_{i,p_i} - \widetilde{\beta}_{i,p_i-1}\right) & \text{if } \widetilde{\beta}_{i,p_i-1} < \varphi_i \le \widetilde{\beta}_{i,p_i} \\ -1/\left(\widetilde{\beta}_{i,p_i+1} - \widetilde{\beta}_{i,p_i}\right) & \text{if } \widetilde{\beta}_{i,p_i} < \varphi_i \le \widetilde{\beta}_{i,p_i+1} \\ 0 & \text{otherwise} \end{cases}.$$

(4.117)

Notice that the gradient (4.117) is a piecewise constant. Thus, the considered Sugeno model can be seen as a piecewise multi-linear system and the interpolation properties of the system are particularly well defined.

**Example 30 (Fuzzy PI controller: continued)** Assume that at present a plant operates under a linear PI controller (or that this has been designed using, *e.g.*, the Ziegler–Nichols rules), and that this PI controller is to be improved by designing a fuzzy PI controller. Thus, the parameters $K_P$ and $K_I$ of a linear PI controller $\Delta u(k) = K_P \Delta e(k) + K_I e(k)$ are *a priori* known. In order to use the nominal system as a starting point in the design of a fuzzy PI control, the equivalent fuzzy representation is needed.

First, the input space needs to be partitioned. Assume that reasonable bounds $[e_{\min}, e_{\max}]$ and $[\Delta e_{\min}, \Delta e_{\max}]$ can be set. Initially, we can place the centers of add-1 triangular fuzzy sets, *e.g.*, at equidistant intervals (using $P_1 = 5$, $P_2 = 3$):

$$\widetilde{\beta}_1 = [e_{\min}, e_{\min} + z_e, e_{\min} + 2z_e, e_{\min} + 3z_e, e_{\max}]$$

(4.118)

$$\widetilde{\beta}_2 = [\Delta e_{\min}, \Delta e_{\min} + z_{\Delta e}, \Delta e_{\max}]$$

(4.119)

where $z_e = \frac{(e_{\max} - e_{\min})}{4}$ and $z_{\Delta e} = \frac{(\Delta e_{\max} - \Delta e_{\min})}{2}$. The $\widetilde{\alpha}_{p_1,p_2}$ remain to be specified. As the nominal system is given, we can set the $\widetilde{\alpha}_{1,1}, \widetilde{\alpha}_{2,1}, \cdots, \widetilde{\alpha}_{5,3}$ to correspond to the system output at points $\left(\widetilde{\beta}_{1,1}, \widetilde{\beta}_{2,1}\right)$, $\left(\widetilde{\beta}_{1,2}, \widetilde{\beta}_{2,1}\right)$, $\cdots$,

$\left(\widetilde{\beta}_{1,5}, \widetilde{\beta}_{2,3}\right)$ by assigning

$$
\begin{aligned}
\widetilde{\alpha}_{1,1} &= K_P \widetilde{\beta}_{2,1} + K_I \widetilde{\beta}_{1,1} \qquad\qquad (4.120) \\
\widetilde{\alpha}_{2,1} &= K_P \widetilde{\beta}_{2,1} + K_I \widetilde{\beta}_{1,2} \\
&\ \ \vdots \\
\widetilde{\alpha}_{5,3} &= K_P \widetilde{\beta}_{2,3} + K_I \widetilde{\beta}_{1,5}
\end{aligned}
$$

Since this type of Sugeno model is (piecewise multi-)linear, the resulting fuzzy PI now produces exactly the same function as the nominal linear PI, as long as the inputs are within the given ranges, *i.e.*, $e \in [e_{\min}, e_{\max}]$ and $\Delta e \in [\Delta e_{\min}, \Delta e_{\max}]$.

Let us conclude this section by making a few remarks.

**Remark 4 (Extension principle)** The *extension principle*, or *compositional rule of inference*, is a means for extending any mapping of a fuzzy set from one space to another. Let $A$ be a fuzzy set defined in $X$, and f be a mapping from $X$ to $Y$, f:$X \to Y$. Then a mapping of $A$ via f is a fuzzy set $\mu_B(A)$ defined in $Y$. The membership function is computed according to:

$$
[\mu_B(A)](y) = \sup_{\text{all } x \in X \text{ for which } y=f(x)} [\mu_A(x)] \qquad\qquad (4.121)
$$

assuming that $\sup \emptyset = 0$ (when no element of $X$ is mapped to $y$). In the MISO case, we have $[\mu_B(A)](y) = \sup_{\text{all } x \in X, \text{ for which } y=f(x)} [\mu_A(\mathbf{x})]$ where $\mu_A(\mathbf{x}) = T[\mu_{A_1}(x_1), \cdots, \mu_{A_I}(x_I)]$.

**Example 31 (Extension principle)** Figure 4.7 illustrates the extension principle for mapping a fuzzy set $A$ (characterized by $\mu_A$) through a function f. The result is a fuzzy set $B$ (characterized by $\mu_B$).

**Example 32 (0-order Sugeno fuzzy system)** Figure 4.8 shows an illustration of the extension principle for a fuzzy input $A'$ and a function given by sampled data points $\{y\} = f\{x\}$. The output is a fuzzy set on a discrete domain $Y$. Note that the fuzzy input $A'$ may be a result of fuzzification of a non-fuzzy input $x_0$. The 'fuzziness' in $A'$ together with a defuzzification method then determines the interpolation/smoothing properties of the system.

**Example 33 (0-order Sugeno fuzzy system: continued)** Figure 4.9 shows an illustration of the extension principle for a crisp input $x_0$ and a

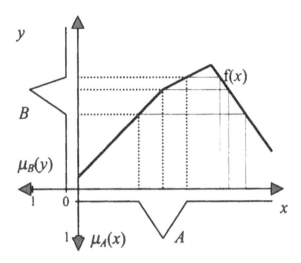

Figure 4.7: Mapping a fuzzy set $A$ through a function f.

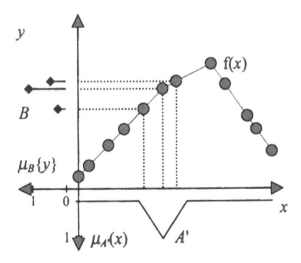

Figure 4.8: Mapping a fuzzy input $A$ through a 'function' given by sampled data points.

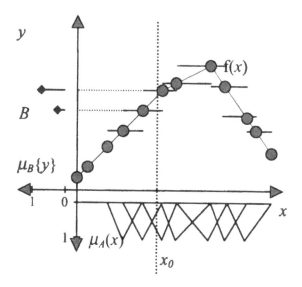

Figure 4.9: Mapping a crisp input through a 'function' given by fuzzy rules.

function given by 'fuzzy rules' (constant local models). The outputs are fuzzy singletons on a discrete domain $Y$. Note that the input is crisp (or point fuzzification is used). The fuzziness in rule antecedents, as well as the defuzzification method, determines the interpolation properties of the system.

**Remark 5 (Fuzzy neural networks)** During the past few years, the close connections between fuzzy and neural systems have been recognized (see, *e.g.*, [86][34]). *Fuzzy neural networks* try to benefit from the advantages of both neural and fuzzy approaches. Functional equivalence of some neural and fuzzy paradigms has been established, and common frameworks, such as the (generalized) basis function network, have been introduced. The links to the 'old' methods of parameter estimation have become apparent, which has enabled the application of efficient parameter estimation methods.

Fuzzy neural networks emphasize that the model contents can be presented as linguistic rules or as numerical parameters. The former allows the use of human experimental knowledge in initializing model parameters, complementing missing data, and validating the identified model. The latter enables, *e.g.*, the application of efficient optimization methods for parameter estimation. In most fuzzy neural network approaches found in the literature, the learning abilities of neural networks are applied to structures sharing the transparent logical interpretability of fuzzy systems.

# Chapter 5

# Non-linear Dynamic Structures

The best approach for describing non-linear dynamic systems is to consider the *a priori* physical information about the system to be characterized. In many cases, suitable information is not available and the designer needs to turn into semi-empirical or black-box methods. In nonlinear dynamic systems, the output of the system depends, often in a complex way, on the past outputs, inputs or internal components of the system. The main problems in system identification are in structure selection, whereas efficient gradient-based or guided random search methods are available for solving the associated parameter estimation problems, even if the model is not linear with respect to the parameters.

A direct extension of linear dynamic models is the *Volterra series* representation [92]. The Volterra representation is very general. In practice, however, a finite truncation of the series must be used, and a discrete approximation of the series made. For a SISO system, a Volterra model can be given by

$$y(k) = w_0 + \sum_{n=0}^{N} w_n u(k-n) + \dots \tag{5.1}$$

$$\sum_{n_1=0}^{N} \sum_{n_2=n_1}^{N} \cdots \sum_{n_P=n_{P-1}}^{N} w_{n_1,n_2,\cdots,n_P} u(k-n_1) u(k-n_2) \cdots u(k-n_P)$$

where $y(k)$ and $u(k)$ are the system output and input at discrete time instant $k$. The system parameters are given by $w_0$, $w_n$, $\cdots$, $w_{n_1,n_2,\cdots,n_P}$ ($n, n_i = 1, 2, ..., N$; $i = 1, 2, ..., P$). $N$ and $P$ are the orders of the system, respectively. The order $N$ is related to the length of the time window (zeros of the polynomials), and the order $P$ is related to the non-linearity of the mapping. The model output is linear with respect to its parameters, which makes the parameter estimation simple. Extension into the MISO case is straightforward. However, there are theoretical and practical drawbacks associated with the Volterra series [92]. In particular, the system may contain a large amount of parameters and suffer from the curse of dimensionality. Due to this, practical applications of Volterra series are often limited to first and second order terms.

The static non-linearity in the Volterra models can be approximated by alternative structures, providing more convenient means for

- including *a priori* knowledge,

- handling of incomplete and noisy data sets,

- more efficient parameterization,

- increasing the transparency of the model,

- improved data compression, *etc.*

In what follows, two types of nonlinear black-box dynamic structures are considered:

- Non-linear time-series, and

- Wiener and Hammerstein models.

In both structures, the non-linear function is a static one. The capability to characterize dynamical process behavior is obtained using delayed inputs and external feedback (non-linear time-series), or internal feedback using linear dynamic filters (Wiener and Hammerstein models).

## 5.1   Non-linear time-series models

There are a large number of different black-box approaches for describing non-linear dynamic systems. In process identification, non-linear dynamic black-box *time-series* structures are common. The ability to characterize dynamical process behavior is obtained by using delayed inputs and external feedback. For most practical purposes in process identification, MISO non-linear dynamic systems can be described with sufficient accuracy using the NARX, NOE and NARMAX time-series structures. The structure determines the inputs to the model, where only externally recurrent feedback connections are allowed. For modelling very complex non-linear dynamic systems, fully recurrent systems can also be considered.

Denote a non-linear static function by f, a function of some parameters $\mathbf{w}$. In the NOE time-series structure the predictor input consists of past inputs of the process and the past predictions of the process output:

$$\widehat{y}(k) = \mathrm{f}\left(u\left(k-d\right),...,u\left(k-d-n_B\right),\widehat{y}\left(k-1\right),...,\widehat{y}\left(k-n_A\right),\mathbf{w}\right) \quad (5.2)$$

In the NARX structure the input consists of past inputs and outputs of the process:

$$\widehat{y}(k) = \mathrm{f}\left(u\left(k-d\right),...,u\left(k-d-n_B\right),y\left(k-1\right),...,y\left(k-n_A\right),\mathbf{w}\right) \quad (5.3)$$

In the NARMAX structure the input consists of past inputs and outputs of the process, as well as past predictions:

$$\begin{aligned}\widehat{y}(k) \;=\; & \mathrm{f}(u\left(k-d\right),...,u\left(k-d-n_B\right), \qquad\qquad (5.4)\\ & y\left(k-1\right),...,y\left(k-n_A\right),\\ & \widehat{y}\left(k-1\right),...,\widehat{y}\left(k-n_C\right),\mathbf{w})\end{aligned}$$

The NARMAX structure is shown in Fig. 5.1. Notice, that the NOE and NARX structures can be seen as special cases of the NARMAX structure.

The structure of the mapping f between the inputs and the output is not determined. If no *a priori* information about the structure of the process is available, it is common to choose some black-box structure: power series, sigmoid neural networks, or 0-order Sugeno fuzzy system (among many others, see Chapter 4). In practice, process modelling using NOE, NARX and NARMAX structures can give accurate predictions on a fixed data set. Most importantly, it is possible to model a wide class of non-linearities. If some non-linear black-box structure is chosen for the static function f, practically all reasonable dynamic functions can be approximated (provided that the input data windows are long enough, and the size (number of parameters) of the black-box model is sufficiently large). The approach is simple, as it extends the linear dynamic time-series structures to non-linear combinations of the inputs. If the mapping f is a linear one, ARX, OE and ARMAX structures result (see Chapter 3).

The main problem with these structures concerns the identification of the static non-linear function f. The complexity (degrees of freedom) of the mapping depends on the structure chosen for the non-linear (parameterized) function. In nonlinear black-box structures, the degree of freedom is usually large since the restriction of linearity of the mapping is removed. Technically, it is simple to apply some gradient-based optimization method with the NARX structure; with NOE and NARMAX the need to take into account the dynamics when computing the gradients increases slightly the

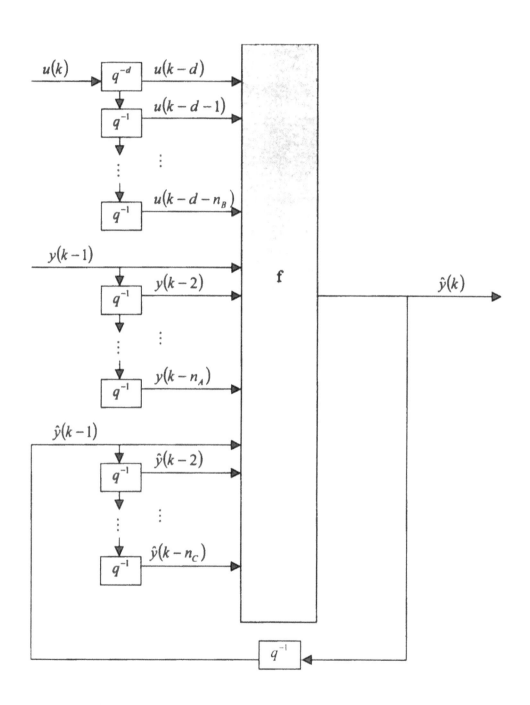

Figure 5.1: NARMAX time-series predictor.

need of computations. However, too many degrees of freedom in f make the parameters **w** sensitive to noise in data, and poor interpolation can be expected if the data set does not contain enough information (covering the whole operating range and all dynamic situations of interest). In general, the extrapolation properties of non-linear time-series models are always poor.

These problems can be tackled in parameter estimation by using optimization under constraints, where constraints can be posed on the structure (regularization), based on *a priori* known properties of the process, or, *e.g.*, deviation from a nominal model [63]. Alternatively, the degrees of freedom in the mapping can be reduced.

Let us next consider the gradients of the general nonlinear black-box time-series models. These are required by gradient-based parameter estimation techniques (Chapter 6).

## 5.1.1 Gradients of non-linear time-series models

For simplicity of notation, let us restrict to SISO systems (extending to MISO is straightforward.) Consider a non-linear time-series NARMAX predictor (5.4), see also Fig. 5.1. Let us calculate the gradient $\partial \widehat{y} / \partial w_j$ of the system output $\widehat{y}$ with respect to its parameters $w_j$, $\mathbf{w} = [w_1, ..., w_j, ...w_J]^T$, $j = 1.2, ..., J$.

For simplicity of notation, denote $f(\mathbf{u}(k-d), \mathbf{y}(k-1), \widehat{\mathbf{y}}(k-1), \mathbf{w})$ by $f(k, \mathbf{w})$. Let us linearize the function f (5.4) around an operating point $\overline{\mathbf{x}} = \left\{ \overline{\mathbf{u}}, \overline{\mathbf{y}}, \overline{\widehat{\mathbf{y}}}, \overline{\mathbf{w}} \right\}$. Using Taylor series, we have that

$$f(k, \mathbf{w}) \approx \widetilde{f}(k, \mathbf{w}) \tag{5.5}$$

$$= \left[ \frac{\partial f}{\partial \mathbf{u}}(k, \mathbf{w}) \right]_{\mathbf{x}=\overline{\mathbf{x}}} \widetilde{\mathbf{u}}(k-d) \tag{5.6}$$

$$+ \left[ \frac{\partial f}{\partial \mathbf{y}}(k, \mathbf{w}) \right]_{\mathbf{x}=\overline{\mathbf{x}}} \widetilde{\mathbf{y}}(k-1)$$

$$+ \left[ \frac{\partial f}{\partial \widehat{\mathbf{y}}}(k, \mathbf{w}) \right]_{\mathbf{x}=\overline{\mathbf{x}}} \widetilde{\widehat{\mathbf{y}}}(k-1, \mathbf{w})$$

$$+ c(k, \mathbf{w})$$

where $c$ is a constant ($c(k, \mathbf{w}) = [f(k, \mathbf{w})]_{\mathbf{x}=\overline{\mathbf{x}}}$) and the tilded variables denote the deviation from the point of linearization ($\mathbf{u} = \overline{\mathbf{u}} + \widetilde{\mathbf{u}}$, *etc.*). The notation $[\cdot]_{\mathbf{x}=\overline{\mathbf{x}}}$ indicates that the expression is evaluated at the point of linearization.

We then have that

$$
\begin{aligned}
\frac{\partial \widehat{y}}{\partial w_j}(k) \approx{} & \left[\frac{\partial}{\partial w_j}\left[\frac{\partial \mathbf{f}}{\partial \mathbf{u}}(k,\mathbf{w})\right]_{\mathbf{x}=\overline{\mathbf{x}}}\right]\widetilde{\mathbf{u}}(k-d) \\
& + \left[\frac{\partial}{\partial w_j}\left[\frac{\partial \mathbf{f}}{\partial \mathbf{y}}(k,\mathbf{w})\right]_{\mathbf{x}=\overline{\mathbf{x}}}\right]\widetilde{\mathbf{y}}(k-1) \\
& + \frac{\partial}{\partial w_j}\left[\left[\frac{\partial \mathbf{f}}{\partial \widehat{\mathbf{y}}}(k,\mathbf{w})\right]_{\mathbf{x}=\overline{\mathbf{x}}}\widetilde{\widehat{\mathbf{y}}}(k-1,\mathbf{w})\right] \\
& + \frac{\partial}{\partial w_j}c(k,\mathbf{w})
\end{aligned}
\tag{5.7}
$$

since the $\widetilde{\mathbf{u}}$ and $\widetilde{\mathbf{y}}$ do not depend on $w_j$, $j=1,2,...,J$, whereas $\mathbf{f}$ and $\widehat{\widetilde{\mathbf{y}}}$ do. For the third term on the right hand side we have that

$$
\begin{aligned}
& \frac{\partial}{\partial w_j}\left[\left[\frac{\partial \mathbf{f}}{\partial \widehat{\mathbf{y}}}(k,\mathbf{w})\right]_{\mathbf{x}=\overline{\mathbf{x}}}\widetilde{\widehat{\mathbf{y}}}(k-1,\mathbf{w})\right] \\
& = \left[\frac{\partial \mathbf{f}}{\partial \widehat{\mathbf{y}}}(k,\mathbf{w})\right]_{\mathbf{x}=\overline{\mathbf{x}}}\left[\frac{\partial}{\partial w_j}\widetilde{\widehat{\mathbf{y}}}(k-1,\mathbf{w})\right] \\
& + \left[\frac{\partial}{\partial w_j}\left[\frac{\partial \mathbf{f}}{\partial \widehat{\mathbf{y}}}(k,\mathbf{w})\right]_{\mathbf{x}=\overline{\mathbf{x}}}\right]\widetilde{\widehat{\mathbf{y}}}(k-1,\mathbf{w})
\end{aligned}
\tag{5.8}
$$

Substituting (5.8) to (5.7) and reorganizing gives

$$
\begin{aligned}
\frac{\partial \widehat{y}}{\partial w_j}(k) \approx{} & \frac{\partial}{\partial w_j}\left[\left[\frac{\partial \mathbf{f}}{\partial \mathbf{u}}(k,\mathbf{w})\right]_{\mathbf{x}=\overline{\mathbf{x}}}\widetilde{\mathbf{u}}(k-d) + \left[\frac{\partial \mathbf{f}}{\partial \mathbf{y}}(k,\mathbf{w})\right]_{\mathbf{x}=\overline{\mathbf{x}}}\widetilde{\mathbf{y}}(k-1)\right. \\
& \left. + \left[\frac{\partial \mathbf{f}}{\partial \widehat{\mathbf{y}}}(k,\mathbf{w})\right]_{\mathbf{x}=\overline{\mathbf{x}}}\widetilde{\widehat{\mathbf{y}}}(k-1) + c(k,\mathbf{w})\right] \\
& + \left[\frac{\partial \mathbf{f}}{\partial \widehat{\mathbf{y}}}(k,\mathbf{w})\right]_{\mathbf{x}=\overline{\mathbf{x}}}\left[\frac{\partial}{\partial w_j}\widetilde{\widehat{\mathbf{y}}}(k-1)\right] \\
\approx{} & \left[\frac{\partial \mathbf{f}(k,\mathbf{w})}{\partial w_j}\right]_{\mathbf{x}=\overline{\mathbf{x}}} + \left[\frac{\partial \mathbf{f}}{\partial \widehat{\mathbf{y}}}(k,\mathbf{w})\right]_{\mathbf{x}=\overline{\mathbf{x}}}\left[\frac{\partial \widehat{y}}{\partial w_j}(k-1)\right]
\end{aligned}
\tag{5.9}\tag{5.10}
$$

since $\frac{\partial}{\partial w_j}\widetilde{\widehat{\mathbf{y}}}(k-1,\mathbf{w})=\frac{\partial}{\partial w_j}\widehat{\mathbf{y}}(k-1,\mathbf{w})$. Thus, the gradient is composed of two terms: the static gradient (first term on the right) and the dynamic effect of the gradient (second term).

Let us summarize the results by writing the above in a more convenient form.

**Algorithm 19 (Gradients of NARMAX predictor)** The gradients for a NARMAX time-series model

$$
\begin{aligned}
\widehat{y}(k) \;=\; & f(u\,(k-d)\,,...,u\,(k-d-n_B)\,, \\
& y\,(k-1)\,,...,y\,(k-n_A)\,, \\
& \widehat{y}\,(k-1)\,,...,\widehat{y}\,(k-n_C)\,,\mathbf{w})
\end{aligned}
\tag{5.11}
$$

with parameters $\mathbf{w} = [w_1,...,w_j,...w_J]^T$ are obtained from

$$
\Psi_j\,(k) = \frac{\partial f}{\partial w_j}\,(k,\mathbf{w}) + \sum_{m=1}^{n_C} \Phi_{\widehat{y}(k-m)}\,(k,\mathbf{w})\,\Psi_j\,(k-m,\mathbf{w})
\tag{5.12}
$$

where $\Psi_j$ denotes the gradient of the model output with respect to its parameters:

$$
\frac{\partial \widehat{y}}{\partial w_j}\,(k) \leftarrow \Psi_j\,(k,\mathbf{w})
\tag{5.13}
$$

$\frac{\partial f}{\partial w_j}\,(k,\mathbf{w})$ $(j = 1,2,...,J)$ are the static gradients of the non-linear function with respect to its parameters. The second term gives the dynamic effect of the feedback in the network to the gradients, a correction by the linearized gain:

$$
\Phi_{\widehat{y}(k-m)}\,(k,\mathbf{w}) \leftarrow \frac{\partial f\,(k,\mathbf{w})}{\partial \widehat{y}\,(k-m)}
\tag{5.14}
$$

**Example 34 (ARMAX structure)** Let us illustrate the above using a simple linear dynamic system

$$
\widehat{y}\,(k) = a y\,(k) + b u\,(k-1) + c \widehat{y}\,(k-1)
\tag{5.15}
$$

The system has three parameters, $\mathbf{w} = [a,b,c]^T$, the function f is linear. The gradients of the static (linear) function are given by:

$$
\frac{\partial f}{\partial w_j}\,(k,\mathbf{w}) = \begin{bmatrix} y\,(k) \\ u\,(k-1) \\ \widehat{y}\,(k-1,\mathbf{w}) \end{bmatrix}
\tag{5.16}
$$

$$
\frac{\partial f\,(k,\mathbf{w})}{\partial \widehat{y}\,(k-1)} = c
\tag{5.17}
$$

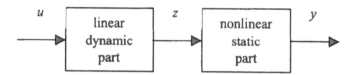

Figure 5.2: A Wiener system. The system input $u$ is put through a linear filter, and a nonlinear mapping of the intermediate signal $z$ gives the system output $y$.

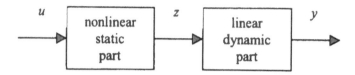

Figure 5.3: A Hammerstein system. A nonlinear mapping of the input signal $u$ gives the intermediate signal $z$. The system output $y$ is the output of a linear filter.

The gradient of the system output with respect to its parameters is given by

$$\Psi(k) = \begin{bmatrix} \frac{\partial \hat{y}(k)}{\partial a} \\ \frac{\partial \hat{y}(k)}{\partial b} \\ \frac{\partial \hat{y}(k)}{\partial c} \end{bmatrix} = \begin{bmatrix} y(k) \\ u(k-1) \\ \hat{y}(k-1, \mathbf{w}) \end{bmatrix} + c\Psi(k-1) \qquad (5.18)$$

Notice that although f is linear, the system output is not linear since the gradients depend also on past data. A similar result was derived in section 3.3.8.

## 5.2  Linear dynamics and static non-linearities

In many cases, dynamics of the non-linear process can be approximated using linear transfer functions for describing the system dynamics. Wiener and Hammerstein structures are typical examples of such structures. A restricted class of Wiener and Hammerstein systems will be considered next.

Wiener and Hammerstein structures consist of a linear dynamic part and a non-linear static part. In a Wiener structure (see Fig. 5.2), the linear dynamic part is followed by the non-linear part. In a Hammerstein structure (see Fig. 5.3), the non-linear part precedes the linear dynamic part.

## 5.2.1 Wiener systems

Assume a SISO Wiener system given by

$$y(k) = f(z(k)) \tag{5.19}$$

where

$$z(k) = \frac{B(q^{-1})}{A(q^{-1})}u(k-d) \tag{5.20}$$

$y(k)$ is the output of the Wiener system, f is a non-linear static SISO function, $z(k)$ is an intermediate variable, $u(k)$ is the input to the system and $d$ is the time delay. $A(q^{-1})$ and $B(q^{-1})$ are polynomials in the backward shift operator $q^{-1}$ :

$$
\begin{aligned}
A(q^{-1}) &= 1 + a_1 q^{-1} + \ldots + a_{n_A} q^{-n_A} \tag{5.21} \\
B(q^{-1}) &= b_0 + b_1 q^{-1} + \ldots + b_{n_B} q^{-n_B} \tag{5.22}
\end{aligned}
$$

Obviously, a predictor for the above deterministic system is given by

$$\widehat{y}(k) = f(\widehat{z}(k)) \tag{5.23}$$

where $A(q^{-1})\widehat{z}(k) = B(q^{-1})u(k-d)$. But this is also the predictor for an OE-system. This leads to consider the following stochastic process:

$$y(k) = f\left[\frac{B(q^{-1})}{A(q^{-1})}u(k-d) + e(k)\right] \tag{5.24}$$

Let us rewrite the noise term:

$$
\begin{aligned}
y(k) &= f[\widehat{z}(k) + e(k)] \tag{5.25} \\
&= f[\widehat{z}(k)] + \{f[\widehat{z}(k) + e(k)] - f[\widehat{z}(k)]\} \tag{5.26} \\
&= f[\widehat{z}(k)] + e_y(k) \tag{5.27}
\end{aligned}
$$

where the prediction error now appears at the output of the Wiener system:

$$e_y(k) = y(k) - \widehat{y}(k) \tag{5.28}$$

Although this may seem nice, the transition from $e(k)$ to $e_y(k)$ is critical. From a statistical point of view, in linear systems the properties of $\{e(k)\}$ convey to $\{e_y(k)\}$ (*e.g.*, if $e(k)$ has Gaussian distribution, then $e_y(k)$ remains Gaussian, too). For nonlinear systems, this is not the case, and even as an approximation it is valid only locally around an operating point provided that the function f is smooth enough[1].

Let us consider the following stochastic Wiener system

$$y\left(k\right) = \mathbf{f}\left(\frac{B\left(q^{-1}\right)}{A\left(q^{-1}\right)}u\left(k-d\right)\right) + e\left(k\right) \tag{5.29}$$

where $\{e\left(k\right)\}$ is a sequence of independent random variables with zero mean and finite variance. The predictor for such a system is given by

$$\widehat{y}\left(k\right) = \mathbf{f}\left(\frac{B\left(q^{-1}\right)}{A\left(q^{-1}\right)}u\left(k-d\right)\right) \tag{5.30}$$

and minimizes the expectation of the squared prediction error[2]. In general, the non-linear system is a function of some parameters $\mathbf{w}$. Hence we have the expression for a SISO Wiener predictor:

$$\widehat{y}\left(k\right) = \mathbf{f}\left(\frac{B\left(q^{-1}\right)}{A\left(q^{-1}\right)}u\left(k-d\right), \mathbf{w}\right) \tag{5.31}$$

It is straightforward to extend these results for MISO systems with multiple linear dynamic systems (one for each input). Let us first define a Wiener system.

---

[1] Under some conditions related to the non-linear mapping f and its inverse $\mathbf{f}^{-1}$ (continuity, differentiability, *etc.*), the density function of the output of the nonlinear system can be expressed as a function of the density function of the input (see [39], p. 34, see also [71]).

[2] Let us find

$$\widehat{y}\left(k\right) = \arg\min_{\widehat{y}} E\left\{\left[y\left(k\right) - \widehat{y}\right]^2\right\}$$

Substituting (5.29) we have that

$$
\begin{aligned}
E\left\{\left[y\left(k\right) - \widehat{y}\right]^2\right\} &= E\left\{\left[\mathbf{f}\left(\frac{B\left(q^{-1}\right)}{A\left(q^{-1}\right)}u\left(k-d\right)\right) + e\left(k\right) - \widehat{y}\right]^2\right\} \\
&= E\left\{\left[\mathbf{f}\left(\frac{B\left(q^{-1}\right)}{A\left(q^{-1}\right)}u\left(k-d\right)\right) - \widehat{y}\right]^2\right\} \\
&\quad + E\left\{\left[\mathbf{f}\left(\frac{B\left(q^{-1}\right)}{A\left(q^{-1}\right)}u\left(k-d\right)\right) - \widehat{y}\right]e\left(k\right)\right\} \\
&\quad + E\left\{e^2\left(k\right)\right\}
\end{aligned}
$$

where the second term is zero (due to the independence of $e\left(k\right)$ with respect to $u\left(k-d\right)$ and $\widehat{y}$ and that $E\left\{e\left(k\right)\right\} = 0$). If the variance is finite, the criterion is minimized by (5.30).

**Definition 12 (Wiener system)** Define a MISO Wiener system by

$$y(k) = f(z(k)) + e(k) \qquad (5.32)$$

where $z(k) = [z_1(k), z_2(k), ..., z_i(k), ..., z_I(k)]^T$ are given by

$$z_i(k) = \frac{B_i(q^{-1})}{A_i(q^{-1})} u_i(k - d_i) \qquad (5.33)$$

$y(k)$ is the output of the Wiener system, f is a non-linear static MISO function, $z_i(k)$ are intermediate variables, $u_i(k)$ are the inputs to the system and $d_i$ are the time delays. $A_i(q^{-1})$ and $B_i(q^{-1})$ are polynomials in the backward shift operator $q^{-1}$ :

$$A_i(q^{-1}) = 1 + a_{i,1}q^{-1} + ... + a_{i,n_{A_i}}q^{-n_{A_i}} \qquad (5.34)$$
$$B_i(q^{-1}) = b_{i,0} + b_{i,1}q^{-1} + ... + b_{i,n_{B_i}}q^{-n_{B_i}} \qquad (5.35)$$

and $i = 1, 2, ..., I$, where $I$ is the number of inputs to the system.

Note, that a general Wiener system may have a single MIMO linear dynamic part. Here we restrict to the case of multiple SISO linear dynamic parts. The MISO predictor can be derived in a way similar to the SISO case.

**Algorithm 20 (Wiener predictor)** A predictor for a MISO Wiener system is given by

$$\widehat{y}(k) = f(\widehat{z}(k), w) \qquad (5.36)$$

where

$$\widehat{z}_i(k) = \frac{B_i(q^{-1})}{A_i(q^{-1})} u_i(k - d_i) \qquad (5.37)$$

$\widehat{y}(k)$ is the predicted output of the Wiener system, f is the non-linear static MISO function of parameters $w$, $u_i(k)$ and $\widehat{z}_i(k)$ are the input to the system and intermediate variables, respectively, and $d_i$ are the time delays. $A_i(q^{-1})$ and $B_i(q^{-1})$ are polynomials in the backward shift operator $q^{-1}$ :

$$A_i(q^{-1}) = 1 + a_{i,1}q^{-1} + ... + a_{i,n_{A_i}}q^{-n_{A_i}} \qquad (5.38)$$
$$B_i(q^{-1}) = b_{i,0} + b_{i,1}q^{-1} + ... + b_{i,n_{B_i}}q^{-n_{B_i}} \qquad (5.39)$$

and $i = 1, 2, ..., I$, where $I$ is the number of inputs to the system.

Let us next consider Hammerstein systems.

## 5.2.2   Hammerstein systems

**Definition 13 (Hammerstein system)** Define a MISO Hammerstein system by

$$y(k) = \frac{B(q^{-1})}{A(q^{-1})} f(u(k), w) + e(k) \tag{5.40}$$

where $u(k) = [u_1(k), u_2(k), ..., u_i(k), ..., u_I(k)]^T$ , $i = 1, 2, ..., I$ where $I$ is the number of inputs to the system, $y(k)$ is the output of the Hammerstein system, and f is a non-linear static MISO function of parameters $w$. $A(q^{-1})$ and $B(q^{-1})$ are polynomials in the backward shift operator $q^{-1}$ :

$$A(q^{-1}) = 1 + a_1 q^{-1} + ... + a_{n_A} q^{-n_A} \tag{5.41}$$
$$B(q^{-1}) = b_0 q^{-d} + b_1 q^{-1-d} + ... + b_{n_B} q^{-n_B - d} \tag{5.42}$$

$d$ is the time delay.

Since the multiple-input non-linearity appears at the input and the system contains just one linear dynamic filter, the prediction is simple to derive.

**Algorithm 21 (Hammerstein predictor)** A predictor for a MISO Hammerstein system is given by

$$\hat{y}(k) = \frac{B(q^{-1})}{A(q^{-1})} f(u(k), w) \tag{5.43}$$

where $u(k) = [u_1(k), u_2(k), ..., u_i(k), ..., u_I(k)]^T$ , $i = 1, 2, ..., I$ where $I$ is the number of inputs to the system. $\hat{y}(k)$ is the predicted output of the Hammerstein system, f is a non-linear static MISO function of parameters $w$. $A(q^{-1})$ and $B(q^{-1})$ are polynomials in the backward shift operator $q^{-1}$ :

$$A(q^{-1}) = 1 + a_1 q^{-1} + ... + a_{n_A} q^{-n_A} \tag{5.44}$$
$$B(q^{-1}) = b_0 q^{-d} + b_1 q^{-1-d} + ... + b_{n_B} q^{-n_B - d} \tag{5.45}$$

$d$ is the time delay.

In the SISO case, the input $u(k)$ is a scalar $u(k)$.

# 5.3 Linear dynamics and steady-state models

The Wiener and Hammerstein systems consist of two parts: the linear dynamic (transfer function) and the static (non-linear) part. In the practice of industrial process engineering, the steady-state characteristics of a process are of main interest, and the dynamic behavior is often poorly known. In fact, often only the control engineers seem to be interested in the modeling of the dynamics of a process, while the system designers and production engineers largely ignore the dynamics. In order to provide models that both parties can understand and in order to employ already existing (steady-state) models of the process –among other reasons (such as increased simplicity in identification and control design)– it is reasonable to consider the case where the non-linear static part in Wiener and Hammerstein models represents the steady-state behavior of a process.

Hence, we assume that the static (non-linear) function is given by the steady-state function of the process

$$y_{ss} = f(u_{ss}) \tag{5.46}$$

where the subscript ss denotes steady-state.

The Wiener system is given by

$$y(k) = f_{ss}(z(k), w) + e(k) \tag{5.47}$$

where $z(k) = [z_1(k), z_2(k), ..., z_i(k), ..., z_I(k)]^T$ are given by

$$z_i(k) = \frac{B_i(q^{-1})}{A_i(q^{-1})} u_i(k - d_i) \tag{5.48}$$

In order to preserve the steady-state function, the steady-state gain of the linear dynamic part has to be equal to one,

$$\lim_{z \to 1} \frac{B_i(z)}{A_i(z)} = 1 \tag{5.49}$$

*i.e.,* in steady-state

$$u_{i,ss} = z_{i,ss} \tag{5.50}$$

for all $i = 1, 2, ..., I$.

Similarly, for the Hammerstein system:

$$y(k) = \frac{B(q^{-1})}{A(q^{-1})} f_{ss}(z(k), w) + e(k) \tag{5.51}$$

we must have

$$\lim_{z \to 1} \frac{B(z)}{A(z)} = 1 \tag{5.52}$$

in order to preserve the steady-state function

$$y_{ss} = f_{ss}(\mathbf{u}_{ss}, \mathbf{w}) \tag{5.53}$$

for all $i = 1, 2, ..., I$.

## 5.3.1   Transfer function with unit steady-state gain

There are several ways to fulfill the requirements (5.49) and (5.52). Let us consider the following constraint on the coefficient of the transfer function, where a substitution for $b_{n_B}^*$ fulfills the requirement.

**Algorithm 22 (TF with unit steady-state gain)** Let the transfer polynomial, $n_B \geq 0$, be given by

$$\frac{B^*(q^{-1})}{A(q^{-1})} = \frac{b_0 + b_1 q^{-1} + ... + b_{n_B-1} q^{-(n_B-1)} + b_{n_B}^* q^{-n_B}}{1 + a_1 q^{-1} + ... + a_{n_A} q^{-n_A}} \tag{5.54}$$

A unit steady-state gain is ensured by

$$b_{n_B}^* = 1 + \sum_{n=1}^{n_A} a_n - \sum_{n=0}^{n_B-1} b_n \tag{5.55}$$

**Proof.** Substituting (5.55) to the z-transform equivalent of (5.54) and letting $z \to 1$ gives

$$\lim_{z \to 1} \frac{B^*(z)}{A(z)} = \frac{b_0 + b_1 + ... + b_{n_B-1}^* + 1 + \sum_{n=1}^{n_A} a_n - \sum_{n=0}^{n_B-1} b_n}{1 + a_1 + ... + a_{n_A}} = 1 \quad (5.56)$$

which shows that the steady-state gain of the dynamic part is one, as desired. ∎

## 5.3.2   Wiener and Hammerstein predictors

Combining the result in section 5.3.1 with the steady-state Wiener structure, we get the following predictor.

**Algorithm 23 (Wiener predictor: continued)** A predictor for a MISO Wiener system with a steady-state non-linear function is given by

$$\widehat{y}(k) = f_{ss}(\widehat{\mathbf{z}}(k), \mathbf{w}) \tag{5.57}$$

where

$$\widehat{z}_i(k) = \frac{B_i^*(q^{-1})}{A_i(q^{-1})} u_i(k - d_i) \tag{5.58}$$

$\widehat{y}(k)$ is the predicted output of the Wiener system, $f_{ss}$ is the non-linear static steady-state MISO function of parameters $\mathbf{w}$, $u_i(k)$ and $\widehat{z}_i(k)$ are the input to the system and intermediate variables, respectively, and $d_i$ are the time delays. $A_i(q^{-1})$ and $B_i^*(q^{-1})$ are polynomials in the backward shift operator $q^{-1}$:

$$A_i(q^{-1}) = 1 + a_{i,1}q^{-1} + \dots + a_{i,n_{A_i}}q^{-n_{A_i}} \tag{5.59}$$

$$B_i^*(q^{-1}) = b_{i,0} + b_{i,1}q^{-1} + \dots + b_{i,n_{B_i}-1}q^{-(n_{B_i}-1)} + b_{i,n_{B_i}}^*q^{-n_{B_i}} \tag{5.60}$$

where

$$b_{i,n_B}^* = 1 + \sum_{n=1}^{n_{A_i}} a_{i,n} - \sum_{n=0}^{n_{B_i}-1} b_{i,n} \tag{5.61}$$

and $i = 1, 2, \dots, I$, where $I$ is the number of inputs to the system.

For the Hammerstein system we get a similar result.

**Algorithm 24 (Hammerstein predictor: continued)** A predictor for a MISO Hammerstein system with a steady-state non-linear function is given by

$$\widehat{y}(k) = \frac{B^*(q^{-1})}{A(q^{-1})} f_{ss}(\mathbf{u}(k), \mathbf{w}) \tag{5.62}$$

where $\widehat{y}(k)$ is the predicted output of the Hammerstein system, and $f_{ss}$ is the non-linear static steady-state MISO function of parameters $\mathbf{w}$. $\mathbf{u}(k) = [u_1(k), \dots, u_i(k), \dots, u_I(k)]^T$, $i = 1, 2, \dots, I$, where $I$ is the number of inputs to the system. $A(q^{-1})$ and $B^*(q^{-1})$ are polynomials in the backward shift operator $q^{-1}$:

$$A(q^{-1}) = 1 + a_1 q^{-1} + \dots + a_{n_A}q^{-n_A} \tag{5.63}$$

$$B^*(q^{-1}) = b_0 q^{-d} + b_1 q^{-1-d} + \dots \tag{5.64}$$
$$+ b_{n_B-1}q^{-(n_B-1)-d} + b_{n_B}^* q^{-n_B-d}$$

where

$$b_{n_B}^* = 1 + \sum_{n=1}^{n_A} a_n - \sum_{n=0}^{n_B-1} b_n \tag{5.65}$$

and $d$ is the time delay.

### 5.3.3  Gradients of the Wiener and Hammerstein predictors

If the parameters of the Wiener or Hammerstein model are unknown, they need to be estimated. Often, the most convenient way is to estimate the parameters from input–output data observed from the process. If gradient-based techniques are used, the gradients with respect to the parameters need to be computed. Assume that all parameters are unknown:

- parameters of the static (steady-state) mapping, $\mathbf{w}$; and

- parameters of the linear transfer function(s), *i.e.*, coefficients of the polynomials $A$ and $B$, as well as delay(s) $d$.

An estimate for the delay(s), is usually obtained by simply looking at the process behavior (step response, *etc.*), whereas parameters $\mathbf{w}$, $A$ and $B$ are estimated by minimizing the prediction error. Sometimes the estimation of $d$ may require several iterative rounds, where a value for $d$ is suggested, parameters in $A$, $B$ and w are estimated, and if the model is unsatisfactory, new value(s) for $d$ are suggested.

Let us compute the gradients for parameters in $A$, $B$, and $\mathbf{w}$ in the Wiener predictor (23). For the static part, denote gradient with respect to parameters by

$$\frac{\partial \widehat{y}}{\partial w_j}(k) \leftarrow \Psi_j(k) \tag{5.66}$$

and the gradients with respect to inputs by

$$\frac{\partial \widehat{y}}{\partial \widehat{z}_i}(k) \leftarrow \Phi_i(k) \tag{5.67}$$

where $\mathbf{w} = [w_1, ..., w_j, ..., w_J]^T$ and $J$ is the number of parameters in the non-linear part. $I$ is the number of inputs to the system, $i = 1, 2, ..., I$. These parameters depend on the structure chosen for the static part.

The chain rule can be applied for calculating the gradients with respect to parameters of the dynamic part:

$$\frac{\partial \widehat{y}}{\partial a_{i,n}}(k) = \Phi_i(k) \frac{\partial \widehat{z}_i}{\partial a_{i,n}}(k) \tag{5.68}$$

and

$$\frac{\partial \widehat{y}}{\partial b_{i,n}}(k) = \Phi_i(k) \frac{\partial \widehat{z}_i}{\partial b_{i,n}}(k) \tag{5.69}$$

Hence, in order to compute the gradients, only the gradients

$$\frac{\partial \widehat{z}_i}{\partial b_{i,n}}(k) \text{ and } \frac{\partial \widehat{z}_i}{\partial a_{i,n}}(k) \tag{5.70}$$

are further needed, where $n = 1, 2, ..., n_{A_i}$ and $n = 0, 1, ..., n_{B_i} - 1$, respectively, $i = 1, 2, ..., I$.

For simplicity, omit the input index $i$ for a moment. The output of the linear part can be written as

$$\begin{aligned}
\widehat{z}(k) = & \left[ b_0 q^{-d} + ... + b_{n_B-1} q^{-(n_B-1)-d} + b_{n_B}^* q^{-n_B-d} \right] u(k) \tag{5.71} \\
& - \left[ \left( a_1 + a_2 q^{-1} + ... + a_{n_A} q^{-(n_A-1)} \right) \right] \widehat{z}(k-1)
\end{aligned}$$

where $b_{n_B}^* = 1 + \sum_{n=1}^{n_A} a_n - \sum_{n=0}^{n_B-1} b_n$. It is now simple to compute the derivatives with respect to the parameters:

$$\frac{\partial \widehat{z}}{\partial b_n}(k) = u(k-n-d) - u(k-n_B-d) - \sum_{m=1}^{n_A} \left[ a_m \frac{\partial \widehat{z}}{\partial b_n}(k-m) \right] \tag{5.72}$$

$$\frac{\partial \widehat{z}}{\partial a_n}(k) = u(k-n_B-d) - \sum_{m=1}^{n_A} \left[ a_m \frac{\partial \widehat{z}}{\partial a_n}(k-m) \right] - \widehat{z}(k-n) \tag{5.73}$$

where $n = 0, 1, ..., n_B-1$ and $n = 1, 2, ..., n_A$, respectively. Assuming that the parameters $n_A$ change slowly, past gradients of $\frac{\partial \widehat{z}}{\partial a_n}$ and $\frac{\partial \widehat{z}}{\partial b_n}$ can be stored and the equations computed recursively, thus avoiding excessive computations.

Let us collect the previous results in what follows.

**Algorithm 25 (Gradients of the Wiener predictor)** The gradients of the Wiener predictor with steady-state non-linear part are given by

$$\frac{\partial \widehat{y}}{\partial w_j}(k) \leftarrow \Psi_j(k) \tag{5.74}$$

$$\frac{\partial \widehat{y}}{\partial a_{i,n}}(k) = \Phi_i(k)\frac{\partial \widehat{z}_i}{\partial a_{i,n}}(k) \tag{5.75}$$

$$\frac{\partial \widehat{y}}{\partial b_{i,n}}(k) = \Phi_i(k)\frac{\partial \widehat{z}_i}{\partial b_{i,n}}(k) \tag{5.76}$$

where

$$\frac{\partial \widehat{y}}{\partial \widehat{z}_i}(k) \leftarrow \Phi_i(k) \tag{5.77}$$

$$\frac{\partial \widehat{z}_i}{\partial b_{i,n}}(k) = u_i(k - n - d_i) - u_i(k - n_{B_i} - d_i) - \sum_{m=1}^{n_{A_i}}\left[a_{i,m}\frac{\partial \widehat{z}_i}{\partial b_{i,n}}(k - m)\right] \tag{5.78}$$

$$\frac{\partial \widehat{z}_i}{\partial a_{i,n}}(k) = u_i(k - n_{B_i} - d_i) - \sum_{m=1}^{n_{A_i}}\left[a_{i,m}\frac{\partial \widehat{z}_i}{\partial a_{i,n}}(k - m)\right] - \widehat{z}_i(k - n) \tag{5.79}$$

where $i = 1, 2, ..., I$ (system inputs), $j = 1, 2, ..., J$ (parameters of the static steady-state part) and $n = 0, 1, ..., n_{B_i} - 1$ and $n = 1, 2, ..., n_{A_i}$, respectively (orders of the polynomials associated with each input).

Owing to the recursion in the computation of the gradients, the polynomial $A_i^*$ needs to be stable. When instability is encountered, the parameters need to be *projected toward a stable region*. A simple method consists of multiplying $A_i^*$ with a constant $\gamma$, $0 \ll \gamma < 1$:

$$A_i^* = 1 + \gamma a_{i,1}q^{-1} + ... + \gamma^{n_{A_i}}a_{i,n_{A_i}}q^{-n_{A_i}} \tag{5.80}$$

until all roots of the polynomial lay inside the unit circle.

Let us next give the gradients for the Hammerstein predictor. The predictor is given by

$$\widehat{y}(k) = \frac{B^*(q^{-1})}{A(q^{-1})}f_{ss}(u(k), w) \tag{5.81}$$

in which the linear dynamic subsystem at the output can be written out

$$\begin{aligned}\widehat{y}(k) =\ & \left[b_0 q^{-d} + b_1 q^{-1-d} + ... + b_{n_B-1}q^{-(n_B-1)-d}\right]\widehat{z}(k) \\ & + \left[\left(1 + \sum_{n=1}^{n_A}a_n - \sum_{n=0}^{n_B-1}b_n\right)q^{-n_B-d}\right]\widehat{z}(k) \\ & - \left[\left(a_1 + a_2 q^{-1} + ... + a_{n_A}q^{-(n_A-1)}\right)\right]\widehat{y}(k - 1)\end{aligned} \tag{5.82}$$

where the output of the static subsystem is given by

$$\widehat{z}(k) = \mathbf{f_{ss}}(\mathbf{u}(k), \mathbf{w}) \tag{5.83}$$

It is simple to calculate the derivatives with respect to the parameters:

$$\frac{\partial \widehat{y}}{\partial b_n}(k) = \widehat{z}(k - n - d) - \widehat{z}(k - n_B - d) - \sum_{m=1}^{n_A}\left[a_m \frac{\partial \widehat{y}}{\partial b_n}(k - m)\right] \tag{5.84}$$

$$\frac{\partial \widehat{y}}{\partial a_n}(k) = \widehat{z}(k - n_B - d) - \sum_{m=1}^{n_A}\left[a_m \frac{\partial \widehat{y}}{\partial a_n}(k - m)\right] - \widehat{y}(k - n) \tag{5.85}$$

where $n = 0, 1, ..., n_B - 1$ and $n = 1, 2, ..., n_A$, respectively.

The gradient of the output of the Hammerstein predictor with respect to the parameters of the static part, denoted by

$$\frac{\partial \widehat{y}}{\partial w_j}(k) \leftarrow \Xi_j(k) \tag{5.86}$$

$(j = 1, 2, ..., J)$ is still required. Denote the gradient of the output of the static part with respect to its parameters by

$$\frac{\partial \widehat{z}}{\partial w_j}(k) \leftarrow \Psi_j(k) \tag{5.87}$$

where $\mathbf{w} = [w_1, ..., w_j, ..., w_J]^T$ contains the parameters of the static mapping. The gradient of the linear subsystem (5.82) is given by

$$\begin{aligned}
\frac{\partial \widehat{y}}{\partial w_j}(k) &= \left(b_0 q^{-d} + b_1 q^{-1-d} + ... + b_{n_B - 1} q^{-(n_B - 1)-d}\right) \frac{\partial \widehat{z}}{\partial w_j}(k) \\
&+ \left(1 + \sum_{n=1}^{n_A} a_n - \sum_{n=0}^{n_B - 1} b_n\right) q^{-n_B - d} \frac{\partial \widehat{z}}{\partial w_j}(k) \\
&- \left[(a_1 + a_2 q^{-1} + ... + a_{n_A} q^{-(n_A - 1)})\right] \frac{\partial \widehat{y}}{\partial w_j}(k - 1)
\end{aligned} \tag{5.88}$$

which is more conveniently expressed as

$$\begin{aligned}
\Xi_j(k) &= \sum_{n=0}^{n_B - 1} b_n \Psi_j(k - n - d) \\
&+ \left(1 + \sum_{n=1}^{n_A} a_n - \sum_{n=0}^{n_B - 1} b_n\right) \Psi_j(k - n_B - d) \\
&- \sum_{n=1}^{n_A} a_n \Xi_j(k - n)
\end{aligned} \tag{5.89}$$

Assuming that the parameters change slowly, past gradients $\Xi_j$ and $\Psi_j$ can be stored and the equations computed recursively, thus avoiding excessive computations.

Let us collect the results for the Hammerstein system.

**Algorithm 26 (Gradients of the Hammerstein predictor)** The gradients of the Hammerstein predictor with steady-state non-linear part are given by

$$\frac{\partial \widehat{y}}{\partial b_n}(k) = \widehat{z}(k-n-d) - \widehat{z}(k-n_B-d) - \sum_{m=1}^{n_A}\left[a_m\frac{\partial \widehat{y}}{\partial b_n}(k-m)\right] \quad (5.90)$$

$$\frac{\partial \widehat{y}}{\partial a_n}(k) = \widehat{z}(k-n_B-d) - \sum_{m=1}^{n_A}\left[a_m\frac{\partial \widehat{y}}{\partial a_n}(k-m)\right] - \widehat{y}(k-n) \quad (5.91)$$

$$\frac{\partial \widehat{y}}{\partial w_j}(k) \leftarrow \Xi_j(k) \quad (5.92)$$

where

$$\Xi_j(k) = \sum_{n=0}^{n_B-1} b_n\Psi_j(k-n-d)$$

$$+ \left(1+\sum_{n=1}^{n_A}a_n - \sum_{n=0}^{n_B-1}b_n\right)\Psi_j(k-n_B-d) \quad (5.93)$$

$$- \sum_{n=1}^{n_A}a_n\Xi_j(k-n)$$

$$\frac{\partial \widehat{z}}{\partial w_j}(k) \leftarrow \Psi_j(k) \quad (5.94)$$

where $j = 1, 2, ..., J$ (parameters of the static steady-state part) and $n = 0, 1, ..., n_B - 1$ and $n = 1, 2, ..., n_A$, respectively (orders of the polynomials associated with the output).

## 5.4   Remarks

Let us conclude this chapter by making a few remarks on the practical use of Wiener and Hammerstein systems. For discussion on the Wiener, Hammerstein, and related structures see, *e.g.*, [72].

## 5.4.1 Inverse of Hammerstein and Wiener systems

Wiener and Hammerstein models are counterparts: the inverse of a Hammerstein structure is a Wiener structure, and *vice versa*. This is important in applications to control.

To show this, let us assume that a system is described by a Hammerstein model

$$y(k) = q^{-d} \frac{B(q^{-1})}{A(q^{-1})} f(\mathbf{u}(k)) \tag{5.95}$$

$d$ is the time delay of the dynamic system, thus we can require that $b_0 \neq 0$.

The linear dynamic part is given by

$$y(k) = q^{-d} \frac{B(q^{-1})}{A(q^{-1})} z(k) \tag{5.96}$$

which can be written out as a difference equation:

$$
\begin{aligned}
y(k) = & -a_1 y(k-1) - \ldots - a_m y(k - n_A) \\
& + b_0 z(k-d) + \ldots + b_{n_B} z(k - d - n_B)
\end{aligned} \tag{5.97}
$$

Its inverse is given by

$$z(k) = q^{d} \frac{A(q^{-1})}{B(q^{-1})} y(k) \tag{5.98}$$

Writing out, we have

$$
\begin{aligned}
z(k) = & \frac{1}{b_0} y(k+d) + \frac{a_1}{b_0} y(k+d-1) + \ldots + \frac{a_{n_A}}{b_0} y(k+d-n_A) \\
& - \frac{b_1}{b_0} z(k-1) - \ldots - \frac{b_{n_B}}{b_0} z(k-1-n_B)
\end{aligned} \tag{5.99}
$$

The non-linear static part is given by

$$z(k) = f(\mathbf{u}(k)) \tag{5.100}$$

Let us assume that the inverse(s) of the non-linear static part exists

$$u_i(k) = f_i^{-1}(z(k)) \tag{5.101}$$

Then, the inverse(s) of a Hammerstein system can be expressed as

$$u_i(k) = f_i^{-1}(z(k)) = f_i^{-1}\left( \frac{A(q^{-1})}{B(q^{-1})} y(k+d) \right) \tag{5.102}$$

which is a Wiener system with input sequence $\{y(k+d)\}$ filtered by $\frac{A}{B}$ and scaled through $f_i^{-1}$, to produce an output sequence $\{u_i(k)\}$.

**Example 35 (Hammerstein inverse)** Let a SISO system be given by a Hammerstein model

$$
\begin{align}
y(k) &= 0.7z(k-1) + 0.3y(k-1) \tag{5.103}\\
z(k) &= u^2(k) \tag{5.104}
\end{align}
$$

with $u \in \Re_+$ (positive real). Thus

$$
A(q^{-1}) = 1 - 0.3q^{-1}; B(q^{-1}) = 0.7; d = 1 \tag{5.105}
$$

The inverse is given by

$$
\begin{align}
z(k) &= \frac{1}{0.7}y(k+1) - \frac{0.3}{0.7}y(k) \tag{5.106}\\
u(k) &= \sqrt{z(k)} \tag{5.107}
\end{align}
$$

which has the structure of a Wiener system. Notice, that for $u \in \Re$ the inverse does not exist (not unique).

## 5.4.2  ARX dynamics

So far, only OE-type of dynamics have been considered. Let us next consider briefly systems with ARX dynamics. An ARX-Hammerstein predictor is given by

$$
\widehat{y}(k) = -A_1(q^{-1})y(k) + B(q^{-1})z(k) \tag{5.108}
$$

where $A = 1 + q^{-1}A_1$, and $z(k) = f(\mathbf{u}(k))$. Since $\{y\}$ is known ($y$ is measurable), a one-step ahead ARX-Hammerstein predictor can be directly implemented. In general, ARX-type of dynamics can not be recommended for the identification of processes where measurements are strongly contaminated by noise.

The gradients are simple to calculate

$$
\begin{align}
\frac{\partial \widehat{y}}{\partial a_n}(k) &= y(k-n) \tag{5.109}\\
\frac{\partial \widehat{y}}{\partial b_n}(k) &= z(k) \tag{5.110}\\
\frac{\partial \widehat{y}}{\partial w_j}(k) &= \sum_{n=0}^{n_B} b_n \Psi_j(k) \tag{5.111}
\end{align}
$$

where $\Psi_j(k) = \frac{\partial \widehat{z}}{\partial w_j}$.

For Wiener systems the intermediate variable $z(k)$ is not available. Thus an ARX implementation is not straightforward. However, if the inverses of the non-linear part exist and are known, we can obtain the intermediate variables using the inverses

$$z_i(k) = f_i^{-1}(y(k), \mathbf{w}) \tag{5.112}$$

Then the gradients of the system can be given as

$$\frac{\partial \widehat{y}}{\partial w_j}(k) = \Psi_j(k) \tag{5.113}$$

$$\frac{\partial \widehat{y}}{\partial b_{i,n}}(k) = \Phi_i(k) u(k-n) \tag{5.114}$$

$$\frac{\partial \widehat{y}}{\partial a_{i,n}}(k) = \Phi_i(k) z_i(k-n) \tag{5.115}$$

where $\Phi_i = \frac{\partial \widehat{y}}{\partial z_i}$. The identification of Wiener systems with ARX type of dynamics requires the identification of the inverse(s) $f_i^{-1}$.

The derivation of the corresponding equations for a system with unit steady-state gain linear dynamics is straightforward and omitted here.

# Chapter 6

# Estimation of Parameters

This chapter considers parameter estimation techniques. These techniques are essential in system identification, as they provide the means for determining (off-line) or adjusting (on-line) the parameters of a chosen model structure, using sampled data (measurements). The least squares method can be applied when the system output is linear with respect to its parameters. This is true for linear static mappings (linear regression models), as well as for some linear dynamic structures (such as ARX structures), and some non-linear systems (*e.g.*, power series and multi-linear systems). However, usually the least squares method can not be directly applied in non-linear or dynamic systems, since these types of systems are, in general, non-linear with respect to their parameters. In the previous chapters we have pointed out structures such as the OE (Chapter 3), the sigmoid neural networks (Chapter 4), or the Wiener structure (Chapter 5), with which the least squares method can not be directly applied.

In this chapter, the least squares method will be extended to the case of non-linear systems. The parameter estimation problem is seen as an optimization problem, minimizing a cost function consisting of a sum of squared prediction errors.

In general, the non-linear parameter estimation techniques are *iterative*: A fixed set of data is used repeatedly; at each iteration the parameters are adjusted towards a smaller cost. This is because usually the criterion will not be quadratic in the parameters, as in the least squares method, and therefore an analytical solution is not available. Note the difference between *recursive* algorithms, such as recursive least squares (RLS). In RLS, new data patterns are added one-by-one, possibly forgetting the older ones (*e.g.*, exponential forgetting), but not used repeatedly. The methods discussed in this chapter are *batch methods*, such as the least squares method, where a fixed data set is used. Also recursive forms can be derived. Unfortunately, on-line identification using non-linear models is less robust, due to several reasons, and can not be recommended.

In practice, gradient-based methods are dominating. Their main drawback is that they can get stuck with the local minima in the cost function. However, these methods have shown to be efficient in practice. For complex systems, suitable methods can be found, *e.g.*, from random search and probabilistics. Unfortunately, the practical implementations are often inefficient.

First, a general overview to prediction error methods is given, and the algorithm for the Levenberg–Marquardt method is given. This is followed by the presentation of the Lagrange multipliers approach for the case of optimization under constraints. The guided random search methods are considered in the next section, with special emphasis on the learning automata. In order to illustrate the feasibility and performance of the methods presented in this chapter, a number of simulation examples on process identification is presented at the end of this chapter.

## 6.1   Prediction error methods

The family of methods that minimize the error between the predicted and the observed values of the output are called prediction error methods[1]. Many other parameter estimation techniques can be regarded as special cases of these methods.

Iterative prediction error minimization methods are based on the following principle [87]: given a function $J(\theta)$ to be minimized

$$J(\theta) = \sum_{k=1}^{K} \frac{1}{2} \left( y(k) - \widehat{y}(k) \right)^2 \qquad (6.1)$$

and an initial state for $\widehat{\theta}(0)$, find a minimization direction $u(l)$ and step size $\eta(l)$, and update the parameters as

$$\widehat{\theta}(l+1) = \widehat{\theta}(l) + \eta(l) u(l) \qquad (6.2)$$

The prediction $\widehat{y}(k) = f(\theta, k)$ is a function of the parameter vector $\theta = [\theta_1, \cdots .\theta_P]^T$, which is computed using the current parameter estimates: $\theta = \widehat{\theta}(l)$; $y(k)$ is the corresponding target observation (measurement); $K$ is the number of data patterns in the training set.

---

[1]Principle of least squares prediction error [3]: Postulate a model that gives the prediction in terms of past data and parameters. Given the observations of past data, adjust the parameters in such a way that the sum of the squares of the prediction errors is as small as possible.

The task of the minimization is to find optimal values for the direction and the step size, when only local information of the function is available. Repeated application of (6.2), each time with the optimal direction and step size, will bring J($\theta$) to a minimum. As a result of the search, a parameter estimate is obtained which minimizes the cost function

$$\widehat{\theta} = \arg \min_{\theta} J(\theta) \tag{6.3}$$

Note, however, that the globality of the minimum can not be guaranteed. The optimization techniques are not necessarily restricted to cost functions of the specific form given by (6.1), as long as the derivatives are computed accordingly.

## 6.1.1 First-order methods

Let us now focus on finding the minimization direction. Let $\overline{\theta}$ be some fixed parameter vector. The cost function (6.1) can be written as a Taylor expansion around $\overline{\theta}$:

$$J(\theta) = J(\overline{\theta}) + \sum_{p=1}^{P} \left[\frac{\partial J}{\partial \theta_p}\right]_{\theta=\overline{\theta}} \widetilde{\theta}_p + \sum_{p=1; p*=1}^{P} \left[\frac{\partial^2 J}{\partial \theta_p \partial \theta_{p*}}\right]_{\theta=\overline{\theta}} \widetilde{\theta}_p \widetilde{\theta}_{p*} + \cdots \tag{6.4}$$

where $\widetilde{\theta}$ is the deviation from $\overline{\theta}$, $\theta = \overline{\theta} + \widetilde{\theta}$. In the first order methods, only the first non-constant term of the Taylor expansion is used:

$$J(\theta) \approx J(\overline{\theta}) + \sum_{p=1}^{P} \left[\frac{\partial J}{\partial \theta_p}\right]_{\theta=\overline{\theta}} \widetilde{\theta}_p \tag{6.5}$$

The derivative of J($\theta$), approximated by (6.5), is given by

$$\frac{\partial}{\partial \widetilde{\theta}_p} J(\theta) = \left[\frac{\partial J}{\partial \theta_p}\right]_{\theta=\overline{\theta}} = \sum_{k=1}^{K} \left([y(k) - \widehat{y}(k)] \left[\frac{\partial \widehat{y}}{\partial \theta_p}(k)\right]_{\theta=\overline{\theta}}\right) \tag{6.6}$$

A natural fixed point $\overline{\theta}$ is the current parameter estimate $\theta(l)$. To compute the new estimate, the minimization direction is given by the negative gradient. The learning rule then becomes

$$\widehat{\theta}(l+1) = \widehat{\theta}(l) - \eta(l) \left[\frac{\partial J}{\partial \theta}\right]_{\theta=\theta(l)} \tag{6.7}$$

The step size $\eta(l)$ is often replaced by a fixed constant, due to lower computational cost. These types of methods are often also referred to as steepest descent, gradient descent, least mean squares, or error backpropagation techniques.

## 6.1.2   Second-order methods

If the second non-constant term from the Taylor expansion is also considered, the cost function becomes

$$J\left(\theta\right) \approx J\left(\overline{\theta}\right) + \sum_{p=1}^{P}\left[\frac{\partial J}{\partial\theta_p}\right]_{\theta=\overline{\theta}}\widetilde{\theta}_p + \sum_{p=1;p*=1}^{P}\left[\frac{\partial^2 J}{\partial\theta_p\partial\theta_{p*}}\right]_{\theta=\overline{\theta}}\widetilde{\theta}_p\widetilde{\theta}_{p*} \qquad (6.8)$$

and can be written as

$$J\left(\theta\right) = J\left(\overline{\theta}\right) - \mathbf{b}^T\widetilde{\theta} + \frac{1}{2}\widetilde{\theta}^T\mathbf{H}^T\widetilde{\theta} \qquad (6.9)$$

where

$$\mathbf{b} = -\left[\frac{\partial J}{\partial\theta}\right]_{\theta=\overline{\theta}} \qquad (6.10)$$

and the elements of the Hessian **H** are given by

$$h_{p,p*} = \left[\frac{\partial^2 J}{\partial\theta_p\partial\theta_{p*}}\right]_{\theta=\overline{\theta}} \qquad (6.11)$$

Minimum of (6.8) is found by setting the derivative to zero, and is located at

$$\mathbf{H}\widetilde{\theta} - \mathbf{b} = 0 \qquad (6.12)$$

from where the optimal $\widetilde{\theta}$ can be obtained. It is given by

$$\widetilde{\theta} = \mathbf{H}^{-1}\mathbf{b} \qquad (6.13)$$

Thus, the optimization reduces to matrix inversion.

Unfortunately, the calculation of the Hessian **H** is computationally prohibitive in practice (analytical solutions are rare (see, *e.g.*, [10]) and approximation methods must be applied. Alternatives include **quasi-Newton** methods such as BFGS and DFP, or the conjugate gradient methods (see, *e.g.*, [7][87][84])).

### Levenberg–Marquardt method

A commonly used second-order method is the *Levenberg–Marquardt method* (see, *e.g.*, [7]).

Define a vector **R** whose $K$ components $r_k$ are the residuals

$$r_k = \widehat{y}\left(k\right) - y\left(k\right) \qquad (6.14)$$

$k = 1, 2, ..., K$, where $K$ is the number of data samples. The cost function and its derivatives can now be expressed as

$$J(\boldsymbol{\theta}) = \frac{1}{2}R(\boldsymbol{\theta})^T R(\boldsymbol{\theta}) \tag{6.15}$$

$$\frac{\partial J(\boldsymbol{\theta})}{\partial \boldsymbol{\theta}} = \sum_{k=1}^{K} r_k \frac{\partial r_k}{\partial \boldsymbol{\theta}} = G(\boldsymbol{\theta})^T R(\boldsymbol{\theta}) \tag{6.16}$$

$$\frac{\partial^2 J(\boldsymbol{\theta})}{\partial \boldsymbol{\theta}^2} = \sum_{k=1}^{K} \left( \frac{\partial r_k}{\partial \boldsymbol{\theta}} \left[ \frac{\partial r_k}{\partial \boldsymbol{\theta}} \right]^T + r_k \frac{\partial^2 r_k}{\partial \boldsymbol{\theta}^2} \right) = G(\boldsymbol{\theta})^T G(\boldsymbol{\theta}) + S(\boldsymbol{\theta}) \tag{6.17}$$

where $G(\boldsymbol{\theta})$ is the Jacobian matrix, whose elements $g_{k,p}$ are given by

$$g_{k,p} = \frac{\partial r_k}{\partial \theta_p} \tag{6.18}$$

(note that $\frac{\partial r_k}{\partial \theta_p} = \frac{\partial \hat{y}}{\partial \theta_p}(k)$) and $S(\boldsymbol{\theta})$ is the part of the Hessian matrix that contains the second derivatives of $r_k$.

The Newton iteration is given by

$$\hat{\boldsymbol{\theta}}(l+1) = \hat{\boldsymbol{\theta}}(l) - \left[ G(\boldsymbol{\theta})^T G(\boldsymbol{\theta}) + S(\boldsymbol{\theta}) \right]^{-1} G(\boldsymbol{\theta})^T R(\boldsymbol{\theta}) \tag{6.19}$$

where the Jacobian $G(\boldsymbol{\theta})$ is easy to calculate, while $S(\boldsymbol{\theta})$ is not. In the Gauss–Newton method, the $S(\boldsymbol{\theta})$ is simply neglected. In the Levenberg–Marquardt method, the step is defined as

$$\hat{\boldsymbol{\theta}}(l+1) = \hat{\boldsymbol{\theta}}(l) - \left[ G(\boldsymbol{\theta})^T G(\boldsymbol{\theta}) + \mu(l) I \right]^{-1} G(\boldsymbol{\theta})^T R(\boldsymbol{\theta}) \tag{6.20}$$

where $\mu(l)$ is increased whenever the step would result to an increased value of J. When a step reduces J, $\mu(l)$ is reduced.

## 6.1.3   Step size

When the minimization direction is available, the problem is to decide how far to go along this line before a new direction is chosen. The step size may be constant or time-varying. Often, the step size parameter is chosen such that it is a decaying function of time, such as $\eta(k) = \frac{1}{k}$, for example. This choice is made due to theoretical requirements (to ensure infinite search

range: $\sum_{l=1}^{\infty} \eta(l) = \infty$ and convergence of the estimates: $\sum_{l=1}^{\infty} \eta^2(l) < \infty$).
Heuristically, the step size should be large when far away from the optimum
(at the beginning of the search), and tend towards zero in the neighborhood
of the optimum point (at the end of search). In practice, however, it may be
difficult to find an efficient step size coefficient fulfilling these requirements.

The following simple procedure was suggested for $\mu$ in the Levenberg–
Marquardt method in [24]: Whenever a step would result in an increased
value of J, $\mu(l)$ is increased by multiplying it by some factor larger than 1;
when a step reduces the value of J, it is divided by some factor larger than
1. Note, that when $\mu$ is large the algorithm becomes steepest descent with a
step size equal to $\frac{1}{\mu}$, while for small $\mu$ the algorithm becomes Gauss–Newton.

Another common method is the *three point method*: Given there are three
values $\theta_{\mathrm{a}} < \theta_{\mathrm{b}} < \theta_{\mathrm{c}}$ such that the function at $\theta_{\mathrm{b}}$, J($\theta_{\mathrm{b}}$), is the lowest, the
sign of the derivative at $\theta_b$ indicates whether a minimum is located in $[\theta_{\mathrm{a}}, \theta_{\mathrm{b}}]$
or in $[\theta_{\mathrm{b}}, \theta_{\mathrm{c}}]$. This section is then linearly interpolated from its endpoints,
and the procedure is repeated.

## 6.1.4  Levenberg–Marquardt algorithm

Let us summarize the Levenberg-Marquardt method in the following algorithm.

**Algorithm 27 (Levenberg–Marquardt)** Given a function J($\theta$) of the
sum of prediction errors

$$J(\theta) = \frac{1}{2} \sum_{k=1}^{K} [y(k) - \widehat{y}(k)]^2 \tag{6.21}$$

the minimizing parameters

$$\widehat{\theta} = \arg \min_{\theta} J(\theta) \tag{6.22}$$

can be found by the following algorithm.

1. Initialize:

   Set iteration index $l = 1$.

   Initialize $\widehat{\theta}(1)$ and $\mu(1)$ and specify $\eta$.

2. Evaluate the model and the residuals:

   Evaluate $\widehat{y}(k)$ and $\frac{\partial \widehat{y}}{\partial \theta_p}(k)$ for all patterns $k$ and parameters $p$.

Compose the residual vector **R**

$$r_k = \widehat{y}(k) - y(k) \tag{6.23}$$

and compute $J\left(\widehat{\boldsymbol{\theta}}(l)\right)$

$$J\left(\widehat{\boldsymbol{\theta}}(l)\right) = \frac{1}{2}\mathbf{R}^T\mathbf{R} \tag{6.24}$$

Compose the Jacobian matrix **G**

$$g_{k,p} = \frac{\partial \widehat{y}}{\partial \theta_p}(k) \tag{6.25}$$

3. Solve the parameter update

$$\Delta\widehat{\boldsymbol{\theta}}(l) = -\left[\mathbf{G}^T\mathbf{G} + \mu(l)\mathbf{I}\right]^{-1}\mathbf{G}^T\mathbf{R} \tag{6.26}$$

4. Repeat Step 2 using $\widehat{\boldsymbol{\theta}}(l) + \Delta\widehat{\boldsymbol{\theta}}(l)$, *i.e.* compute $J\left(\widehat{\boldsymbol{\theta}}(l) + \Delta\widehat{\boldsymbol{\theta}}(l)\right)$.

If $J\left(\widehat{\boldsymbol{\theta}}(l) + \Delta\widehat{\boldsymbol{\theta}}(l)\right) < J\left(\widehat{\boldsymbol{\theta}}(l)\right)$ then increase the step size

$$\mu(l+1) = \mu(l)/\eta \tag{6.27}$$

and update the parameters

$$\widehat{\boldsymbol{\theta}}(l+1) = \widehat{\boldsymbol{\theta}}(l) + \Delta\widehat{\boldsymbol{\theta}}(l) \tag{6.28}$$

otherwise reduce the step size

$$\mu(l+1) = \eta\mu(l) \tag{6.29}$$

5. Set $l = l + 1$ and return to Step 2, or quit.

**Example 36 (pH neutralization)** Let us consider a simple simulated example related to pH neutralization. The process model is given in detail in Chapter 8. Here, we will concentrate on the problem of identifying a SISO steady state titration curve: the effect of base stream to the effluent pH. The data was obtained by solving the steady state model for randomly selected inputs, scaling both variables to $[0,1]$, and adding Gaussian noise $N(0, 0.05^2)$ to the output measurement.

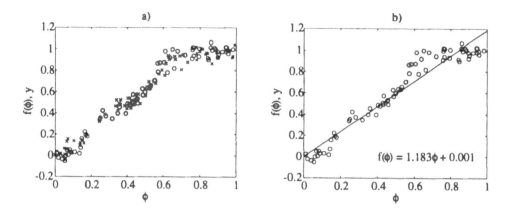

Figure 6.1: Titration curve data. Plot a) shows the observed data. Training data is indicated by circles, test data by crosses. Plot b) shows the estimated linear model.

Let us now examine the identification of a sigmoid neural network model for the process, estimating the parameters with the Levenberg–Marquardt (LM) method.

**Data.** Assume that two sets of data from the process have been measured, both containing 75 input–output observations of the plant behavior. The first set will be used for parameter estimation (training data), while the second is conserved for model validation (test set). In addition, in this simulated example, we can evaluate the true noiseless function. This gives us a third set (500 observations). The data are shown in Fig. 6.1a where the training data is indicated by circles and the test data by crosses.

**Model structure.** Given the data, the next thing to do is to select the model structure. First, a linear regression model was estimated. The prediction is shown in Fig. 6.1b. Clearly, the plot indicates that the process may possess some non-linearities. Since we are not aware of any data transformation for a titration curve that would convert the parameter estimation into a linear problem, we consider a sigmoid neural network (SNN) black-box structure. Let us stick to the simple one-hidden layer network. With SNN the number of nodes $H$ still needs to be set. The non-linearities seem mild, so let us experiment with several moderate network sizes, say, 3, 5, and 10.

**Parameter estimation.** For the LM method, the initial parameter vector $\theta(1)$, initial step size $\mu(1)$ and its adjustment factor $\eta$ need to be specified. For SNN, a reasonable starting point is obtained by initializing $\widehat{\theta}$ with small random values, say $\widehat{\theta}_p \in N(0, 0.01) \; \forall p$. Let us set the initial step size to $\mu(1) = 0.01$ and a relatively moderate adjustment factor $\eta = 1.2$.

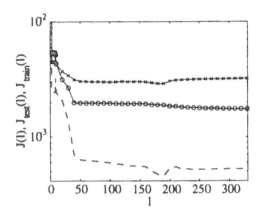

Figure 6.2: Evolution of the criteria $J = \frac{1}{K}\sum_k (\widehat{y}(k) - y(k))^2$ during parameter estimation as a function of the LM iterations $l$. $y(k)$ are the output data in the noiseless (J), test ($J_{test}$) and training ($J_{train}$) data sets and $\widehat{y}(k)$ are the corresponding predictions by the model. $H = 5$.

With these values set, we can proceed in estimating the parameters. Let us start with the medium sized network ($H = 5$). The evolution of several criteria is shown in Fig. 6.2. The LM algorithm performs the task of minimizing $J_{train}$, which is the average sum of residual errors between observations in the training data and the corresponding model predictions. As seen from Fig. 6.2 (the curve with circles), the $J_{train}$ drops rapidly during the first 50 iteration rounds, and for $l > 50$ the evolution is slower. With the stopping criterion of $\|G^T R\| < 10^{-3}$ the model shown in Fig. 6.3a is obtained (predictions connected with a solid line).

**Validation**. At first sight, this seems to be a reasonable fit to the training data (circles in Fig. 6.3a). To have a better view of the model performance, we can compute the linearized parameters. These are shown in Fig. 6.3b. These show peaks at the areas where the prediction grows rapidly. Also, the gain is positive throughout the operating area indicating that the model is monotone. Thus, the model seems to coincide with our expectations of what a titration curve should look like.

It is interesting to look at the estimated basis functions, see Fig. 6.3c, showing the adjusted sigmoid functions. The mapping seems to be composed of a practically linear term (with coefficient 3.3), a slowly increasing term (coefficient 0.4) and two sharper correction terms (coefficients $-0.3$ and $-0.2$). The model output is obtained by multiplying the sigmoids with the associated coefficients, and adding up the results (see Fig 6.3d).

If the network has too many degrees of freedom, the noise in the finite

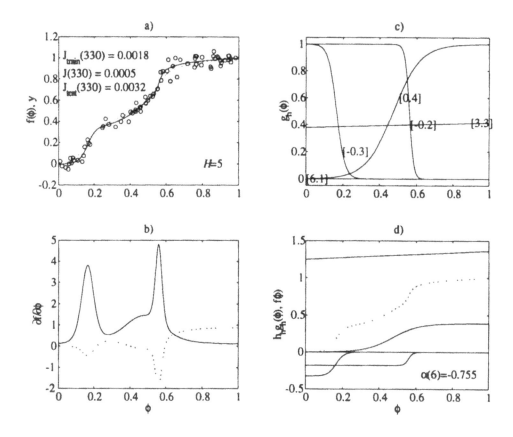

Figure 6.3: Performance of the model ($H = 5$). Plot a) shows the training data (circles), the prediction by the model (solid curve) and the noiseless function (dotted curve). The upper left corner shows the values of the performance criteria (see caption of Fig. 6.2 at the end of training). Plot b) shows the linearized parameters, *i.e.*, the gradients with respect to the system input (solid curve) and the constant (dotted curve). Plot c) shows the adjusted basis functions and the associated coefficients $\alpha_h$. Plot d) shows the basis functions multiplied by their constants (solid curves) and the final mapping (dotted curve) obtained by summing the components and adding the bias term. All plots are shown as a function of the system input $\varphi$.

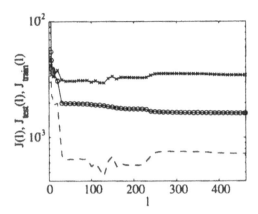

Figure 6.4: Evolution of the criteria as a function of the LM iterations $l$. $H = 10$ (see the legend of Fig. 6.2).

data set will be captured by the model parameters. A rough cross-validation method is to spare a set of observations for model validation. Computing the minimization criterion for the test data can reveal whether this is the case. Fig. 6.2 (the curve with crosses) shows the evolution of the $J_{test}$ during the estimation. It can be seen to drop rapidly during the first 50 iteration rounds and then grow, with a larger increase at $l = 200$. This indicates that our model may have captured some noise in it. Indeed, looking at the criterion J (the dashed curve in Fig. 6.2), computed using the model prediction and the noiseless function (notice that this is not available in practical problems), shows that the minimum occurs at $l = 180$ and then starts to grow. This effect is even more pronounced for the largest network experimented ($H = 10$), as shown in Fig. 6.4. Figure 6.5 shows the prediction by the largest network, which indeed contains some spurious wiggles at $\varphi \in [0, 0.15]$. During the parameter estimation for the smallest network ($H = 3$) such phenomena could not be observed, and the prediction is the smoothest among the SNNs experimented (see Fig. 6.6).

Comparing the final values of $J_{test}$ (see Figs. 6.3a, 6.5 and 6.6) would suggest the smallest network ($H = 3$) as the 'optimal' one. For this simulated example, we can also compute the deviation of the model from the noiseless function (Js in Figs. 6.3a, 6.5 and 6.6), and we find the medium network ($H = 5$) to have the smallest J. This gives a small taste of the difficulties associated with automatic data driven structure selection methods. Taking into account that, in general, different values for initial $\widehat{\theta}$, initial $\mu(1)$, $\eta$ and the stopping criterion result in different parameter estimates, it is easy to see

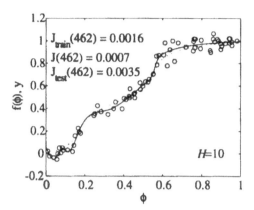

Figure 6.5: Prediction by the largest SNN model (solid curve), training data (circles) and the noiseless function (dotted curve). The upper left corner shows the values of the performance criteria (see legend of Fig. 6.2 at the end of training.)

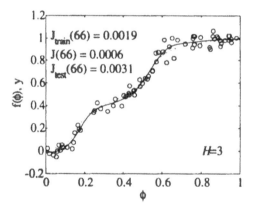

Figure 6.6: Prediction by the smallest SNN model (solid curve), training data (circles) and the noiseless function (dotted curve). The upper left corner shows the values of the performance criteria (see legend of Fig. 6.2 at the end of training.)

that common sense engineering and process knowledge help a lot in assessing the validity of an identified model.

## 6.2 Optimization under constraints

Parameter estimation rarely consists only of minimizing the prediction error on a fixed data set. Instead, more or less vague conditions are imposed on the model, either implicitly or explicitly. One way to include *a priori* knowledge explicitly into the identification is to consider optimization under constraints. A common technique for solving this kind of problem is the Lagrange multipliers approach (see, *e.g.*, [15]).

### 6.2.1 Equality constraints

Consider a function $J(\boldsymbol{\theta})$ of a parameter vector $\boldsymbol{\theta} = [\theta_1, \cdots, \theta_P]^T$ of $P$ parameters to be minimized

$$\min_{\boldsymbol{\theta}} J\left(\boldsymbol{\theta}\right) \tag{6.30}$$

subject to $C$ equality constraints

$$h_1\left(\boldsymbol{\theta}\right) = 0 \tag{6.31}$$
$$\vdots$$
$$h_C\left(\boldsymbol{\theta}\right) = 0$$

Construct a *Lagrange function* $L(\boldsymbol{\theta}, \boldsymbol{\lambda})$

$$L\left(\boldsymbol{\theta}, \boldsymbol{\lambda}\right) = J\left(\boldsymbol{\theta}\right) + \sum_{c=1}^{C} \lambda_c h_c\left(\boldsymbol{\theta}\right) \tag{6.32}$$

which is to be minimized with respect to $\boldsymbol{\theta}$ and maximized with respect to $\boldsymbol{\lambda}$. $\boldsymbol{\lambda} = [\lambda_1, \cdots, \lambda_C]$ are the *Lagrange multipliers*, also referred to as the Kuhn–Tucker parameters.

The Taylor expansion of this Lagrange function is given by

$$L\left(\boldsymbol{\theta}, \boldsymbol{\lambda}\right) \approx L\left(\overline{\boldsymbol{\theta}}, \tilde{\boldsymbol{\lambda}}\right) + \sum_{p=1}^{P} \left[\frac{\partial L\left(\boldsymbol{\theta}, \boldsymbol{\lambda}\right)}{\partial \theta_p}\right]_{\boldsymbol{\theta} = \overline{\boldsymbol{\theta}}; \lambda = \overline{\lambda}} \tilde{\theta}_p \tag{6.33}$$

$$+ \sum_{c=1}^{C} \left[\frac{\partial L\left(\boldsymbol{\theta}, \boldsymbol{\lambda}\right)}{\partial \lambda_c}\right]_{\boldsymbol{\theta} = \overline{\boldsymbol{\theta}}; \lambda = \overline{\lambda}} \tilde{\lambda}_c + \cdots \tag{6.34}$$

The optimality conditions are given by

$$\frac{\partial L\left(\boldsymbol{\theta},\lambda\right)}{\partial\widetilde{\theta}_p} = \frac{\partial L\left(\boldsymbol{\theta},\lambda\right)}{\partial\theta_p} = 0 \tag{6.35}$$

$$\frac{\partial L\left(\boldsymbol{\theta},\lambda\right)}{\partial\widetilde{\lambda}_c} = \frac{\partial L\left(\boldsymbol{\theta},\lambda\right)}{\partial\lambda_c} = 0 \tag{6.36}$$

**Example 37 (Parameter estimation algorithm)** Consider a linear regression model

$$y\left(k\right) = \boldsymbol{\theta}^T\left(k\right)\boldsymbol{\varphi}\left(k-1\right) \tag{6.37}$$

and a criterion

$$J = \frac{1}{2}\left\|\boldsymbol{\theta}\left(k\right) - \boldsymbol{\theta}\left(k-1\right)\right\|^2 \tag{6.38}$$

Determine an identification algorithm for $\boldsymbol{\theta}\left(k\right)$ which minimizes $J$.

Consider the following Lagrange function:

$$L = \frac{1}{2}\sum_{p=1}^{P}\left(\theta_p\left(k\right) - \theta_p\left(k-1\right)\right)^2 + \lambda\left[y\left(k\right) - \boldsymbol{\theta}^T\left(k\right)\boldsymbol{\varphi}\left(k-1\right)\right] \tag{6.39}$$

From the optimality constraints it follows that

$$\begin{aligned}\boldsymbol{\theta}\left(k\right) - \boldsymbol{\theta}\left(k-1\right) - \lambda\boldsymbol{\varphi}\left(k-1\right) &= 0 \tag{6.40}\\ y\left(k\right) - \boldsymbol{\theta}^T\left(k\right)\boldsymbol{\varphi}\left(k-1\right) &= 0 \tag{6.41}\end{aligned}$$

Solving for $\boldsymbol{\theta}\left(k\right)$ we derive

$$y\left(k\right) - \boldsymbol{\varphi}^T\left(k-1\right)\left[\boldsymbol{\theta}\left(k-1\right) + \lambda\boldsymbol{\varphi}\left(k-1\right)\right] = 0 \tag{6.42}$$

which gives for $\lambda$

$$\lambda = \frac{y\left(k\right) - \boldsymbol{\varphi}^T\left(k-1\right)\boldsymbol{\theta}\left(k-1\right)}{\boldsymbol{\varphi}\left(k-1\right)\boldsymbol{\varphi}^T\left(k-1\right)} \tag{6.43}$$

The identification algorithm can be summarized as follows:

$$\widehat{\boldsymbol{\theta}}\left(k\right) = \widehat{\boldsymbol{\theta}}\left(k-1\right) + \frac{\boldsymbol{\varphi}\left(k-1\right)}{\boldsymbol{\varphi}^T\left(k-1\right)\boldsymbol{\varphi}\left(k-1\right)}\left[y\left(k\right) - \boldsymbol{\varphi}^T\left(k-1\right)\widehat{\boldsymbol{\theta}}\left(k-1\right)\right]$$

$$\tag{6.44}$$

## 6.2.2 Inequality constraints

Let us extend the previous results in order to deal with inequality constraints. Assume that a function $J(\boldsymbol{\theta})$ of a parameter vector $\boldsymbol{\theta} = [\theta_1, \cdots, \theta_P]$ of $P$ parameters is to be minimized

$$\min_{\boldsymbol{\theta}} J(\boldsymbol{\theta}) \tag{6.45}$$

subject to $C$ inequality constraints

$$q_1(\boldsymbol{\theta}) \leq 0 \tag{6.46}$$

$$\vdots$$

$$q_C(\boldsymbol{\theta}) \leq 0$$

Note, that an equality constraint can be constructed using two inequality constraints.

Again, a Lagrange function $L(\boldsymbol{\theta}, \boldsymbol{\lambda})$ can be constructed

$$L(\boldsymbol{\theta}, \boldsymbol{\lambda}) = J(\boldsymbol{\theta}) + \sum_{c=1}^{C} \lambda_c q_c(\boldsymbol{\theta}) \tag{6.47}$$

where now $\lambda_c \geq 0$ (for more sophisticated approaches see, *e.g.*, [15], pp. 302-319, pp. 334-342). The Lagrange function is simultaneously minimized with respect to the parameter vector $\boldsymbol{\theta}$, and maximized with respect to the multipliers $\boldsymbol{\lambda}$.

A simple recursive algorithm solving this problem can be written as follows

$$\widehat{\boldsymbol{\theta}}(l+1) = \widehat{\boldsymbol{\theta}}(l) - \eta(l) \frac{\partial L(\boldsymbol{\theta}, \boldsymbol{\lambda})}{\partial \boldsymbol{\theta}} \tag{6.48}$$

$$\boldsymbol{\lambda}(l+1) = \max\left(0, \boldsymbol{\lambda}(l) + \eta(l) \frac{\partial L(\boldsymbol{\theta}, \boldsymbol{\lambda})}{\partial \boldsymbol{\lambda}}\right) \tag{6.49}$$

where $\eta(k)$ is the learning rate (step size).

The derivatives with respect to Lagrange multipliers are given directly by the constraints

$$\frac{\partial L(\boldsymbol{\theta}, \boldsymbol{\lambda})}{\partial \lambda_c}(l) = q_c\left(\widehat{\boldsymbol{\theta}}(l)\right) \tag{6.50}$$

where $c = 1, 2, ..., C$. In order to calculate the gradients with respect to the parameters

$$\frac{\partial L(\boldsymbol{\theta}, \boldsymbol{\lambda})}{\partial \theta_p}(l) = \frac{\partial}{\partial \theta_p} J\left(\widehat{\boldsymbol{\theta}}(l)\right) + \sum_{c=1}^{C} \lambda_c \frac{\partial}{\partial \theta_p} q_c\left(\widehat{\boldsymbol{\theta}}(l)\right) \tag{6.51}$$

$(p = 1, 2, ..., P)$, the gradients

$$\frac{\partial}{\partial \theta_p} \mathrm{J} \left( \widehat{\boldsymbol{\theta}} \, (l) \right) \text{ and } \frac{\partial}{\partial \theta_p} \mathrm{q}_c \left( \widehat{\boldsymbol{\theta}} \, (l) \right) \qquad (6.52)$$

need to be available. In general, they can be approximated by finite differences

$$\frac{\partial}{\partial \theta_p} \mathrm{J} \left( \widehat{\boldsymbol{\theta}} \, (l) \right) \approx \frac{\mathrm{J} \left( \boldsymbol{\theta} + \mathbf{e}_p \Delta \theta_p \right) - \mathrm{J} \left( \boldsymbol{\theta} - \mathbf{e}_p \Delta \theta_p \right)}{2 \Delta \theta_p} \qquad (6.53)$$

$$\frac{\partial}{\partial \theta_p} \mathrm{q}_c \left( \widehat{\boldsymbol{\theta}} \, (l) \right) \approx \frac{\mathrm{q}_c \left( \boldsymbol{\theta} + \mathbf{e}_p \Delta \theta_p \right) - \mathrm{q}_c \left( \boldsymbol{\theta} - \mathbf{e}_p \Delta \theta_p \right)}{2 \Delta \theta_p} \qquad (6.54)$$

where $\Delta \theta_p$, $p = 1, 2, ..., P$, are some small variations of the parameters; $\mathbf{e}_p$ is a column vector with 1 as the $p$'th element, zeros elsewhere.

In some cases, an analytical form can be given. In the case of parameter estimation, the gradients $\frac{\partial}{\partial \theta_p} \mathrm{J}(\boldsymbol{\theta})$ $(p = 1, 2, ..., P)$ are usually available. In prediction error methods, the cost function is given by

$$J\left( \boldsymbol{\theta} \right) = \frac{1}{2} \sum_{k=1}^{K} \left( y\left( k \right) - \widehat{y}\left( k \right) \right)^2 \qquad (6.55)$$

where $y$ is the target output and $\widehat{y}$ is the output predicted by model. Thus,

$$\frac{\partial}{\partial \theta_p} J\left( \boldsymbol{\theta} \right) = \sum_{k=1}^{K} \left[ \left( y\left( k \right) - \widehat{y}\left( k \right) \right) \frac{\partial \widehat{y}}{\partial \theta_p} \left( k \right) \right] \qquad (6.56)$$

In black-box models, the gradient $\frac{\partial \widehat{y}}{\partial \theta_p}$ is required by (unconstrained) parameter estimation methods, and is usually easily available.

The availability of the gradients of the constraints, $\frac{\partial}{\partial \theta_p} \mathrm{q}_c \left( \boldsymbol{\theta} \right)$, depends on the type of constraints. For simple constraints (upper and lower bounds on the output or parameters, fixed points), analytic forms of the gradients are easy to obtain. For other typical constraints, such as constraints on gains, poles, deviation from a nominal model etc., these may be difficult to obtain.

Let us collect the results to an algorithm.

**Algorithm 28 (Lagrange multipliers)** Problem: minimize $\mathrm{J}(\boldsymbol{\theta})$ subject to $\mathrm{q}_c\left( \boldsymbol{\theta} \right) \leq 0$, $c = 1, 2, ..., C$, with respect to $\boldsymbol{\theta} = [\theta_1, \cdots, \theta_P]$.

1. Initialize:

   Set iteration index $l = 1$.

Initialize $\widehat{\boldsymbol{\theta}}\,(l)$ and $\boldsymbol{\lambda}\,(l) = \mathbf{0}$.

2. Evaluate model and constraints:

   Evaluate $\widehat{y}\,(k)$ and $\frac{\partial \widehat{y}}{\partial \theta_p}\,(k)$ for all patterns $k$.

   Evaluate $q_c$ and $\frac{\partial q_c}{\partial \theta_p}$ for all constraints $c$.

   Evaluate $\frac{\partial J}{\partial \theta_p}$ for all parameters $p$.

3. Compose gradients of the Lagrangian, $\frac{\partial L}{\partial \theta}$ and $\frac{\partial L}{\partial \lambda}$:

$$\frac{\partial L}{\partial \lambda_c}\,(l) = q_c\,(l) \tag{6.57}$$

$$\frac{\partial L}{\partial \theta_p}\,(l) = \frac{\partial J}{\partial \theta_p}\,(l) + \sum_{c=1}^{C} \lambda_c\,(l)\,\frac{\partial q_c}{\partial \theta_p}\,(l) \tag{6.58}$$

4. Update the parameters and Lagrange multipliers:

$$\widehat{\boldsymbol{\theta}}\,(l+1) = \widehat{\boldsymbol{\theta}}\,(l) - \eta\,(l)\,\frac{\partial L}{\partial \theta}\,(l) \tag{6.59}$$

$$\boldsymbol{\lambda}\,(l+1) = \max\left(0, \boldsymbol{\lambda}\,(l) + \eta\,(l)\,\frac{\partial L}{\partial \lambda}\,(l)\right) \tag{6.60}$$

5. Quit, or set $l = l + 1$ and return to Step 2.

## 6.3 Guided random search methods

For solving optimization tasks, such as in parameter estimation, the most popular approaches are gradient-based. However, a cost function may have several local optima, and it is well known that gradient-based estimation routines may converge to a local optimum instead of a global one.

A common solution is to repeat the gradient-based search several times, starting at different (random) initial locations. An alternative is to use some (guided) random search method instead of a gradient-based method, or use both as in hybrid methods.

With random search methods, the computation of the gradients at each iteration (often the most time-consuming phase in the implementation and computation of gradient-based methods) can be avoided. As well, constraints can be easily included. Most importantly, random search methods perform a global search in the search space, and are thus not easily fooled by the local optima. As this has, however, not shown to be a severe problem in many of the practical applications, various gradient methods are commonly used due to their efficacy:

- *Matyas random optimization* method [6] uses the idea of 'contaminating' the current solution in order to explore the search space around the current solution. At each iteration, a Gaussian random vector is added to the current solution. If the new solution improves the model performance, it will be used in the next iteration. If no improvement occurs, the old solution is kept and a new Gaussian random vector is generated. In *simulated annealing* [46], occasional upward steps in the criterion are allowed. The acceptance of upward steps is treated probabilistically, so that as the optimization proceeds, the probability is decreased until the system 'freezes'.

- A different approach was taken by Luus and Jaakola [56], whose method is based on *direct search* and systematic *search region reduction*. In each iteration, a number of random vectors belonging to the search space are generated and evaluated against the criterion. In order to improve the solution, instead of concentrating on the space close to the best solution, the direct search is performed in a much larger space. In each iteration, the search space is, however, slightly reduced until it becomes so small that a desired accuracy has been obtained.

- Where simulated annealing finds background from statistical mechanics and evolutionary processes, *genetic algorithms* [22] are motivated by the mechanisms of natural selection and genetics. From an initial population of solutions, a genetic algorithm chooses the most fitted solutions (in the sense of a given criterion) using a selection operator. In order to generate new solutions, operators such as crossover and mutation are used. They are based on a specific form of coding of the solutions as strings, which allows recombination operators similar to those observed with chromosomes. On the new population of solutions, selection and recombination operators are used repeatedly until the population converges, giving a population of fittest solutions.

Optimization techniques based on *learning automata* [75] also belong to the class of random search techniques. The concept of learning automata was initially introduced in connection of modeling of biological systems. They have been widely used to solve problems for which an analytical solution is not possible, or which are mathematically intractable. They have also attracted interest due to their potential usefulness in engineering problems of optimization and control characterized by non-linearity and high level of uncertainty. In general, the learning automata are very simple machines, and have few and transparent tuning parameters. As an example of the random search paradigms, we will next consider the stochastic learning automata in more detail.

## 6.3.1 Stochastic learning automaton

Optimization techniques based on learning automata (LA) belong to the class of (guided) random search techniques. In general, random search methods have attained fairly little interest in optimization, although they have some very appealing features. Learning automata can be applied to a large class of optimization problems, since there are only few assumptions concerning the function to be optimized. They are simple, transparent and easy to apply, even for complexly structured or constrained systems. The main advantages, if compared to the more popular gradient-based algorithms, are that the gradients need not be computed and that the search for the global minimum is not easily fooled by the local minima.

A learning automaton is a sequential machine characterized by:

- a set of *actions*,

- an *action probability distribution* and

- a *reinforcement learning* scheme.

It is connected in a feedback loop (see Fig. 6.7) to the random environment (the function to be optimized, the process to be controlled, *etc.*). At every sampling instant, the automaton chooses randomly an action from a finite set of actions on the basis of a probability distribution. The selected action causes a reaction of the environment, which in turn is the input signal for the automaton. With a reinforcement scheme, the learning automaton recursively updates its probability distribution, and should be capable of changing its structure and/or parameters to achieve the desired goal or optimal performance in the sense of a given criterion.

To describe an *automaton*, introduce the following [75]:

1. $U$ denotes the set $\{u_1, u_2, \cdots, u_A\}$ of the $A$ actions of the automaton, $A \in [2, \infty[$.

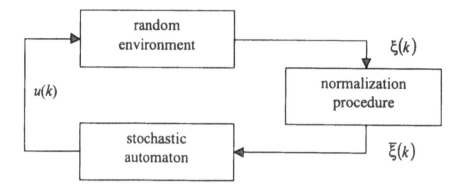

Figure 6.7: A learning automaton with a normalization procedure connected in a feedback loop with the environment. The automaton produces an action, $u(k)$, based on the probabilities of the actions. The environment response, $\xi(k)$, is normalized and fed back to the automaton. The automaton adjusts its action probabilities and produces a new action.

2. $\{u(k)\}$ is a sequence of automaton outputs (actions), $u(k) \in U$.

3. $\mathbf{p}(k) = [p_1(k), \cdots, p_A(k)]^T$ is a vector of action probabilities at iteration $k$, for which

$$\sum_{a=1}^{A} p_a(k) = 1, \forall k \qquad (6.61)$$

4. $\{\overline{\xi}(k)\}$ is a sequence of automaton inputs (environment responses). Automaton inputs are provided by the environment either in a binary (P-model environment) or continuous (S-model environment) form.

5. T represents the reinforcement scheme which changes the probability vector

$$\mathbf{p}(k+1) = \mathbf{p}(k) + \eta(k)\,\mathbf{T}\left(\mathbf{p}(k), \left\{\overline{\xi}(\kappa)\right\}, u(\kappa)\right)_{\kappa=1,2,\cdots,k} \qquad (6.62)$$

$$p_a(1) > 0 \ \forall a, \qquad (6.63)$$

where $\eta(k)$ is a scalar learning rate that may be time-varying. The vector $\mathbf{T}=[T_1(.), ..., T_A(.)]^T$ satisfies the following conditions for preserving the probability measure:

$$\sum_{a=1}^{A} T_a(.) = 0, \ \forall k, \qquad (6.64)$$

$$p_a(k) + \eta(k) \, \mathrm{T}_a(.) \in [0,1], \forall k, \; \forall a. \tag{6.65}$$

The operation of a learning automaton can be summarized as follows (see Fig. 6.7)

1. Select randomly an action $u(k)$ from the action set $U$ according to the probability distribution $p(k)$.

2. Calculate the normalized environment response $\overline{\xi}(k)$.

3. Adjust the probability vector $\mathbf{p}(k)$.

4. Return to Step 1, or quit.

A practical method for choosing an action according to a probability distribution (Step 1) is to generate a uniformly distributed random variable $\varsigma = U(0,1)$. The $a$'th action $u(k) = u_a$ is then chosen such that $a$ is equal to the least value of $i$, satisfying the following constraint:

$$\sum_{i=1}^{a} p_i(k) \geq \varsigma \tag{6.66}$$

In the S-model environment, the continuous environment responses (Step 2) need to be in the range of $\xi(k) \in [0,1]$. To achieve this, a normalization procedure can be applied, *e.g.*,

$$\overline{\xi}(k) = \frac{s_a(k) - \min_{i=1,\ldots,A} s_i(k)}{\max_{i=1,\ldots,A} s_i(k) - \min_{i=1,\ldots,A} s_i(k)} \tag{6.67}$$

where $u_a$ is the chosen action, $s_a$ is the expectation of the environment response $\xi$ for action $a$, and $\overline{\xi}(k)$ denotes the normalized environment response, $\overline{\xi}(k) \in [0,1]$.

A number of reinforcement schemes (Step 3) have been described in the literature [75]. A general non-linear reinforcement scheme is of the form [67]:

• if $u(k) = u_a$:

$$p_a(k+1) = p_a(k) + \left(1 - \overline{\xi}(k)\right) \sum_{j=1, \, j\neq a}^{A} g_i(\mathbf{p}(k)) \tag{6.68}$$

$$-\overline{\xi}(k) \sum_{j=1, \, j\neq a}^{A} h_i(\mathbf{p}(k))$$

- if $u\left(k\right) \neq u_a$:

$$p_a\left(k+1\right) = p_a\left(k\right) - \left(1 - \overline{\xi}\left(k\right)\right) g_a\left(\mathbf{p}\left(k\right)\right) + \overline{\xi}\left(k\right) h_a\left(\mathbf{p}\left(k\right)\right) \quad (6.69)$$

where the functions $g_a$ and $h_a$ are associated with reward and penalty, respectively.

A simple reinforcement scheme, the *linear reward-penalty* ($L_{R\text{-}P}$) *scheme* [11], for an automaton of $A$ actions operating in an S-model environment is obtained by selecting

$$g_a\left(\mathbf{p}\left(k\right)\right) = \eta p_a\left(k\right) \qquad (6.70)$$

$$h_a\left(\mathbf{p}\left(k\right)\right) = \frac{1}{A-1} - \eta p_a\left(k\right) \qquad (6.71)$$

Substituting the preceding equations into (6.68)–(6.69) we have the following $L_{R\text{-}P}$ learning scheme:

- if $u\left(k\right) = u_a$:

$$p_a\left(k+1\right) = p_a\left(k\right) + \eta\left(1 - p_a\left(k\right) - \overline{\xi}\left(k\right)\right) \qquad (6.72)$$

- if $u\left(k\right) \neq u_a$:

$$p_a\left(k+1\right) = p_a\left(k\right) - \eta\left(p_a\left(k\right) - \frac{\overline{\xi}\left(k\right)}{A-1}\right) \qquad (6.73)$$

Learning automata can be applied to a large variety of complex optimization problems, since there are only few assumptions concerning the function to be optimized. Typical applications include multimodal function optimization problems, see Fig. 6.8. What makes the automata approach particularly interesting is the existence of theoretical proofs ($\varepsilon$-accuracy, $\varepsilon$-optimality, convergence with probability one, convergence in the mean square sense, rate of convergence) (see, *e.g.*, [75]). There are almost no conditions concerning the function to be optimized (continuity, unimodality, differentiability, convexity, *etc.*). Learning automata perform a global search on the search space (action space), and they are not easily fooled by the local minima. In general, learning automata are simple machines, and have few and transparent tuning parameters. The main drawback is in the lack of efficiency (slow convergence rates), in particular for large action spaces.

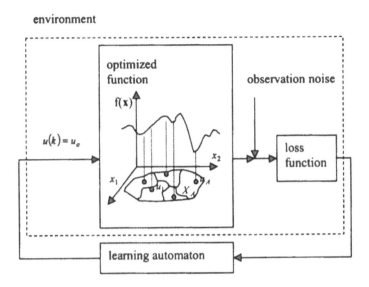

Figure 6.8: Multimodal function optimization using a learning automaton. The search region is quantified using $X_a, X_b \subset X$; $X_a \cap_{a \neq b} X_b = 0$; $\cup_a X_a = X$, where $u(k) \in \{u_1, \cdots, u_A\}$; and $u_a \in X_a$ are fixed points.

## 6.4 Simulation examples

This section will concern three applications related to process engineering. First, a SISO Wiener model for a simulated pneumatic valve is identified. Both grey-box and black-box approaches for modeling the static part are considered. The results are compared with those from the literature [95], and the estimated parameters are examined. The second example considers the estimation of the parameters in a MISO Hammerstein model. The data is drawn from a binary distillation column model, considered also in [17]. Parameter estimation under constraints posed on the properties of the static part is illustrated. The third example illustrates identification of a two-tank system under constraints using a Wiener model.

All simulations were performed on a Pentium PC (450 MHz) and Matlab$^{\text{TM}}$ 5.2. In the parameter estimation, the functions `leastsq.m` (Levenberg – Marquardt method with a mixed quadratic and cubic line search) and `constr.m` (mechanization of the Lagrange multipliers approach) from the Matlab optimization toolbox were used. The differential equations were solved using the `ode23.m` function, inverse problems with `fsolve.m`.

## 6.4.1   Pneumatic valve: identification of a Wiener system

Let us first consider a simple example on the identification of a Wiener system. A simple model for a pneumatic valve for fluid flow control is given in [95], where also a Wiener model for the system is identified. Some of the simulation results can also be found from [36].

### Process and data

Pneumatic as well as electrical valves are commonly used for fluid flow control. The static characteristics of a valve for fluid flow control vary with operating conditions. The input of the model represents the pneumatic control signal applied to the stem, while the internal variable represents the stem position, or equivalently, the position of the valve plug. Linear dynamics describe the dynamic balance between the control signal and the stem position:

$$\frac{z(k)}{u(k)} = \frac{0.1044q^{-1} + 0.0883q^{-2}}{1 - 1.4138q^{-1} + 0.6065q^{-2}} \tag{6.74}$$

The flow through the valve is given by a non-linear function of the stem position:

$$y(k) = \frac{z(k)}{\sqrt{c_1 + c_2 z^2(k)}}; c_1 = 0.1; c_2 = 0.9 \tag{6.75}$$

Based on the above model, data sets of 1000 input-output pairs were generated in the same fashion as in [95] (see Fig. 6.9): a pseudo random sequence (PRS) was used as input. In practice it is impossible to obtain perfect measurements; to make the simulations more realistic in the case of noisy observations a Gaussian noise was introduced at the output measurement. For reference purposes, also a noiseless data set was generated, as well as two test sets of 200 input–output patterns.

### Model structure and parameter estimation

The structure of the linear dynamic SISO model was assumed to be known $(n_B = 1, n_A = 2, d = 1)$

$$z(k) = \frac{b_0 + b_1^* q^{-1}}{1 - a_1 q^{-1} + a_2 q^{-2}} u(k-1) \tag{6.76}$$

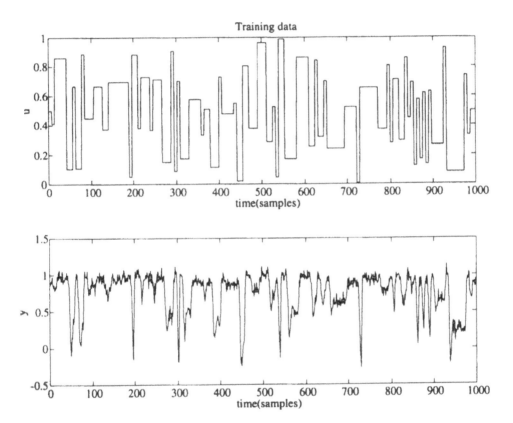

Figure 6.9: Training data for the pneumatic valve model. The system input (upper plot) was a PRS signal with a basic clock period of seven samples. The system output (lower plot) was corrupted by Gaussian noise $N(0, 0.05^2)$.

Thus the linear dynamic part contained three parameters to be estimated from data (remember that $b_1^*$ is fixed, see Algorithm 23).

Two model structures were considered for the static part. In the first case, the SQRT structure of (6.75) with the two parameters $c_1$ and $c_2$ to be estimated, was considered. The required derivatives are simple to compute, resulting in

$$\frac{\partial y}{\partial z} = \frac{c_1}{(c_1 + c_2 z^2)^{\frac{3}{2}}} \tag{6.77}$$

$$\frac{\partial y}{\partial c_1} = \frac{-z}{2(c_1 + c_2 z^2)^{\frac{3}{2}}} \tag{6.78}$$

$$\frac{\partial y}{\partial c_2} = \frac{-z^3}{2(c_1 + c_2 z^2)^{\frac{3}{2}}} \tag{6.79}$$

In the second case, a 0-order Sugeno fuzzy model was used (see Algorithm 18). Let us assume crisp system input(s), triangular fuzzy sets, add-one partition, product t-norm, and weighted average defuzzification. Then, a 0-order Sugeno model can be expressed as

$$\hat{y} = \sum_{p_1=1}^{P_1} \sum_{p_2=1}^{P_2} \cdots \sum_{p_I=1}^{P_I} \alpha_{p_1,p_2,\cdots,p_I} \tilde{\mu}_{1,p_1}(z_1,\cdot) \tilde{\mu}_{2,p_2}(z_2,\cdot) \cdots \tilde{\mu}_{I,p_I}(z_I,\cdot) \tag{6.80}$$

where the input degrees of membership are given by

$$\tilde{\mu}_{i,p_i}\left(z_i,\tilde{\beta}_i\right) = \max\left(\min\left(\frac{z_i - \tilde{\beta}_{i,p_i-1}}{\tilde{\beta}_{i,p_i} - \tilde{\beta}_{i,p_i-1}}, \frac{\tilde{\beta}_{i,p_i+1} - z_i}{\tilde{\beta}_{i,p_i+1} - \tilde{\beta}_{i,p_i}}\right),0\right) \tag{6.81}$$

where $\tilde{\mu}_{i,p_i}(\varphi_i,\cdot)$ is the membership function associated with the $p_i$'th fuzzy set partitioning the $i$'th input ($p_i = 1, 2, \cdots, P_i$). The input partition is given by $\tilde{\beta}_i\left(\tilde{\beta}_{i,p_i-1} < \tilde{\beta}_{i,p_i}\right)$. The derivatives with respect to the inputs are given by

$$\frac{\partial \hat{y}}{\partial z} = \sum_{p_1=1}^{P_1} \sum_{p_2=1}^{P_2} \cdots \sum_{p_I=1}^{P_I} \alpha_{p_1,p_2,\cdots,p_I} \prod_{j=1;j\neq i}^{I} \tilde{\mu}_{j,p_j}(z_j,\cdot) \frac{\partial}{\partial z_i}\tilde{\mu}_{i,p_i}(z_i,\cdot) \tag{6.82}$$

where

$$\frac{\partial}{\partial z_i}\tilde{\mu}_{i,p_i}\left(z_i,\tilde{\beta}_i\right) = \begin{cases} 1/\left(\tilde{\beta}_{i,p_i} - \tilde{\beta}_{i,p_i-1}\right) & \text{if } \tilde{\beta}_{i,p_i-1} < z_i \leq \tilde{\beta}_{i,p_i} \\ -1/\left(\tilde{\beta}_{i,p_i+1} - \tilde{\beta}_{i,p_i}\right) & \text{if } \tilde{\beta}_{i,p_i} < z_i \leq \tilde{\beta}_{i,p_i+1} \\ 0 & \text{otherwise} \end{cases} \tag{6.83}$$

and with respect to the consequent parameters by

$$\frac{\partial \widehat{y}}{\partial \alpha_{p_1, p_2, \cdots, p_I}} = \prod_{i=1}^{I} \widetilde{\mu}_{i, p_i} \left( z_i, \cdot \right) \tag{6.84}$$

In the SISO problem $(I = 1)$ considered here, five triangular membership functions were used $(P_1 = 5)$. This results in a piece-wise linear structure which is functionally equivalent to that used in [95]. The antecedent parameters (knots) were set to $\widetilde{\beta}_1 = [0, 0.2, 0.4, 0.7, 1]$ as in [95], and the consequent parameters $\alpha$ were to be estimated from data.

The parameters of the two model structures were estimated using the Levenberg–Marquardt method, using both noiseless and noisy data resulting in four simulations. In addition, the results were compared with those reported in [95].

**Analysis**

Table 6.1 shows the number of iterations required in the four cases. The training was fast, completed in a few minutes. Note, that the recursive prediction error method used in [95] is not a batch method, and thus the results are not directly comparable with [95] (bottom line of Table 6.1).

The root-mean-squared errors (RMSE)

$$RMSE = \sqrt{\frac{1}{K} \sum_{k=1}^{K} \left( y\left(k\right) - \widehat{y}\left(k\right) \right)^2} \tag{6.85}$$

on the corresponding training set and on noiseless test data are given in Table 6.2. When the model structure was exactly known (the SQRT case), the match was perfect; the RMSE on training and test set were zero up to four digits in the noiseless case, and close to the standard deviation of the noise in the case of noisy data. Also in the case of a more general black-box model (0-Sugeno) the accuracy of all the identified models was good.

Tables 6.3, 6.4 and 6.5 show the true values of the parameters, estimated values, and the parameter values estimated in [95]. Table 6.3 shows the coefficients of the polynomials $A$ and $B$, *i.e* linear dynamics. The maximum deviation from true zero at $q = -0.8458$ was less than 1% in the case of SQRT model, as well as in the case of 0-Sugeno model identified from noiseless data. For the 0-Sugeno models identified from noisy data, however, the deviation was more important, yet less than 20%. In all cases, deviation from the true poles located at $q = 0.7069 \pm 0.3268i$ was less than 2%.

|                      | training time | training epochs |
|----------------------|:-------------:|:---------------:|
| SQRT–no noise        |               | 47              |
| SQRT–noise           |               | 64              |
| 0-Sugeno–no noise    |               | 37              |
| 0-Sugeno–noise       |               | 37              |
| 0-Sugeno–noise[95]   |               | (1)             |

Table 6.1: Number of iterations required by the Levenberg–Marquardt method.

Table 6.4 gives the parameters of the SQRT model, and Table 6.5 the consequent parameters of the 0-Sugeno model. Note, that the redundancy in gains is removed by fixing the gain of the dynamic part; thus parameter $b_1^*$ is not independently estimated. In [95], the problem of redundancy was solved by fixing the gain of the static part on an interval $u \in [0.2, 0.4]$ to 1.5; thus only the bias for the interval was identified (0.2402). The 'non-identified' parameters are indicated by the parentheses in the tables. However, as the steady-state gain of the transfer function of the true system (6.74)–(6.75) is one, the parameters are directly comparable. Table 6.4 shows that the parameters were correctly estimated up to two digits in the case where the correct form of the non-linearity was known. Table 6.5 indicates that the difference between the consequent parameters estimated here and the knot parameters estimated in [95] was small, the largest difference appeared in the first parameter ($|\Delta w_1| < 0.08$), for which region the training set contained only a few data points.

The performance of the identified 0-Sugeno model on test set data (previously unseen to the model) is illustrated in Fig. 6.10. The upper part of the figure shows the steps in the system input $u$, and the filtered intermediate variable $z$. Note that the steady-state gain of the linear dynamic filter is one. The lower part in Fig. 6.10 shows the output of the true system $y$ and the prediction by the model $\widehat{y}$, which is obtained by putting the intermediate variable $z$ through the static (non-linear) part. The fit between the desired and obtained signals is close, although some deviation can be seen at lower values of $y$. This mismatch is due to the noise in the few data samples from that operating area.

## Results

In this example we found that the correct parameter values were estimated using the Wiener structure and the Levenberg–Marquardt method in pa-

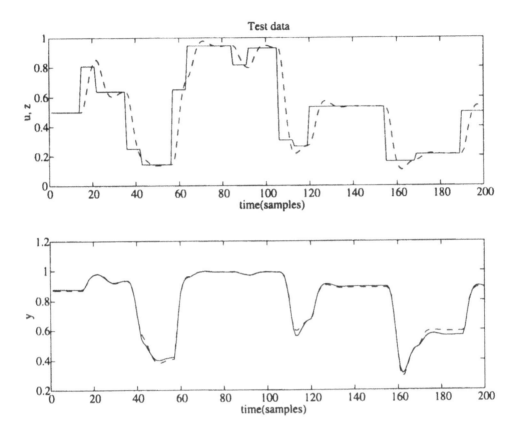

Figure 6.10: Prediction by the 0-Sugeno model identified on noisy data. The intermediate signal $z$ (upper plot – dashed line) is obtained by filtering the input sigal $u$ (upper plot – solid line). The model output (lower plot – dashed line) is computed by putting the intermediate signal through the non-linear static function. The true output is shown by a solid line.

| RMSE | training data | test data |
|---|---|---|
| true model–no noise | 0 | 0 |
| SQRT–no noise | 0.0000 | 0.0000 |
| SQRT–noise | 0.0488 | 0.0014 |
| 0-Sugeno–no noise | 0.0297 | 0.0315 |
| 0-Sugeno–noise | 0.0525 | 0.0126 |
| 0-Sugeno–noise[95] | 0.0554 | 0.0111 |

Table 6.2: Root-mean-squared error on training and test data.

| Linear dynamics | $b_0$ | $b_1 (b_1^*)$ | $a_1$ | $a_2$ |
|---|---|---|---|---|
| true parameters–no noise | 0.1044 | 0.0883 | −1.4138 | 0.6065 |
| SQRT–no noise | 0.1044 | (0.0883) | −1.4138 | 0.6065 |
| SQRT–noise | 0.1046 | (0.0902) | −1.4104 | 0.6052 |
| 0-Sugeno–no noise | 0.1058 | (0.0884) | −1.3983 | 0.5925 |
| 0-Sugeno–noise | 0.0997 | (0.0991) | −1.3926 | 0.5914 |
| 0-Sugeno–noise[95] | 0.0980 | 0.0984 | −1.4026 | 0.5990 |

Table 6.3: Estimated parameters of the linear dynamic block.

| SQRT-parameters | $c_1$ | $c_2$ |
|---|---|---|
| true parameters–no noise | 0.1 | 0.9 |
| SQRT–no noise | 0.1000 | 0.9000 |
| SQRT–noise | 0.0992 | 0.8999 |

Table 6.4: Estimated parameters of the static SQRT-block.

| 0-Sugeno parameters | $w_1$ | $w_2$ | $w_3$ | $w_4$ | $w_5$ |
|---|---|---|---|---|---|
| 0-Sugeno–no noise | −0.0545 | 0.5743 | 0.8253 | 0.9626 | 1.0021 |
| 0-Sugeno–noise | −0.0397 | 0.5770 | 0.8252 | 0.9608 | 1.0030 |
| 0-Sugeno–noise[95] | 0.0221 | (0.5402) | (0.8402) | 0.9607 | 0.9852 |

Table 6.5: Estimated consequent parameters of the static 0-Sugeno block.

rameter estimation. Both noisy and noiseless cases were experimented. As expected, the performance of the approach was similar to that in [95], when comparable. However, the approach suggested here is not restricted to any particular form of the static mapping. As illustrated, the suggested identification procedure does not restrict the type of the non-linear static model, as long as the gradients can be computed (or approximated). In the example, it was shown how a grey-box SQRT-model with a structure justified by physical background could also be used, as well as a fuzzy black-box model.

In the next example, we will consider the identification of a more complicated MISO system.

## 6.4.2 Binary distillation column: identification of Hammerstein model under constraints

In a second example, identification of a binary distillation column was studied. Hammerstein modeling of this process was considered in [17], see also [37]. In what follows, special emphasis is focused on the role of process identification under constraints.

### Process and data

Distillation is a complex chemical operation for separation of components of liquid mixtures and purification of products. It is widely used in the petroleum, chemical and pharmaceutical industries. In a typical distillation column, the feed enters near the center of the column and flows down. Vapors that are released by heating are condensed and can be removed as overhead product or distillate. Any liquid that is returned to the column is called reflux. The reflux flows down the column and joins the feed stream. The reflux rate has a major influence on the separation process. Too much reflux makes the product excessively pure, but wastes energy because more reflux liquid has to be vaporized, while too little reflux causes an impure product.

A simple model for a binary distillation column was given in [90] (pp. 70–74). The model is described by a set of differential equations. The composition dynamics at the bottom, $1^{st}$ tray, feed tray, top trays and condenser are given by

$$\frac{dx_b}{dt} = \frac{1}{M_b}\left(L_1 x_1 - B x_b - V y_b\right) \tag{6.86}$$

$$\frac{dx_1}{dt} = \frac{1}{M_1}\left(V\left(y_b - y_1\right) + L_2 x_2 - L_1 x_1\right) \tag{6.87}$$

$$\frac{dx_1}{dt} = \frac{1}{M_{n_f}} \left( V \left( y_{n_f-1} - y_{n_f} \right) + L_{n_f+1} x_{n_f+1} - L_{n_f} x_{n_f} + F z_t \right) \tag{6.88}$$

$$\frac{dx_N}{dt} = \frac{1}{M_N} \left( V \left( y_{N-1} - y_N \right) + R x_d - L_N x_N \right) \tag{6.89}$$

$$\frac{dx_d}{dt} = \frac{1}{M_d} \left( V y_N - R x_d - D x_d \right) \tag{6.90}$$

respectively. At other trays, the composition dynamics are given by

$$\frac{dx_n}{dt} = \frac{1}{M_n} \left( V \left( y_{n-1} - y_n \right) + L_{n+1} x_{n+1} - L_n x_n \right) \tag{6.91}$$

The vapor composition at trays $n = 1, 2, ..., N$ and at the bottom are computed using a constant relative volatility:

$$y_n = \frac{\alpha x_n}{1 + (\alpha - 1) x_n} \tag{6.92}$$

$$y_b = \frac{\alpha x_b}{1 + (\alpha - 1) x_b} \tag{6.93}$$

The relations between the changes in flows are assumed to be immediate:

$$B = F - V + R \tag{6.94}$$

$$L_n = \begin{cases} R + F & \text{if } n \le n_f \\ R & \text{if } n > n_f \end{cases} \tag{6.95}$$

$$D = V - R \tag{6.96}$$

where $B$, $D$, and $L_n$ are flows of the bottom, distillate and liquid at tray $n$, respectively. The steady-state operating parameters of the distillation column model are given in Table 6.6. The model is distinguished by an open-book like steady-state non-linearity between reflux and top composition and a strong variation of the apparent time constant with change in reflux.

Using the model, a data set of 900 data patterns was generated by varying the values of reflux flow $R$ and distillate flow $V$ (PRS with a maximum amplitude of 5%, sampling time 10 min). The top composition $x_d$ was observed and a model for it was to be identified.

## Model structure and parameter estimation

In [17], a Hammerstein model for the process was considered: $u_1 \leftarrow R$, $u_2 \leftarrow V$, $y \leftarrow x_d$, with first order dynamics, $n_B = 0$, $n_A = 1$, $d = 1$. The same setting for the dynamic part was used here.

| Parameter | Value |
|---|---|
| reflux $R$ | 1.477 |
| vapour boilup $V$ | 1.977 |
| feed flow $F$ | 1 |
| feed composition $z_f$ | 0.5 |
| number of trays $N$ | 25 |
| feed tray $n_f$ | 12 |
| relative volatility $\alpha$ | 2 |
| holdups $M_b = M_n = M_d$ | 0.5 |
| bottom composition $x_b$ | 0.005 |
| top composition $x_d$ | 0.995 |

Table 6.6: Steady-state operating parameters of the binary distillation column model.

The steady state mapping was identified using a sigmoid neural network (SNN) structure. The output of a one-hidden-layer SNN with $H$ hidden nodes (see Algorithm 17) is given by

$$\widehat{z} = \sum_{h=1}^{H} \alpha_h g_h (\mathbf{u}, \boldsymbol{\beta}_h) + \alpha_{H+1} \qquad (6.97)$$

$$g_h (\mathbf{u}, \boldsymbol{\beta}_h) = \cfrac{1}{1 + \exp\left( -\sum_{i=1}^{I} \beta_{h,i} u_i - \beta_{h,I+1} \right)} \qquad (6.98)$$

where $\mathbf{u}$ is the $I$ dimensional input vector. The parameters of the network are contained in an $H+1$ dimensional vector $\boldsymbol{\alpha}$ and $H \times (I+1)$ dimensional matrix $\boldsymbol{\beta}$. The gradients with respect to the parameters are given by

$$\frac{\partial \widehat{z}}{\partial u_i} = \sum_{h=1}^{H} \alpha_h g_h (\mathbf{u}, \boldsymbol{\beta}_h) [1 - g_h (\mathbf{u}, \boldsymbol{\beta}_h)] \beta_{h,i+1} \qquad (6.99)$$

$$\frac{\partial \widehat{z}}{\partial \alpha_h} = g_h (\mathbf{u}, \boldsymbol{\beta}_h) ; \frac{\partial \widehat{z}}{\partial \alpha_{H+1}} = 1 \qquad (6.100)$$

$$\frac{\partial \widehat{z}}{\partial \beta_{h,i}} = \alpha_h g_h (\mathbf{u}, \boldsymbol{\beta}_h) [1 - g_h (\mathbf{u}, \boldsymbol{\beta}_h)] \varphi_i \qquad (6.101)$$

$$\frac{\partial \widehat{z}}{\partial \beta_{h,I+1}} = \alpha_h g_h (\mathbf{u}, \boldsymbol{\beta}_h) [1 - g_h (\mathbf{u}, \boldsymbol{\beta}_h)] \qquad (6.102)$$

where $h = 1, 2, ..., H$ and $i = 1, 2, ..., I$. In the distillation column example, $H = 6$ was used.

In the parameter estimation, optimization under constraints was considered. Constraints were posed on the gain of the static mapping so that

$$\frac{\partial \widehat{z}}{\partial u_1}(k) \geq 0 \qquad (6.103)$$

$$\frac{\partial \widehat{z}}{\partial u_2}(k) \leq 0 \qquad (6.104)$$

was required. The constraints were evaluated at 625 points, forming a grid with regular intervals on the input space spanned by $R$ and $V$: $u_1 \in \{0.95u_1^{ss}, ..., 1.05u_1^{ss}\}$, $u_2 \in \{0.95u_2^{ss}, ..., 1.05u_2^{ss}\}$, $u_1^{ss} = 1.477$, $u_2^{ss} = 1.977$. This results in 1250 constraint evaluations at each iteration. The sum of squared errors on the training set was then to be minimized under these constraints. The parameters were estimated using the Lagrange multipliers approach.

### Analysis

The training data and the prediction of the identified model on training data are illustrated in Fig. 6.11. The RMSE on training data was 0.6731. For reference purposes, the parameters were also estimated using the Levenberg–Marquardt method (no constraints). This resulted in a RMSE of 0.2048 on training data. Hence, a more accurate description of training data points was obtained using the Levenberg–Marquardt method.

However, the examination of the static mapping shows a significant problem with the unconstrained model. Fig. 6.12 shows the mappings obtained in the constrained and unconstrained cases. In the unconstrained case, the static mapping is non-monotonic. This is due to the small amount of data and the mismatch in the structure of the plant and the model. The constrained case corresponds better to the *a priori* knowledge of the process behavior (monotonic increasing with respect to $R$, monotonic decreasing with respect to $V$).

Compared with [17], visual inspection of model predictions reveals that more accurate descriptions of the process were identified with the approach suggested here. This can be attributed mainly to the more flexible structure used for the static part (a power series was used in [17]). As pointed out in [17], however, a logarithmic transformation of the output measurement would provide a more reasonable resolution for real applications.

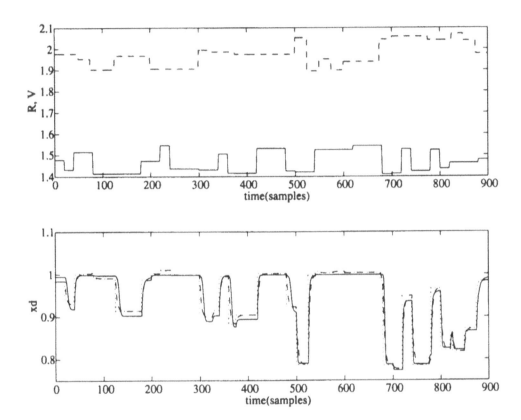

Figure 6.11: Prediction by the Hammerstein model identified under constraints on static gains. The upper plot shows the model inputs; the lower plot shows the plant response (solid line) and model responses: the intermediate variable (dotted line) and the prediction by the model (dash-dot line).

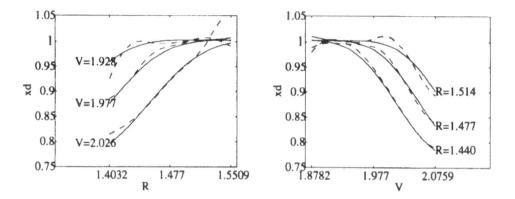

Figure 6.12: Static mapping in the constrained (solid lines) and unconstrained (dashed lines) cases.

### Results

In this example, a Hammerstein model for a MISO process was identified. Parameters were estimated under constraints, where constraints were posed on the static mapping. The suggested approach enables to pose constraints directly based on the *a priori* knowledge on steady-state behavior. Typically, information such as minimum and maximum bounds of plant output, knowledge on sign or bounds of the gains, fixed equilibrium points, *etc.*, is available. With linear dynamics, it is simple to pose constraints also on the dynamical part, such as bounds on the location of poles and zeros. This applies both for the Hammerstein and Wiener approaches. For purposes such as process control, clearly the constrained model can be expected to give better performance.

### 6.4.3   Two-tank system:  Wiener modeling under constraints

As a final example, let us illustrate the identification of a two-tank process under constraints, using a Wiener structure.

**Process**   Consider a two-tank system [64], see Fig. 6.13. Mass-balance considerations lead to the following non-linear model:

$$\frac{dY_1(t)}{dt} = \frac{1}{A_1}\left(Q(t) - k_1\sqrt{Y_1(t)}\right) \qquad (6.105)$$

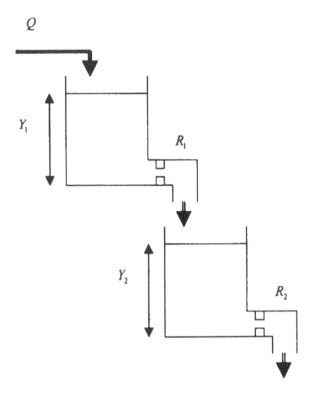

Figure 6.13: A two-tank system.

$$\frac{dY_2\left(t\right)}{dt} = \frac{1}{A_2}\left(k_1\sqrt{Y_1\left(t\right)} - k_2\sqrt{Y_2\left(t\right)}\right) \qquad (6.106)$$

where $Y_1$ and $Y_2$ are the levels, $A_1$ and $A_2$ are the cross-surfaces, and $k_1$ and $k_2$ are the coefficients involved in the modeling of the restrictions and of the two tanks, respectively. The following values were adopted: $A_1 = A_2 = 1$, $k_1 = 1$, $k_2 = 0.9$.

**Experiment design**

The system was simulated using an input consisting of a pseudo random sequence. The output measurement was corrupted with a normally distributed random sequence with a variance equal to 0.04. From the simulations, a set of 398 measurements describing the behavior of the system were sampled using $T = 1$.

**Model structure**

A SISO, $I = 1$, Wiener model was constructed from the input flow, $Q(t)$, to the level of the second tank, $Y_2(t)$. Second order linear dynamics, $N = 0$, $M = 2$, with delay of one sample, $d = 1$, were considered. A sigmoid neural network with six hidden nodes, $H = 6$, was used to model the non-linear static behavior of the system.

**Parameter estimation (under constraints)**

A number of constraints were considered for the static part. These constraints were evaluated in $C = 56$ points $Q_c = \{0.0, 0.02, 0.04, ..., 1.1\}$, $c = 1, 2, ..., C$.

- Constraints on the output:

$$J_c(\theta) = f(Q_c, \theta) - Y_{max}; Y_{max} = 1.203 \qquad (6.107)$$

$$J_{C+c}(\theta) = Y_{min} - f(Q_c, \theta); Y_{min} = 0 \qquad (6.108)$$

- Constraints on the static gain:

$$J_{2C+c}(\theta) = \frac{\partial f(Q_c, \theta)}{\partial Q_c} - K_{max}; K_{max} = 2.5 \qquad (6.109)$$

$$J_{3C+c}(\theta) = K_{min} - \frac{\partial f(Q_c, \theta)}{\partial Q_c}; K_{min} = 0 \qquad (6.110)$$

- Fixed point in the origin: $f(0, \theta) = 0$

$$J_{4C+1}(\theta) = f(0, \theta) \qquad (6.111)$$

$$J_{4C+2}(\theta) = -f(0, \theta) \qquad (6.112)$$

In addition, the poles $p_1$ and $p_2$ were restricted to belong to the circle centered at the origin with radius $p_s$:

$$J_{4C+3}(\theta) = |p_1(\theta)| - p_s; p_s = 0.95 \qquad (6.113)$$

$$J_{4C+4}(\theta) = |p_2(\theta)| - p_s; p_s = 0.95 \qquad (6.114)$$

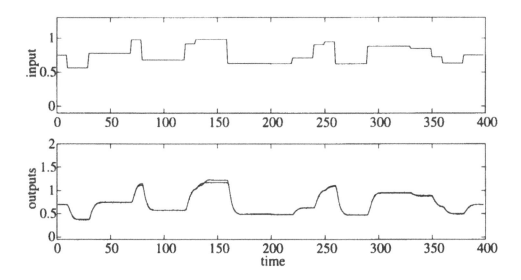

Figure 6.14: Performance on training data. Upper plot shows the input flow $Q(t)$. Lower plot shows the level of the second tank, $Y_2(t)$. Dots indicate the points contained in the training set. Solid lines show the corresponding predictions given by the constrained and unconstrained Wiener models.

Hence, a total number of 228 constraints were posed to the model.

Using the training data, the parameters were estimated under the constraints given by Eqs. (6.107)-(6.114). For comparison, the same data set was used for training a Wiener model with the same structure using the Levenberg–Marquardt method (without constraints). Figure 6.14 shows the performance of the Wiener models after training (8000 iterations). The results indicate that the information contained in the training set was well captured in both cases.

Figure 6.15 shows the static mappings provided by the two models. In both cases, the static mapping is accurate on the operating area for which measurements were provided in the training set. However, extrapolation outside the operating area contained in the training set gives poor results with the unconstrained model.

It is simple to include additional *a priori* information using the constraints. Figure 6.15 shows that the constraints posed on the output of the model, on the gain, and on the fixed point are satisfied by the Wiener model identified under constraints. At the same time, the prediction error on measured data is minimized. For the dynamic part, the following linear model

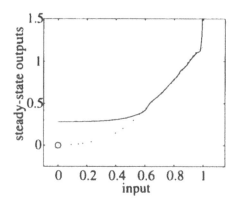

Figure 6.15: Non-linear static mappings identified by the Wiener models. Solid line shows the response for the static part of the Wiener model in the unconstrained case. Dotted line shows the behavior of the model identified under constraints. The circle indicates the equality constraint.

was identified:

$$\widehat{z}(k) = 0.14Q(k) + 1.30\widehat{z}(k-1) - 0.44\widehat{z}(k-2) \qquad (6.115)$$

with poles $p_1 = 0.6476 + i0.143$, $p_2 = 0.6476 - i0.143$, $|p_1| = |p_1| = 0.6632$. Thus the constraints on the dynamic part were fulfilled, too.

Figure 6.16 depicts the performance of the Wiener models on test set data. Note that in the test data the input varies in a wider range than in the training data. The performance of the unconstrained Wiener model is poor, whereas for the constrained Wiener model the performance is much better. All the prior information was captured by the constrained Wiener model. Note that the static output of the Wiener model was constrained to be always less than (height of the second tank), which was not taken into account in the simulation of the plant, Eqs. (6.105)-(6.106), as shown in Fig. 6.16.

### 6.4.4   Conclusions

The application of Wiener and Hammerstein structures in the identification of industrial processes was considered. Structures and associated parameter estimation methods were proposed, which resulted in a non-linear steady-state description of the process with dynamics identified as linear OE-type filters.

In many cases, the dynamics of a non-linear process can be approximated using linear transfer functions, and the system non-linearities can be pre-

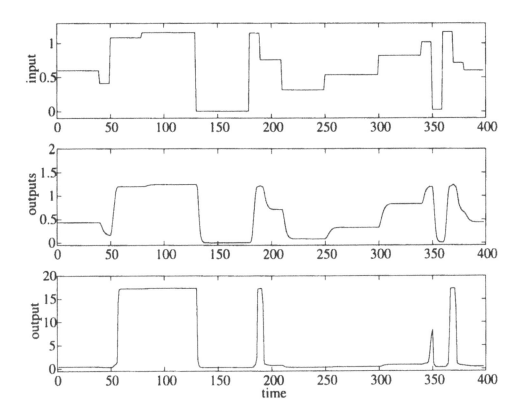

Figure 6.16: Performance on test data. Upper plot shows the input flow $Q(t)$. Middle and lower plots show the level of the second tank, $Y_2(t)$. Solid lines show the corresponding predictions given by the constrained Wiener model (middle plot) and unconstrained Wiener model (lower plot).

sented by a non-linear gain only. This provides many benefits in the form of robustness in dealing with the bias–variance dilemma, availability of the well-developed tools for handling both linear dynamic and non-linear static systems, and increased transparency of the plant description. In this section, examples of identifying a steady-state static plant model were presented, thus emphasizing the transparency aspects.

In industrial practice, it is common that the steady-state behavior of a process is much better known than its dynamic characteristics. With the approach considered in the examples, it is simple to use this knowledge in the initialization and validation of a black-box model. If a reliable steady-state model is available, it can be used as a white-box or grey-box static mapping in the Wiener or Hammerstein structure. Furthermore, there were few restrictions posed on the form of the static mapping; no specific properties of a certain paradigm were used. This enables a non-linear structure to be chosen depending on the application requirements (good transparency–fuzzy systems, high accuracy–neural networks, efficiency and speed–power series, expectable interpolation–piecewise linear systems, *etc.*). These properties are important from the practical point of view of process modeling. In addition, the identification of OE-type of linear dynamics was considered. This type of model is more robust towards noisy measurements, and particularly suitable for long-range simulation purposes.

# Part II

# Control

# Chapter 7

# Predictive Control

## 7.1 Introduction to model-based control

Models are a basic tool in modern process control. Explicit models are required by many of the modern control methods, or models are required during control design. In the control of non-linear processes the role of models is even more emphasized. In the model-based approaches, the controller can be seen as an algorithm operating on a model of the process (subject to disturbances), and optimized in order to reach given control design objectives.

In modeling, the choice of both the model structure and the associated parameter estimation techniques are constrained by the function approximation and interpolation capabilities (*e.g.*, linear approximations, smoothness of non-linearities, *a priori* information). From the control design point of view, the need for convenient ways to characterize a desired closed-loop performance gives additional restrictions (*e.g.*, existence of derivatives and analytic solutions). In addition, many other properties may be of importance (handling of uncertainties, non-ideal sampling, data fusion, tuning, transparency, *etc.*). Clearly, the choice of a modeling method is of essential importance, and therefore a large part of this book has been consecrated for explaining the various approaches.

In some cases, the behavior of the process operator is modeled (common, *e.g.*, in fuzzy control), or a model of a control-oriented cost function is directly desired (*e.g.*, in some passivity-based control approaches). Usually, however, the characterization of the input–output behavior of the process (or the closed-loop control relevant characteristics) is the target of modeling (on/off-line, in open/closed-loop, *etc.*).

The theory of modeling and control of linear systems is well-developed. In the control of non-linear systems, a common approach has been to consider a non-linear model, to linearize it around an operating point, and design a controller based on the linear description. This is simple and efficient, fits well to most regulation problems, and can be seen as gain scheduling or indirect adaptive control. In particular, linear approaches are difficult to beat in the analysis of dynamical systems. For servo problems, fully non-linear approaches have been considered, based on the properties of known non-linearities or on the exploitation of raw computing power (*e.g.*, non-linear predictive control).

Predictive control is a model-based control approach that uses explicitly a process model in order to determine the control actions. In this chapter, the predictive control approach will be discussed for the case of linear SISO models.

## 7.2   The basic idea

Predictive controllers are based on a guess, a prediction, of the future behavior of the process, forecasted using a model of the process. There exists a multitude of predictive control schemes, which all have four major features in common:

1. A *model* of the process to be controlled. The model is used to predict the process output, with given inputs, over the prediction horizon.

2. A *criterion* function (usually quadratic) that is minimized in order to obtain the optimal controller output sequence over the predicted horizon.

3. A *reference* trajectory for the process output, *i.e.* a sequence of desired future outputs.

4. A *minimization procedure.*

The basic concept of predictive control is simple. A predictive controller calculates such *future controller sequence* that the predicted output of the process is close to the desired process output. Predictive controllers use the *receding horizon principle*: Only the first element of the controller output sequence is applied to control the process, and the whole procedure is repeated at the next sample.

Any model that describes the relationship between the input and the output of the process can be used, including disturbance models, non-linear models, or constrained models. The approach can also be extended for multivariable control. Calculation of the controller output sequence is an optimization (minimization) problem. In general, solving requires an iterative procedure. Although many types of models can be considered, a major problem in deriving predictive controllers for non-linear process models is the non-linear optimization problem that must be solved at every sample. The way this problem is solved depends on the type of non-linearity of the process model. However, if:

- the criterion is quadratic,

- the model is linear, and

- there are no constraints,

then an analytical solution is available. The resulting controller is linear and time-invariant if the model is time-invariant. This appealing case will be considered in the following sections.

**Example 38 (Car driver)** Consider the process of driving a car. This process can be assimilated to a SISO system where the input is the variation of the position of the steering wheel towards a given fixed point of dash board. The output is the position of the car with respect to the direction of the road ahead. At each sampling instant the driver of the car calculates the variation of the control variable and implements it, based on his observations of the road and the traffic ahead (to see further than the end of one's nose) and his prediction of the behavior of the car. This procedure is repeated at each sampling period which depends on the driver.

# 7.3 Linear quadratic predictive control

In this section, the state space formulation is adopted (see, *e.g.*, [69][83][96]). Remember, that a transfer function model can always be converted into a state space form; in fact, for each transfer function, there is an infinite number of state space representations (see Appendix A for a brief recap on state space models). First, the state space model and the principle of certainty equivalence control are introduced. The *i*-step-ahead predictors for the model in state space form will be derived. A simple quadratic cost function is then formulated and the optimal solution minimizing the cost function is derived. Finally, the issues of control horizon, integral control action, state estimation and closed-loop behavior are briefly discussed.

## 7.3.1 Plant and model

Let a SISO system (plant, process) be described by a state-space model

$$\mathbf{x}(k+1) = \mathbf{A}\mathbf{x}(k) + \mathbf{B}u(k) \qquad (7.1)$$
$$y(k) = \mathbf{C}\mathbf{x}(k) \qquad (7.2)$$

where

$\mathbf{x}$ is the state vector $(n \times 1)$,

$u$ is the system input (controller output) $(1 \times 1)$

$y$ is the system output (measured) $(1 \times 1)$

$\mathbf{A}$ is the state transition matrix $(n \times n)$

$\mathbf{B}$ is the input transition vector $(n \times 1)$

$\mathbf{C}$ is the state observer vector $(1 \times n)$

Let us assume that a model (approximation) for the system is known and given by $\widehat{\mathbf{A}}$, $\widehat{\mathbf{B}}$ and $\widehat{\mathbf{C}}$, and that the states $\mathbf{x}$ and output $y$ are measurable. In the *certainty equivalence* control, the uncertainty in the parameters is not considered; the estimated parameters are used as if they were the true ones $(\mathbf{A} \leftarrow \widehat{\mathbf{A}}, \mathbf{B} \leftarrow \widehat{\mathbf{B}}, \mathbf{C} \leftarrow \widehat{\mathbf{C}})$. Thus, in what follows, we allow ourselves to simplify the notation by dropping out the 'hats'.

The target is to find the control input $u(k)$ so that the desired control objectives are fulfilled. The objectives concern the future behavior of the process, from the next-to-current state up to the prediction horizon, $H_p$. The prediction horizon is generally chosen to be at least equal to the equivalent time delay (the maximum time delay augmented by the number of unstable zeros). Let the cost function (to be minimized) be given by

$$J = \sum_{i=1}^{H_p} (w(k+i) - \widehat{y}(k+i))^2 + ru(k+i-1)^2 \qquad (7.3)$$

where $w(k+i)$ is the desired system output at instant $k+i$. $r$ is a scalar which can be used for balancing the relative importance of the two squared terms in (7.3). The minimization

$$\{u(k), ..., u(k+H_p-1)\} = \arg \min_{u(k),...,u(k+H_p-1)} J \qquad (7.4)$$

gives a sequence of future controls $\{u(k), u(k+1), ..., u(k+H_p-1)\}$. The first value of the sequence $(u(k))$ is applied to control the system, at next control instant the optimization is repeated (receding horizon control).

## 7.3.2 $i$-step ahead predictions

Let us consider the $i$-step ahead predictions. At instant $k$, the measured state vector $\mathbf{x}(k)$ is available. For future values of $\mathbf{x}$, the model has to be used. The prediction for $y(k+1)$, based on information at $k$, is given by

$$\widehat{y}(k+1) = \mathbf{C}\left[\mathbf{A}\mathbf{x}(k) + \mathbf{B}u(k)\right] \tag{7.5}$$

For $y(k+2)$ we have

$$\widehat{y}(k+2) = \mathbf{C}\left[\mathbf{A}\mathbf{x}(k+1) + \mathbf{B}u(k+1)\right] \tag{7.6}$$

where the estimate for $\mathbf{x}(k+1)$ can be obtained using the model, $\mathbf{x}(k+1) = \mathbf{A}\mathbf{x}(k) + \mathbf{B}u(k)$. Substituting this gives

$$\begin{aligned}
\widehat{y}(k+2) &= \mathbf{C}\left[\mathbf{A}\left[\mathbf{A}\mathbf{x}(k) + \mathbf{B}u(k)\right] + \mathbf{B}u(k+1)\right] & (7.7) \\
&= \mathbf{C}\mathbf{A}^2\mathbf{x}(k) + \mathbf{C}\mathbf{A}\mathbf{B}u(k) + \mathbf{C}\mathbf{B}u(k+1) & (7.8)
\end{aligned}$$

In a similar way we have that

$$\widehat{y}(k+3) = \mathbf{C}\mathbf{A}^3\mathbf{x}(k) + \mathbf{C}\mathbf{A}^2\mathbf{B}u(k) + \mathbf{C}\mathbf{A}\mathbf{B}u(k+1) + \mathbf{C}\mathbf{B}u(k+2) \tag{7.9}$$

and, by induction, for the $i$-step ahead prediction

$$\widehat{y}(k+i) = \mathbf{C}\mathbf{A}^i\mathbf{x}(k) + \sum_{j=1}^{i}\mathbf{C}\mathbf{A}^{i-j}\mathbf{B}u(k+j-1) \tag{7.10}$$

Let us use a more compact matrix notation. Collect the predicted system outputs, the system inputs, and the desired future outputs at instant $k$ into vectors of size $(H_{\mathrm{p}} \times 1)$:

$$\begin{aligned}
\widehat{\mathbf{y}}(k+1) &= \left[\widehat{y}(k+1), \cdots, \widehat{y}(k+H_{\mathrm{p}})\right]^T & (7.11) \\
\mathbf{u}(k) &= \left[u(k), \cdots, u(k+H_{\mathrm{p}}-1)\right]^T & (7.12) \\
\mathbf{w}(k+1) &= \left[w(k+1), \cdots, w(k+H_{\mathrm{p}})\right]^T & (7.13)
\end{aligned}$$

The future predictions can be calculated from

$$\widehat{\mathbf{y}}(k+1) = \mathbf{K}_{\mathrm{CA}}\mathbf{x}(k) + \mathbf{K}_{\mathrm{CAB}}\mathbf{u}(k) \tag{7.14}$$

where

$$\mathbf{K}_{\mathrm{CA}} = \begin{bmatrix} \mathbf{C}\mathbf{A} \\ \vdots \\ \mathbf{C}\mathbf{A}^{H_{\mathrm{p}}} \end{bmatrix} \tag{7.15}$$

$$\mathbf{K}_{\mathrm{CAB}} = \begin{bmatrix} \mathbf{C}\mathbf{B} & \cdots & 0 \\ \vdots & \ddots & \vdots \\ \mathbf{C}\mathbf{A}^{H_{\mathrm{p}}-1}\mathbf{B} & \cdots & \mathbf{C}\mathbf{B} \end{bmatrix} \tag{7.16}$$

### 7.3.3   Cost function

The cost function (7.3) can be expressed in a vector form

$$J = (\mathbf{w}(k+1) - \hat{\mathbf{y}}(k+1))^T (\mathbf{w}(k+1) - \hat{\mathbf{y}}(k+1)) \qquad (7.17)$$
$$+ \mathbf{u}^T(k)\, \mathbf{R}\mathbf{u}(k)$$

where $\mathbf{R} = r\mathbf{I}$. The solution for $\mathbf{u}$ minimizing $J$ is given by

$$\mathbf{u}(k) = \left[\mathbf{R} + \mathbf{K}_{\text{CAB}}^T \mathbf{K}_{\text{CAB}}\right]^{-1} \mathbf{K}_{\text{CAB}}^T (\mathbf{w}(k+1) - \mathbf{K}_{\text{CA}}\mathbf{x}(k)) \qquad (7.18)$$

**Proof.** Let us simplify the notations by dropping out the sample indexes $k$ related to time. Minimization can be done analytically by setting the derivative $\frac{\partial J}{\partial \mathbf{u}} = \mathbf{0}$. The derivative is given by

$$\frac{\partial J}{\partial \mathbf{u}} = \left[(\mathbf{w} - \hat{\mathbf{y}})^T \frac{\partial}{\partial \mathbf{u}}(\mathbf{w} - \hat{\mathbf{y}})\right]^T$$
$$+ \left[\frac{\partial}{\partial \mathbf{u}}(\mathbf{w} - \hat{\mathbf{y}})\right]^T (\mathbf{w} - \hat{\mathbf{y}}) \qquad (7.19)$$
$$+ \left[\mathbf{u}^T \frac{\partial}{\partial \mathbf{u}}\mathbf{R}\mathbf{u}\right]^T + \left[\frac{\partial}{\partial \mathbf{u}}\mathbf{u}^T\right]^T \mathbf{R}\mathbf{u}$$

For the partial derivatives we get

$$\frac{\partial}{\partial \mathbf{u}}(\mathbf{w} - \hat{\mathbf{y}}) = -\frac{\partial}{\partial \mathbf{u}}\hat{\mathbf{y}} = -\mathbf{K}_{\text{CAB}} \qquad (7.20)$$

$$\frac{\partial}{\partial \mathbf{u}}\mathbf{R}\mathbf{u} = \mathbf{R}; \quad \frac{\partial}{\partial \mathbf{u}}\mathbf{u}^T = \mathbf{I} \qquad (7.21)$$

Thus, the derivative (7.19) can be written as

$$\frac{\partial J}{\partial \mathbf{u}} = -2\mathbf{K}_{\text{CAB}}^T (\mathbf{w} - \hat{\mathbf{y}}) + 2\mathbf{R}^T \mathbf{u} \qquad (7.22)$$

Setting the derivative to zero and substituting the vector of future predictions from (7.14) we have

$$\mathbf{K}_{\text{CAB}}^T (\mathbf{w} - \mathbf{K}_{\text{CA}}\mathbf{x} - \mathbf{K}_{\text{CAB}}\mathbf{u}) = \mathbf{R}^T \mathbf{u} \qquad (7.23)$$

Solving for $\mathbf{u}$ gives the optimal control sequence (7.18). ∎
Let us introduce a gain matrix $\mathbf{K}$:

$$\mathbf{K} = \left[\mathbf{R} + \mathbf{K}_{\text{CAB}}^T \mathbf{K}_{\text{CAB}}\right]^{-1} \mathbf{K}_{\text{CAB}}^T \qquad (7.24)$$

Denote the first row of $\mathbf{K}$ by $\mathbf{K}_1$. Since only the first element of the optimal sequence is applied to the process, the on-line control computations are reduced to

$$u(k) = \mathbf{K}_1 (\mathbf{w}(k+1) - \mathbf{K}_{CA}\mathbf{x}(k)) \qquad (7.25)$$

If the system parameters, $\mathbf{A}$, $\mathbf{B}$ and $\mathbf{C}$, are constant, the gain matrices $\mathbf{K}_1$ and $\mathbf{K}_{CA}$ can be computed beforehand.

## 7.3.4 Remarks

In many cases, it is useful to consider an additional parameter in the tuning of the predictive controller, the *control horizon*. The control horizon $H_c$ specifies the allowed number of changes in the control signal during optimization, *i.e.*

$$\Delta u(k+i) = 0 \text{ for } i \geq H_c \qquad (7.26)$$

where $\Delta = 1 - q^{-1}$.

A simple way to implement the control horizon is to modify the $\mathbf{K}_{CAB}$ matrix. Let us decompose the matrix in two parts. The first part, $\mathbf{K}_{CAB}^a$, contains the first $H_c - 1$ columns from the left of the $\mathbf{K}_{CAB}$ matrix. The second part, vector $\mathbf{K}_{CAB}^b$, sums row-wise the remaining elements of the $\mathbf{K}_{CAB}$ matrix, *i.e.*

$$k_i^b = \sum_{j=H_c}^{H_p} k_{i,j} \qquad (7.27)$$

where $k_i^b$ and $k_{i,j}$ are the elements ($i^{th}$ row and $j^{th}$ column) of the $\mathbf{K}_{CAB}^b$ and $\mathbf{K}_{CAB}$ matrices. The new $K_{CAB}$ matrix is then formed by

$$\mathbf{K}_{CAB} = \begin{bmatrix} \mathbf{K}_{CAB}^a & \mathbf{K}_{CAB}^b \end{bmatrix} \qquad (7.28)$$

In practice, it is useful to introduce also a *minimum horizon*, which specifies the beginning of the horizon to be used in the cost function, *i.e.* $J = \sum_{i=H_m}^{H_p} (\cdot)$ in (7.3). A simple implementation can be done by removing the first $H_m - 1$ rows from $\mathbf{K}_{CA}$ and $\mathbf{K}_{CAB}$ in (7.15) and (7.16), respectively.

Notice, that there is no *integral action* present. Thus, in the case of modeling errors, a steady state error may occur. A simple way to include an integral term to the controller is to use an augmented state space model, with an additional state constructed of the integral-of-error, $x_I(k) = x_I(k-1) + y(k) - \widehat{y}(k)$. This state then has a gain $k_I$ from the augmented state $x_I$ to the controller output $u$.

In general, the states $\mathbf{x}$ are not directly measurable. When noise is not present an *observer* is used for state "recovering". In the presence of noise, a Kalman filter can be used to estimate the states (see Chapter 3). Provided that the covariances of the input and output noises are available or can be estimated, a *state estimate* minimizing the variance of the state estimation error can then be constructed. The Kalman filter uses both the system model $(\mathbf{A}, \mathbf{B}, \mathbf{C})$ and system input–output measurements $u$, $y$ in order to provide an optimal state estimate.

The behavior of this dynamic system under the feedback, that is simply a function which maps the state space into the space of control variables, is analyzed in the next subsection.

### 7.3.5   Closed-loop behavior

In order to analyze the behavior of the closed-loop system, let us derive its characteristic function. Taking into account the control strategy (7.25), from the state-space model (7.1)–(7.2) we derive the relation between the output $y(k)$ and the desired system output $\mathbf{w}(k) = \begin{bmatrix} 1 & \cdots & 1 \end{bmatrix}^T w(k)$: Substitute (7.25) to $\mathbf{x}(k+1)$ in (7.1) with $k \leftarrow k+1$:

$$\mathbf{x}(k) = \mathbf{A}\mathbf{x}(k-1) + \mathbf{B}\mathbf{K}_1\left(\mathbf{w}(k) - \mathbf{K}_{\mathrm{CA}}\mathbf{x}(k-1)\right) \tag{7.29}$$

Reorganizing gives

$$\mathbf{x}(k) = \left[\mathbf{I} - q^{-1}\left(\mathbf{A} - \mathbf{B}\mathbf{K}_1\mathbf{K}_{\mathrm{CA}}\right)\right]^{-1}\mathbf{B}\mathbf{K}_1\mathbf{w}(k) \tag{7.30}$$

Substituting to (7.2) gives the relation between $y(k)$ and $w(k)$:

$$y(k) = \mathbf{C}\left[\mathbf{I} - q^{-1}\left(\mathbf{A} - \mathbf{B}\mathbf{K}_1\mathbf{K}_{\mathrm{CA}}\right)\right]^{-1}\mathbf{B}\mathbf{K}_1 \begin{bmatrix} 1 \\ \vdots \\ 1 \end{bmatrix} w(k) \tag{7.31}$$

and the characteristic polynomial

$$\det\left[\mathbf{I} - q^{-1}\left(\mathbf{A} - \mathbf{B}\mathbf{K}_1\mathbf{K}_{\mathrm{CA}}\right)\right] \tag{7.32}$$

**Example 39 (Characteristic polynomial)** Let a process be described by the following transfer function

$$y(k) = \frac{0.1989q^{-3}}{1 - 0.9732q^{-1}}u(k) \tag{7.33}$$

(this example is discussed in more detail at the end of this chapter). The equivalent control canonical state-space presentation is given by

$$\mathbf{A} = \begin{bmatrix} 0.97 & 0 & 0 \\ 1 & 0 & 0 \\ 0 & 1 & 0 \end{bmatrix} ; \mathbf{B} = \begin{bmatrix} 1 \\ 0 \\ 0 \end{bmatrix} ; \mathbf{C} = \begin{bmatrix} 0 & 0 & 0.1989 \end{bmatrix} \qquad (7.34)$$

Let us design a predictive controller using $H_p = 5$ and $r = 1$. This results to a gain vector

$$\mathbf{K}_1 = \begin{bmatrix} 0 & 0 & 0.1799 & 0.1623 & 0.1514 \end{bmatrix} \qquad (7.35)$$

and

$$\mathbf{K}_{CA} = \begin{bmatrix} 0 & 0.1989 & 0 \\ 0.1989 & 0 & 0 \\ 0.1929 & 0 & 0 \\ 0.1871 & 0 & 0 \\ 0.1815 & 0 & 0 \end{bmatrix} \qquad (7.36)$$

The matrix $\mathbf{A} - \mathbf{B}\mathbf{K}_1\mathbf{K}_{CA}$ is given by

$$\begin{bmatrix} 0.8774 & 0 & 0 \\ 1 & 0 & 0 \\ 0 & 1 & 0 \end{bmatrix} \qquad (7.37)$$

and the characteristic polynomial will be $1 - 0.8774q^{-1}$. For $r = 0.01$, which penalizes less the control actions, the characteristic polynomial will be $1 - 0.1692q^{-1}$, a much faster response.

Note that the control strategy (7.18) associated with the cost function (7.17) is linear towards the system input, output and the desired output. It can be easily expressed in the R-S-T-form:

$$R\left(q^{-1}\right) u\left(k\right) = S\left(q^{-1}\right) y\left(k\right) + T\left(q^{-1}\right) w\left(k\right). \qquad (7.38)$$

In the next section, the approach of generalized predictive control is considered, where a disturbance model is included in the plant description.

## 7.4 Generalized predictive control

An appealing formulation called generalized predictive control (GPC) of long-range predictive control was derived by Clarke and co-workers [13]. It represents a unification of many long-range predictive control algorithms (ID-COM [79], DMC [14]) and a computationally simple approach. In the GPC, an ARMAX/ARIMAX representation of the plant is used. In what follows, $i$-step-ahead predictors for the ARMAX/ARIMAX model in state space form will be derived, a cost function formulated and the optimal solution minimizing the cost function derived. In the next section, a simulation example illustrates the performance and tuning of the GPC controller.

## 7.4.1  ARMAX/ARIMAX model

Recall the ARMAX and ARIMAX structures from Chapter 3. An AR-MAX/ARIMAX model in the polynomial form is given by:

$$F\left(q^{-1}\right) y\left(k\right) = B\left(q^{-1}\right) v\left(k\right) + C\left(q^{-1}\right) e\left(k\right) \tag{7.39}$$

where $f_j$, $b_j$ and $c_j$ are the coefficients of the polynomials $F\left(q^{-1}\right)$, $B\left(q^{-1}\right)$ and $C\left(q^{-1}\right)$, $j = 1, 2, ..., n$. For notational convenience, without loss of generality, we assume that the polynomials are all of order $n$; $F\left(q^{-1}\right)$ and $C\left(q^{-1}\right)$ are monic, and $b_0 = 0$. Substituting $v\left(k\right) \longleftarrow u\left(k\right)$ and $F\left(q^{-1}\right) \longleftarrow A\left(q^{-1}\right)$ in (7.39) gives the ARMAX model, and substituting $v\left(k\right) \longleftarrow \Delta u\left(k\right)$ and $F\left(q^{-1}\right) \longleftarrow \Delta A\left(q^{-1}\right)$ gives the ARIMAX model structure. In what follows, we denote the controller output by $v\left(k\right)$. In the ARIMAX case, the final controller output to be applied to the plant will be $u\left(k\right) = u\left(k-1\right) + \Delta u\left(k\right)$.

The ARMAX/ARIMAX model can be represented in the state-space form[1] as

$$
\begin{aligned}
\mathbf{x}\left(k+1\right) &= \mathbf{A}\mathbf{x}\left(k\right) + \mathbf{B}v\left(k\right) + \mathbf{G}e\left(k\right) \tag{7.40}\\
y\left(k\right) &= \mathbf{C}\mathbf{x}\left(k\right) + e\left(k\right) \tag{7.41}
\end{aligned}
$$

---

[1]The relation between the state-space description and input–output description is given by

$$\frac{B\left(q\right)}{F\left(q\right)} = \mathbf{C}^T \left[q\mathbf{I} - \mathbf{A}\right]^{-1} \mathbf{B} \; ; \; \frac{C\left(q\right)}{F\left(q\right)} = \mathbf{C}^T \left[q\mathbf{I} - \mathbf{A}\right]^{-1} \mathbf{G} + 1$$

and

$$
\begin{aligned}
F\left(q\right) &= \det\left[q\mathbf{I} - \mathbf{A}\right] \; ; \; B\left(q\right) = \mathbf{C}^T \mathrm{adj}\left[q\mathbf{I} - \mathbf{A}\right]\mathbf{B}\\
C\left(q\right) &= \mathbf{C}^T \mathrm{adj}\left[q\mathbf{I} - \mathbf{A}\right]\mathbf{G} + \det\left[q\mathbf{I} - \mathbf{A}\right]
\end{aligned}
$$

Note that the polynomials are given in terms of the feedforward operator $q$.

where

$$
\mathbf{A} = \begin{bmatrix} -f_1 & 1 & 0 & \cdots & 0 \\ -f_2 & 0 & 1 & & 0 \\ \vdots & \vdots & & \ddots & \\ -f_{n-1} & 0 & 0 & & 1 \\ -f_n & 0 & 0 & \cdots & 0 \end{bmatrix} \tag{7.42}
$$

$$
\mathbf{B} = \begin{bmatrix} b_1 & b_2 & \cdots & b_{n-1} & b_n \end{bmatrix}^T \tag{7.43}
$$

$$
\mathbf{G} = \begin{bmatrix} c_1 - f_1 & c_2 - f_2 & \cdots & c_{n-1} - f_{n-1} & c_n - f_n \end{bmatrix}^T \tag{7.44}
$$

$$
\mathbf{C} = \begin{bmatrix} 1 & 0 & \cdots & 0 \end{bmatrix} \tag{7.45}
$$

If the coefficients of the polynomials $F(q^{-1})$ and $B(q^{-1})$ are unknown, they can be obtained through identification (see previous chapters). An estimate of $C(q^{-1})$ may also be identified. On can also consider estimating the matrices $\mathbf{A}$, $\mathbf{B}$ and $\mathbf{C}$ (and $\mathbf{G}$) directly from input–output data using subspace methods [48][54].

## 7.4.2   $i$-step-ahead predictions

The prediction is simple to derive. Let us consider a 1-step-ahead prediction

$$
\begin{aligned}
y(k+1) &= \mathbf{C}\mathbf{x}(k+1) + e(k+1) && (7.46) \\
&= \mathbf{C}[\mathbf{A}\mathbf{x}(k) + \mathbf{B}v(k) + \mathbf{G}e(k)] + e(k+1) && (7.47) \\
&= \mathbf{C}(\mathbf{A} - \mathbf{G}\mathbf{C})\mathbf{x}(k) + \mathbf{C}\mathbf{B}v(k) + \mathbf{C}\mathbf{G}y(k) && (7.48) \\
&\quad + e(k+1)
\end{aligned}
$$

where the last equality is obtained by substituting $e(k) = y(k) - \mathbf{C}\mathbf{x}(k)$ from (7.41) and future noise is not known but assumed zero mean. The 1-step-ahead predictor becomes[2]

$$
\widehat{y}(k+1) = \mathbf{C}(\mathbf{A} - \mathbf{G}\mathbf{C})\mathbf{x}(k) + \mathbf{C}\mathbf{B}v(k) + \mathbf{C}\mathbf{G}y(k) \tag{7.49}
$$

---

[2]The task is to find $\widehat{y}(k+1) = \arg\min_{\widehat{y}} E\left\{[y(k+1) - \widehat{y}]^2\right\}$

Similarly, for the 2-step-ahead prediction, we have

$$
\begin{aligned}
y(k+2) &= \mathbf{C}\mathbf{x}(k+2) + e(k+2) & (7.50)\\
&= \mathbf{C}\left[\mathbf{A}\mathbf{x}(k+1) + \mathbf{B}v(k+1) + \mathbf{G}e(k+1)\right] & (7.51)\\
&\quad + e(k+2)\\
&= \mathbf{C}\left[\mathbf{A}\left[\mathbf{A}\mathbf{x}(k) + \mathbf{B}v(k) + \mathbf{G}e(k)\right] + \mathbf{B}v(k+1)\right] & (7.52)\\
&\quad + \mathbf{C}\mathbf{G}e(k+1) + e(k+2)
\end{aligned}
$$

and the 2-step ahead predictor becomes[3]

$$
\begin{aligned}
\widehat{y}(k+2) &= \mathbf{C}\mathbf{A}\left[\mathbf{A} - \mathbf{G}\mathbf{C}\right]\mathbf{x}(k) + \mathbf{C}\mathbf{A}\mathbf{B}v(k) + \mathbf{C}\mathbf{B}v(k+1) & (7.53)\\
&\quad + \mathbf{C}\mathbf{A}\mathbf{G}y(k)
\end{aligned}
$$

By induction, we have the following formula for an $i$-step-ahead prediction

$$
\begin{aligned}
\widehat{y}(k+i) &= \left[\sum_{j=1}^{i} \mathbf{C}\mathbf{A}^{i-j}\mathbf{B}v(k+j-1)\right] & (7.54)\\
&\quad + \mathbf{C}\mathbf{A}^{i-1}\left[\mathbf{A} - \mathbf{G}\mathbf{C}\right]\mathbf{x}(k)\\
&\quad + \mathbf{C}\mathbf{A}^{i-1}\mathbf{G}y(k)
\end{aligned}
$$

---

$$
\begin{aligned}
& E\left\{[y(k+1) - \widehat{y}]^2\right\}\\
&= E\left\{[\mathbf{C}(\mathbf{A} - \mathbf{G}\mathbf{C})\mathbf{x}(k) + \mathbf{C}\mathbf{B}v(k) + \mathbf{C}\mathbf{G}y(k) + e(k+1) - \widehat{y}]^2\right\}\\
&= E\left\{[\mathbf{C}(\mathbf{A} - \mathbf{G}\mathbf{C})\mathbf{x}(k) + \mathbf{C}\mathbf{B}v(k) + \mathbf{C}\mathbf{G}y(k) - \widehat{y}]^2\right.\\
&\quad + 2\left[\mathbf{C}(\mathbf{A} - \mathbf{G}\mathbf{C})\mathbf{x}(k) + \mathbf{C}\mathbf{B}v(k) + \mathbf{C}\mathbf{G}y(k) - \widehat{y}\right]e(k+1)\\
&\quad \left. + e^2(k+1)\right\}\\
&= E\left\{[\mathbf{C}(\mathbf{A} - \mathbf{G}\mathbf{C})\mathbf{x}(k) + \mathbf{C}\mathbf{B}v(k) + \mathbf{C}\mathbf{G}y(k) - \widehat{y}]^2\right\} + E\left\{e^2(k+1)\right\}
\end{aligned}
$$

since $e(k+1)$ does not correlate with $\mathbf{x}(k)$, $v(k)$, $y(k)$ or $\widehat{y}$. The minimum is obtained when the first term is zero, *i.e.* (7.49).

[3]Proceeding in the same way as with the 1-step ahead predictor, we have

$$
\begin{aligned}
& E\left\{[y(k+2) - \widehat{y}]^2\right\}\\
&= E\left\{[\mathbf{C}\mathbf{A}(\mathbf{A} - \mathbf{G}\mathbf{C})\mathbf{x}(k) + \mathbf{C}\mathbf{A}\mathbf{B}v(k) + \mathbf{C}\mathbf{B}v(k+1) + \mathbf{C}\mathbf{A}\mathbf{G}y(k) - \widehat{y}]^2\right\}\\
&\quad + E\left\{[\mathbf{C}\mathbf{G}e(k+1)]^2\right\} + E\left\{[e(k+2)]^2\right\}
\end{aligned}
$$

since $e(k+1)$ and $e(k+2)$ do not correlate with $\mathbf{x}(k)$, $v(k)$, $y(k)$, $\widehat{y}$ or with each other. The variance is is minimized when (7.53) holds.

Let us use a more compact matrix notation. Collect the predicted system outputs, the system inputs, and the desired future outputs at instant $k$ into vectors of size $(H_p \times 1)$:

$$\hat{\mathbf{y}}(k+1) = [\hat{y}(k+1), \cdots, \hat{y}(k+H_p)]^T \tag{7.55}$$

$$\mathbf{v}(k) = [v(k), \cdots, v(k+H_p-1)]^T \tag{7.56}$$

$$\mathbf{w}(k+1) = [w(k+1), \cdots, w(k+H_p)]^T \tag{7.57}$$

The future predictions can be calculated from

$$\hat{\mathbf{y}}(k+1) = \mathbf{K}_{\text{CAGC}}\mathbf{x}(k) + \mathbf{K}_{\text{CAB}}\mathbf{v}(k) + \mathbf{K}_{\text{CAG}}y(k) \tag{7.58}$$

where

$$\mathbf{K}_{\text{CAGC}} = \begin{bmatrix} \mathbf{C}[\mathbf{A}-\mathbf{GC}] \\ \vdots \\ \mathbf{CA}^{H_p-1}[\mathbf{A}-\mathbf{GC}] \end{bmatrix} \tag{7.59}$$

$$\mathbf{K}_{\text{CAB}} = \begin{bmatrix} \mathbf{CB} & \cdots & 0 \\ \vdots & \ddots & \vdots \\ \mathbf{CA}^{H_p-1}\mathbf{B} & \cdots & \mathbf{CB} \end{bmatrix} \tag{7.60}$$

$$\mathbf{K}_{\text{CAG}} = \begin{bmatrix} \mathbf{CG} \\ \vdots \\ \mathbf{CA}^{H_p-1}\mathbf{G} \end{bmatrix}^T \tag{7.61}$$

## 7.4.3  Cost function

Let us minimize the following cost function, expressed in a vector form

$$\begin{aligned} J &= (\mathbf{w}(k+1) - \hat{\mathbf{y}}(k+1))^T \mathbf{Q}(\mathbf{w}(k+1) - \hat{\mathbf{y}}(k+1)) \\ &\quad + \mathbf{v}^T(k)\mathbf{R}\mathbf{v}(k) \end{aligned} \tag{7.62}$$

where $\mathbf{Q} = \text{diag}[q_1, \cdots, q_{H_p}]$ and $\mathbf{R} = \text{diag}[r_1, \cdots, r_{H_p}]$. Notice that if $v(k) \longleftarrow \Delta u(k)$, the control costs are taken on the increments of the control action, whereas if $v(k) \longleftarrow u(k)$, the costs are on the absolute values of the control, as in (7.17). The introduction of diagonal weighting matrices $\mathbf{Q}$ and $\mathbf{R}$ enables the weighting of the terms in the cost function also with respect to their appearance in time.

The optimal sequence is given by

$$\begin{aligned} \mathbf{v}(k) &= [\mathbf{R} + \mathbf{K}_{\text{CAB}}^T \mathbf{Q}\mathbf{K}_{\text{CAB}}]^{-1} \mathbf{K}_{\text{CAB}}^T \mathbf{Q} \times \\ &\quad (\mathbf{w}(k+1) - \mathbf{K}_{\text{CAGC}}\mathbf{x}(k) - \mathbf{K}_{\text{CAG}}y(k)) \end{aligned} \tag{7.63}$$

**Proof.** Let us simplify the notations by dropping out the sample indexes $k$. Minimization can be done analytically by setting the derivative $\frac{\partial J}{\partial u} = 0$. The derivative is given by

$$
\begin{aligned}
\frac{\partial J}{\partial \mathbf{v}} &= \left[ (\mathbf{w} - \widehat{\mathbf{y}})^T \mathbf{Q} \frac{\partial}{\partial \mathbf{v}} (\mathbf{w} - \widehat{\mathbf{y}}) \right]^T \\
&\quad + \left[ \frac{\partial}{\partial \mathbf{v}} (\mathbf{w} - \widehat{\mathbf{y}}) \right]^T \mathbf{Q} (\mathbf{w} - \widehat{\mathbf{y}}) \\
&\quad + \left[ \mathbf{v}^T \frac{\partial}{\partial \Delta \mathbf{u}} \mathbf{R} \mathbf{v} \right]^T + \left[ \frac{\partial}{\partial \mathbf{v}} \mathbf{v}^T \right]^T \mathbf{R} \mathbf{v}
\end{aligned} \tag{7.64}
$$

For the partial derivatives we get

$$
\frac{\partial}{\partial \mathbf{v}} (\mathbf{w} - \widehat{\mathbf{y}}) = -\frac{\partial}{\partial \mathbf{v}} \widehat{\mathbf{y}} = -\mathbf{K}_{\text{CAB}} \tag{7.65}
$$

$$
\frac{\partial}{\partial \mathbf{v}} \mathbf{R} \mathbf{v} = \mathbf{R}; \; \frac{\partial}{\partial \mathbf{v}} \mathbf{v}^T = \mathbf{I} \tag{7.66}
$$

Thus, the derivative can be written as

$$
\frac{\partial J}{\partial \mathbf{v}} = -2\mathbf{K}_{\text{CAB}}^T \mathbf{Q} (\mathbf{w} - \widehat{\mathbf{y}}) + 2\mathbf{R}^T \mathbf{v} \tag{7.67}
$$

Setting the derivative to zero and substituting the vector of future predictions from (7.58), we have

$$
\mathbf{K}_{\text{CAB}}^T \mathbf{Q} (\mathbf{w} - \mathbf{K}_{\text{CAGC}} \mathbf{x} - \mathbf{K}_{\text{CAB}} \mathbf{v} - \mathbf{K}_{\text{CAG}} y) = \mathbf{R}^T \mathbf{v} \tag{7.68}
$$

Solving for $\mathbf{v}$ gives the optimal control sequence (7.63). ■

Let us introduce a gain matrix $\mathbf{K}$:

$$
\mathbf{K} = \left[ \mathbf{R} + \mathbf{K}_{\text{CAB}}^T \mathbf{Q} \mathbf{K}_{\text{CAB}} \right]^{-1} \mathbf{K}_{\text{CAB}}^T \mathbf{Q} \tag{7.69}
$$

and denote the first row of $\mathbf{K}$ by $\mathbf{K}_1$. Since only the first element of the optimal sequence is applied to the process, the on-line control computations are reduced to

$$
v(k) = \mathbf{K}_1 \left[ \mathbf{w}(k+1) - \mathbf{K}_{\text{CAGC}} \mathbf{x}(k) - \mathbf{K}_{\text{CAG}} y(k) \right] \tag{7.70}
$$

If the system parameters, $\mathbf{A}$, $\mathbf{B}$, $\mathbf{G}$, and $\mathbf{C}$, are constant, the gain matrices $\mathbf{K}_1$, $\mathbf{K}_{\text{CAGC}}$ and $\mathbf{K}_{\text{CAG}}$, can be computed beforehand.

## 7.4.4 Remarks

The *disturbance model* in the ARIMAX structure

$$y(k) = \frac{B(q^{-1})}{A(q^{-1})}u(k-d) + \frac{C(q^{-1})}{\Delta(q^{-1})A(q^{-1})}e(k) \qquad (7.71)$$

allows a versatile design of disturbance control in predictive control. In particular:

- with $C(q^{-1}) = \Delta(q^{-1})A(q^{-1})$, the approach reduces to that of section 7.3, with no integral action;

- with $C(q^{-1}) = A(q^{-1})$, a pure integral control of disturbances is obtained (noise characteristics $\frac{1}{\Delta}$);

- with $C(q^{-1}) = C_1(q^{-1})$, an ARIMAX model with noise characteristics $\frac{C_1}{\Delta A}$ is obtained;

- with $C(q^{-1}) = \Delta(q^{-1})C_1(q^{-1})$, an ARMAX model with noise characteristics $\frac{C_1}{A}$ is obtained;

- with $C(q^{-1}) = \Delta(q^{-1})A(q^{-1})C_1(q^{-1})$, an arbitrary FIR filter can be designed for the noise (no integral action); *etc.*

Since the controller is operating on $\Delta\mathbf{u}$, the *control horizon* is simple to implement. A control horizon $H_c$ is obtained when only the first $H_c$ columns of the matrix $\mathbf{K}_{\mathrm{CAB}}$ in (7.63) are used. Accordingly, the control weighting matrix $\mathbf{R}$, associated with future $vs$, has to be adjusted by specifying only the first $H_c$ rows and columns. The future control increments: $v(k + H_c)$, $v(k + H_c + 1)$, ... are then assumed to be equal to zero. $H_c = 1$ results in *mean-level control*, where the optimization seeks for a constant control input (only one change in $u$ allowed), which minimizes the difference between targets $\mathbf{w}$ and predictions $\hat{\mathbf{y}}$ in the given horizon. With large $H_p$, the plant is driven to a constant reference trajectory (in the absence of disturbances) with the same dynamics as the open-loop plant.

A *minimum horizon* specifies the beginning of the horizon to be used in the cost function. If the plant model has a dead time of $d$ [assuming that $b_0$ is nonzero in (7.71)], then only the predicted outputs at $k + d$, $k + d + 1$, ... are affected by a change in $u(k)$. Thus, the calculation of earlier predictions would be unnecessary. If $d$ is not known, or is variable, $H_m$ can be set to 1. A simple implementation can be done by removing the first $H_m - 1$ rows from $K_{\mathrm{CAGC}}$, $K_{\mathrm{CAB}}$ and $K_{\mathrm{CAG}}$ in (7.59)–(7.61). The corresponding (first $H_m - 1$) rows and columns of the weighting matrix $\mathbf{Q}$ need to be removed, too. With

$H_c = n_A + 1$, $H_p = n_A + n_B + 1$, $H_m = n_B + 1$ a *dead-beat control* [8] results, where the output of the process is driven to a constant reference trajectory in $n_B + 1$ samples, $n_A + 1$ controller outputs are required to do so. The GPC represents an unification of many long-range predictive control algorithms, as well as a computationally simple approach. For example the *generalized minimum variance controller* corresponds to the GPC in which both the $H_m$ and $H_p$ are set equal to time delay and only one control signal is weighted.

In some cases it is more relevant to consider a *cost function with weights on the non-incremental control* input

$$
\begin{aligned}
J &= (\mathbf{w}(k+1) - \widehat{\mathbf{y}}(k+1))^T \mathbf{Q} (\mathbf{w}(k+1) - \widehat{\mathbf{y}}(k+1)) \quad (7.72) \\
&+ \mathbf{u}(k)^T \mathbf{R}\mathbf{u}(k)
\end{aligned}
$$

The above equations are still valid with substitutions $F(q^{-1}) \longleftarrow A(q^{-1})$ and $v(k) \longleftarrow u(k)$ (ARMAX structure). This is a good choice, *e.g.*, if the process already includes an integrator in itself. Note, that the control horizon is then implemented as by (7.27) and (7.28).

The ARMAX/ARIMAX model can be seen as a *fixed gain state observer*. For the noise, we always have $e(k) = y(k) - \mathbf{C}\mathbf{x}(k)$. In general, the states $\mathbf{x}$ are not known (not measured, or there is noise in the measurements). Using the state-space model (7.40)-(7.41), a prediction $\widehat{\mathbf{x}}(k)$ of the state $\mathbf{x}(k)$, given $y$ and $u$ up to and including instant $k-1$, can be written as

$$
\widehat{\mathbf{x}}(k) = [\mathbf{A} - \mathbf{GC}]\widehat{\mathbf{x}}(k-1) + \mathbf{B}v(k-1) + \mathbf{G}y(k-1) \quad (7.73)
$$

or, equivalently,

$$
\widehat{\mathbf{x}}(k) = \mathbf{A}\widehat{\mathbf{x}}(k-1) + \mathbf{B}v(k-1) + \mathbf{G}[y(k-1) - \mathbf{C}\widehat{\mathbf{x}}(k-1)] \quad (7.74)
$$

The prediction $\widehat{\mathbf{x}}(k)$ is then used for $\mathbf{x}(k)$ in the GPC equations.

The above observer is also called an asymptotic state estimate [69], an estimate where the optimal estimate tends to when time tends to infinity. An optimal estimate of the state can be obtained from a *Kalman filter*:

$$
\begin{aligned}
\widehat{\mathbf{x}}(k) &= [\mathbf{A} - \mathbf{GC}]\widehat{\mathbf{x}}(k-1) + \mathbf{B}v(k-1) + \mathbf{G}y(k-1) \quad (7.75) \\
&+ \mathbf{K}(k)[y(k) - \widehat{y}(k)]
\end{aligned}
$$

where

$$
\widehat{y}(k) = \mathbf{C}(\mathbf{A} - \mathbf{GC})\widehat{\mathbf{x}}(k-1) + \mathbf{CB}v(k-1) + \mathbf{CG}y(k-1) \quad (7.76)
$$

and the Kalman filter gain vector is obtained from the following recursive equations

$$\mathbf{K}(k) = (\mathbf{A} - \mathbf{GC})\mathbf{P}(k-1)(\mathbf{A} - \mathbf{GC})^T \times \qquad (7.77)$$
$$\left[ Y + \mathbf{C}(\mathbf{A} - \mathbf{GC})\mathbf{P}(k-1)(\mathbf{A} - \mathbf{GC})^T \mathbf{C}^T \right]^{-1} \mathbf{C}$$

$$\mathbf{P}(k) = (\mathbf{A} - \mathbf{GC})\mathbf{P}(k-1)(\mathbf{A} - \mathbf{GC})^T \qquad (7.78)$$
$$-\mathbf{K}(k)\mathbf{C}(\mathbf{A} - \mathbf{GC})\mathbf{P}(k-1)(\mathbf{A} - \mathbf{GC})^T$$

where the initial condition is $\mathbf{P}(0)$, the covariance matrix of the initial state estimation error: $\mathbf{P}(0) = E\left\{ (\mathbf{x}(0) - \widehat{\mathbf{x}}(0))(\mathbf{x}(0) - \widehat{\mathbf{x}}(0))^T \right\}$ and $Y$ is the variance of $e(k)$. The asymptotic estimate is obtained when $\lim_{k \to \infty} \mathbf{K}(k+1) = 0$, which is true if the eigenvalues of the matrix $(\mathbf{A} - \mathbf{GC})$ are less than one.

### 7.4.5 Closed-loop behavior

The GPC control strategy is a linear combination of the system input, output and the desired output. It can be expressed in the R-S-T-form. As for the linear quadratic predictive controller, the characteristic function can be derived, proceeding in a similar way as in section 7.3.5. The controller is given by (7.70). Substituting (7.41) for $y(k)$ in (7.70), substituting the result to (7.40), regrouping and solving for $\mathbf{x}(k)$ and using (7.41) again, we have

$$y(k) = \left\{ \mathbf{I} - q^{-1}[\mathbf{A} - \mathbf{BK}_1(\mathbf{K}_{\text{CAGC}} + \mathbf{K}_{\text{CAG}}\mathbf{C})] \right\}^{-1} \qquad (7.79)$$
$$\times \left\{ \mathbf{BK}_1 \begin{bmatrix} 1 \\ \vdots \\ 1 \end{bmatrix} w(k) + (\mathbf{BK}_1\mathbf{K}_{\text{CAG}} + \mathbf{G})e(k-1) + e(k) \right\}$$

and the characteristic polynomial is given by

$$\det\left\{ \mathbf{I} - q^{-1}[\mathbf{A} - \mathbf{BK}_1(\mathbf{K}_{\text{CAGC}} + \mathbf{K}_{\text{CAG}}\mathbf{C})] \right\} \qquad (7.80)$$

The next subsection is dedicated to a control problem originating from an industrial process.

## 7.5 Simulation example

Let us consider an example of the control of a fluidized-bed combustor (see Appendix B).

Consider a nominal steady-state point given by $Q_C = 2.6 \frac{kg}{s}$ (fuel feed rate), $F_1 = 3.1 \frac{Nm^3}{s}$ (primary air flow) and $F_2 = 8.4 \frac{Nm^3}{s}$ (secondary air flow). The following linearized and discretized description between combustion power and fuel feed is obtained from the plant model using a sampling time of 10 seconds:

$$P(k) = \frac{0.1989 q^{-3}}{1 - 0.9732 q^{-1}} Q_C(k) \tag{7.81}$$

Assuming an ARIMAX-model structure with $C(q^{-1}) = A(q^{-1})$ (integrating output measurement noise) we have the following state-space model for the system

$$\begin{align}
x(k+1) &= Ax(k) + B\Delta u(k) + Ge(k) \tag{7.82} \\
y(k) &= Cx(k) + e(k) \tag{7.83}
\end{align}$$

where $y \leftarrow P$, $u \leftarrow Q_C$, and the matrices are given by

$$A = \begin{bmatrix} 1.9732 & 1 & 0 \\ -0.9732 & 0 & 1 \\ 0.0000 & 0 & 0 \end{bmatrix}, B = \begin{bmatrix} 0.0000 \\ 0.0000 \\ 0.1989 \end{bmatrix} \tag{7.84}$$

$$C = \begin{bmatrix} 1 & 0 & 0 \end{bmatrix}, G = \begin{bmatrix} 1 \\ -0.9732 \\ 0.0000 \end{bmatrix} \tag{7.85}$$

Let us first design a mean-level controller: $H_c = 1$ (control horizon), $H_p = 360$ (large prediction horizon corresponding to 1 hour of operation). The gain matrices are then given by

$$K_{CAB} = \begin{bmatrix} 0 \\ 0 \\ 0.1989 \\ 0.3925 \\ \vdots \\ 7.4212 \end{bmatrix}, K_{CAG} = \begin{bmatrix} 1 \\ 1 \\ 1 \\ 1 \\ \vdots \\ 1 \end{bmatrix} \tag{7.86}$$

$$K_{CAGC} = \begin{bmatrix} 0.9732 & 1 & 0 \\ 1.9203 & 1.9732 & 1 \\ 2.8421 & 2.9203 & 1.9732 \\ 3.7393 & 3.8421 & 2.9203 \\ \vdots & \vdots & \vdots \\ 36.3114 & 37.3113 & 37.3113 \end{bmatrix} \tag{7.87}$$

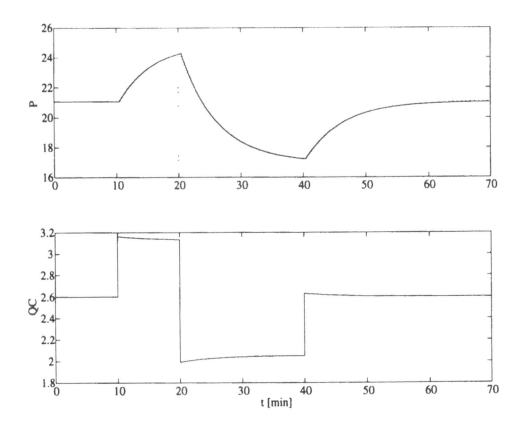

Figure 7.1: Mean-level control. $H_c = 1$, $H_m = 3$, $H_p = 360$, $\mathbf{R} = 0$, $\mathbf{Q} = \mathbf{I}$. The upper plot shows the combustion power, $P$ [MW], controlled by the fuel feed rate $Q_C$ $[\frac{kg}{s}]$.

$H_m$ can be given as equal to the time delay, $H_m = 3$.

The 'ideal' mean-level control result (using weighting matrices $\mathbf{Q} = \mathbf{I}$ and $\mathbf{R} = 0$) is shown in Fig. 7.1, where the linear model (7.81) is used as the process to be controlled. In mean-level control, the plant has open loop dynamics, the closed loop characteristic polynomial, (7.80), is $1 - 0.97q^{-1}$. A tighter control can be obtained by reducing the length of the prediction horizon ($H_p = 30$ in Fig. 7.2,) and/or increasing the control horizon ($H_p = 30$, $H_c = 2$ in Fig. 7.3). The characteristic polynomials are given by $1 - 0.93q^{-1}$ and 1, respectively. Notice, however, that in the latter simulation the control signal is bounded, whereas the computation of the characteristic polynomial was based on an (unconstrained) linear model.

Figure 7.4 shows a more realistic simulation, where the differential equation model was used for simulating the plant. Measurement noise with a

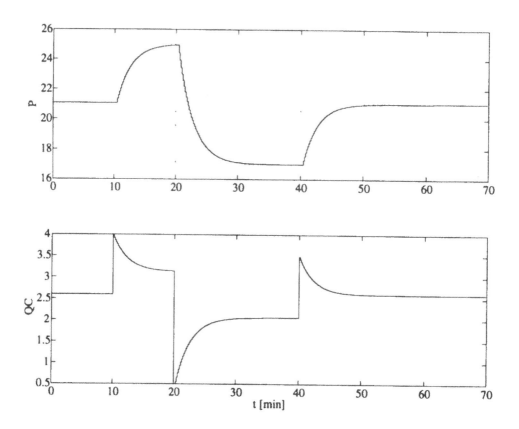

Figure 7.2: A typical GPC setting. $H_\mathrm{p} = 30$, see Fig. 7.1 for other details.

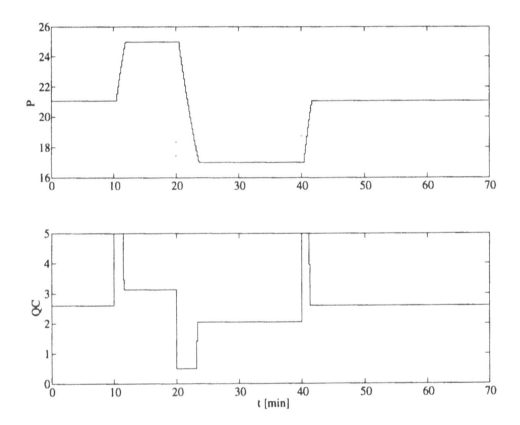

Figure 7.3: Dead-beat type of setting. $H_p = 30$, $H_c = 2$ (see Fig. 7.1 for other details). Note that the input was constrained on the range $[0.5, 5]$.

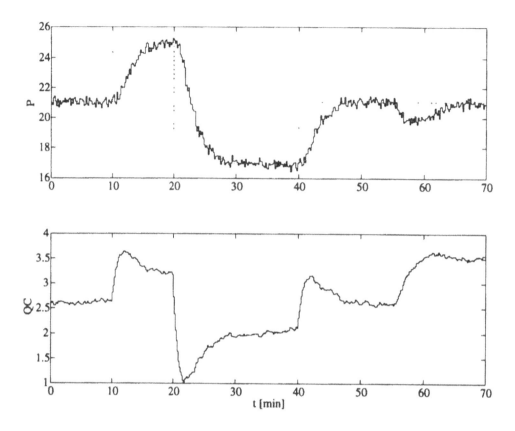

Figure 7.4: GPC control. $H_c = 1$, $H_m = 3$, $H_p = 30$, $\mathbf{R} = 100\mathbf{I}$, $\mathbf{Q} = \mathbf{I}$. The upper plot shows the combustion power, $P$ [MW], controlled by the fuel feed rate $Q_C$ [$\frac{kg}{s}$]. The plant was simulated using the differential equation model, with output noise $N(0, 0.21)$. An unmeasured 25% heat value loss affects the process at $t = 55$ min.

standard deviation of 1% of the nominal value was added to the output. In addition, an unmeasured disturbance (25% step-wise drop in fuel power) affects the simulated process at $t = 55$ min. An ARIMAX model with $C(q^{-1}) = 1 - 0.9q^{-1}$ was designed for disturbance rejection. In addition, a nonzero control weight was used, $\mathbf{R} = 100\mathbf{I}$ to reduce jitter in the controller output.

# Chapter 8

# Multivariable Systems

In this chapter, the control of linear multivariable systems is considered. First, the design of a MIMO control system is reduced to several SISO design problems. The relative gain array (RGA) method aims at helping to choose suitable pairs of control and controlled variables. If the interactions between the variables are strong, the system may not be satisfactorily controlled by SISO controllers only. In this case the interactions can be actively reduced by decouplers and the control of the decoupled system can then be designed using SISO methods. Decoupling is considered in the second section, and a simple multivariable PI controller (MPI) based on decoupling on both low and high frequencies is presented. The third approach considered in this section is a 'true' multivariable control approach. The design of a multivariable generalized predictive controller (MGPC) is considered, which solves the MIMO control design problem by minimizing a quadratic cost function. Simulation examples conclude this chapter.

All methods are based on models of the system. However, only steady-state gains are required by the RGA method; steady-state and high frequency gains by the MPI approach. These can be determined experimentally by using relatively simple plant experiments. The MGPC approach requires a dynamic model of the MIMO system, the identification of which may be a more laborious task and require more extensive experimenting with the plant.

For MIMO systems, the state-space formulation is simpler than, *e.g.*, that of polynomial matrices. Therefore, state-space models are assumed in what follows. In the case of MGPC, the conversion of a system model from a polynomial matrix form to a state-space form is also considered.

## 8.1  Relative gain array method

For processes with $N$ controlled outputs and $N$ manipulated variables, there
are $N!$ different ways to select input–output pairs for SISO control loops. One
way to select the 'best' possible SISO controllers among the configurations, is
to consider all the $N!$ loops and select those input–output pairs that minimize
the amount of interaction between the SISO controllers. This is the relative
gain array (RGA) method, also known as *Bristol's method* (see, *e.g.*, [90],
pp. 494-503).

The RGA method tries to *minimize the interactions between SISO loops*,
by selecting an *appropriate pairing*. It does not eliminate the interactions,
it merely tries to minimize the effect. It only relies upon steady-state information. If dynamic interactions are more important than those occurring at
steady-state, then clearly RGA is not a good method for such systems.

### 8.1.1  The basic idea

Consider a stable $N$-input $N$-output process. Let us define a *relative gain*
between an output $y_o$ ($o = 1, 2, ..., O$) and a manipulated variable $u_i$ ($i =
1, 2, ..., I$) ($O = I = N$) by:

$$\lambda_{o,i} = \frac{\left[\frac{\Delta y_o}{\Delta u_i}\right]_{u_k \text{ constant } \forall k \neq i}}{\left[\frac{\Delta y_o}{\Delta u_i}\right]_{y_k \text{ constant } \forall k \neq o}} \qquad (8.1)$$

where the notation '$u_k$ constant $\forall k \neq i$' denotes that the values of the manipulated variables other than $u_i$ are kept constant. Similarly '$y_k$ constant
$\forall k \neq o$' denotes that all outputs except the $o$'th one are kept constant by
some control loops. Thus, the numerator in (8.1) is the *open-loop* steady-
state gain of the system (the difference between initial and final steady-states
in output $o$, divided by the amplitude of the step change in input $i$). The
denominator in (8.1) is the *closed-loop* steady-state gain, where all other outputs except the $o$'th one are controlled using a controller which eliminates
steady-state error (*e.g.*, a PI-controller). The ratio of the two gains defines
the relative gain $\lambda_{o,i}$.

The value of $\lambda_{o,i}$ is a useful measure of interaction. In particular (see
[52]):

1. If $\lambda_{o,i} = 1$, the output $y_o$ is completely decoupled from all other inputs
   than the $i$'th one. This pair of variables is a perfect choice for SISO
   control.

2. If $0 < \lambda_{o,i} < 1$, there is interaction between the output $y_o$ and input variables other than $u_i$. The smaller the $\lambda_{o,i}$, the smaller the interaction between output $y_o$ and input $u_i$.

3. If $\lambda_{o,i} = 0$, then output $y_o$ does not respond to changes in input $u_i$. Consequently, the input $u_i$ can not be used to control the $o$'th output.

4. If $\lambda_{o,i} < 0$, then the gains of the open- and the closed-loop systems have different signs. This is dangerous, as the system is only conditionally stable[1].

5. If $\lambda_{o,i} > 1$, the open-loop gain is greater than closed-loop gain. This case is also undesirable[2].

A $N \times N$ matrix of relative gains (*Bristol's matrix*) collects all the relative gains into a matrix form.

$$\Lambda = \begin{bmatrix} \lambda_{1,1} & \lambda_{1,2} & \cdots & \lambda_{1,N} \\ \lambda_{2,1} & \lambda_{2,2} & & \lambda_{2,N} \\ \vdots & & \ddots & \vdots \\ \lambda_{N,1} & \lambda_{N,2} & \cdots & \lambda_{N,N} \end{bmatrix} \quad (8.2)$$

The sum of each row and column of the matrix is equal to one.

The RGA method recommends the following way to pair the controlled outputs with the manipulated variables:

**Proposition 1 (Bristol's method)** Select the control loops in such a way that the relative gains $\lambda_{o,i}$ are positive and as close to unity as possible.

In other words, those pairs of input and output variables are selected that minimize the amount of interaction among the resulting loops.

---

[1]Assume, for example, that the system is in open loop, and that the gain between $y_o$ and $u_i$ is positive. This would then fix the gain(s) of the controller (*e.g.*, positive gains in PI-control $\Delta u_i = k_P \Delta e_o + k_I e_o$ ($e_o = w_o - y_o$)). If the other loops are then put to automatic mode (controlled), the sign of the gain between $y_o$ and $u_i$ changes sign (since $\lambda_{o,i} < 0$). Consequently, the gain of the controller designed for the open loop system has gain with a wrong sign, which results in instability.

[2]In most instances the $y_o - u_i$ controller will be tuned with the other control loops in manual mode. When the other control loops are then put into automatic mode, the gain between $y_o$ and $u_i$ will reduce (since $\lambda_{o,i} > 1$) and the control performance for $y_o$ will probably degrade. If the $y_o - u_i$ controller is then re-tuned with a higher gain, a potential problem may arise: If the other loops are put back in manual mode, the gain between $y_o$ and $u_i$ would increase. Coupled with the new high gain controller, instability could result. The greater $\lambda_{o,i}$ is, the more pronounced this effect is.

## 8.1.2   Algorithm

When a model of the system is available, the Bristol's method is simple to compute. Consider a static model of an $N$-input $N$-output process:

$$\mathbf{y} = \mathbf{K_{ss}}\mathbf{u} \tag{8.3}$$

Without loss of generality we can assume for a linear system that the initial state is at $\mathbf{y} = \mathbf{0}$, $\mathbf{u} = \mathbf{0}$. The open loop gains for a unit step are given by the coefficients of the gain matrix $[\mathbf{K_{ss}}]_{o,i} = k_{ss\,o,i}$:

$$\left[\frac{\Delta y_o}{\Delta u_i}\right]_{u_k\ \text{constant}\ \forall k\neq i} = k_{ss\,o,i} \tag{8.4}$$

In order to solve the closed-loop gains let us compute the inverse of the system

$$\mathbf{u} = \mathbf{K_{ss}^{-1}}\mathbf{y} = \mathbf{M}\mathbf{y} \tag{8.5}$$

and denote the inverse matrix by $\mathbf{M}$, $[\mathbf{M}]_{o,i} = m_{o,i}$. In closed loop, all the other outputs are controlled so that the steady-state remains the same, except for the $o$'th one ($\Delta y_j = 0, \forall j \neq o$, $\Delta y_o = \Delta y_o^*$). We can then write the following steady-state relation between the $i$'th input and the $o$'th output:

$$\Delta\mathbf{u} = \mathbf{M}\Delta\mathbf{y} \tag{8.6}$$

$$\begin{bmatrix} \Delta u_1 \\ \vdots \\ \Delta u_{i-1} \\ \Delta u_i \\ \Delta u_{i+1} \\ \vdots \\ \Delta u_N \end{bmatrix} = \mathbf{M} \begin{bmatrix} 0 \\ \vdots \\ 0 \\ \Delta y_o^* \\ 0 \\ \vdots \\ 0 \end{bmatrix} = \begin{bmatrix} m_{1,o} \\ \vdots \\ m_{i,o} \\ \vdots \\ m_{N,o} \end{bmatrix} \Delta y_o^* \tag{8.7}$$

Taking the $i$'th row of the above system of equations gives

$$\Delta u_i = m_{i,o}\Delta y_o^* \tag{8.8}$$

and

$$\left[\frac{\Delta y_o^*}{\Delta u_i}\right]_{y_k\ \text{constant}\ \forall k\neq o} = \frac{1}{m_{i,o}} \tag{8.9}$$

where $m_{i,o}$ is the $(i, o)$'th element of the inverse of the process' steady state gain matrix. Thus, the elements of the Bristol's matrix are given by

$$\lambda_{o,i} = k_{ss\,o,i} m_{i,o} \tag{8.10}$$

Let us give an algorithm for computing the Bristol's matrix, when a linear model for the system is available.

**Algorithm 29 (Bristol's method)** Given a steady-state process model

$$\mathbf{y} = \mathbf{K}_{ss}\mathbf{u} \tag{8.11}$$

the Bristol matrix is given by

$$\Lambda = \mathbf{K}_{ss} \odot \left(\mathbf{K}_{ss}^{-1}\right)^T \tag{8.12}$$

where $\odot$ denotes the element-wise multiplication.

**Example 40 (Bristol's method)** Consider a $2 \times 2$ system

- Let the following steady-state information be available

$$\left[\begin{array}{c} y_1 \\ y_2 \end{array}\right] = \mathbf{K}_{ss} \left[\begin{array}{c} u_1 \\ u_2 \end{array}\right] \tag{8.13}$$

where

$$\mathbf{K}_{ss} = \left[\begin{array}{cc} 1 & 0 \\ 0.15 & -0.2 \end{array}\right] \tag{8.14}$$

This results in the following matrix of relative gains

$$\Lambda = \left[\begin{array}{cc} 1 & 0 \\ 0 & 1 \end{array}\right] \tag{8.15}$$

The Bristol's method then suggests to select SISO controllers for pairs $y_1 - u_1$ and $y_2 - u_2$, which is intuitively clear since the input $u_2$ has no effect on the output $y_1$.

- Let the system be given by

$$\mathbf{K}_{ss} = \left[\begin{array}{cc} 1 & 2 \\ 0.15 & -0.2 \end{array}\right] \tag{8.16}$$

This results in the following matrix of relative gains

$$\Lambda = \left[\begin{array}{cc} 0.4 & 0.6 \\ 0.6 & 0.4 \end{array}\right] \tag{8.17}$$

The Bristol's method then suggests to select SISO controllers for pairs $y_1 - u_2$ and $y_2 - u_1$.

- Let the system be given by

$$\mathbf{K}_{\mathrm{su}} = \begin{bmatrix} 1 & 2 \\ 0.15 & 0.2 \end{bmatrix} \tag{8.18}$$

This results in the following matrix of relative gains

$$\Lambda = \begin{bmatrix} -2 & 3 \\ 3 & -2 \end{bmatrix} \tag{8.19}$$

The Bristol's method then suggests to select SISO controllers for pairs $y_1 - u_2$ and $y_2 - u_1$. There may be problems in switching between automatic and manual modes, but at least the gains in open and closed loop will have same signs.

**Example 41 (Fluidized bed combustion)** A steady-state model for an FBC plant (see Appendix B) in the neighborhood of an operating point is given by

$$\begin{bmatrix} C_{\mathrm{F}} \\ T_{\mathrm{B}} \\ T_{\mathrm{F}} \\ P \end{bmatrix} = \begin{bmatrix} -0.0688 & 0.0155 & 0.0155 \\ 212.29 & -93.73 & 0 \\ 162.72 & -5.87 & -18.29 \\ 8.103 & 0 & 0 \end{bmatrix} \begin{bmatrix} Q_{\mathrm{C}} \\ F_1 \\ F_2 \end{bmatrix} \tag{8.20}$$

- Let us first consider that the outputs $C_{\mathrm{F}}$ (flue gas $O_2$), $T_{\mathrm{B}}$ (bed temperatures) and $P$ (combustion power) are controlled by the three inputs (fuel feed, primary and secondary airs). The Bristol's matrix becomes

$$\Lambda = \begin{bmatrix} 0 & 0 & 1 \\ 0 & 1 & 0 \\ 1 & 0 & 0 \end{bmatrix} \tag{8.21}$$

Thus the suggestion is to control oxygen with secondary air, power with fuel feed, and bed temperatures with primary air. For the first two, this is common practice in reality; the bed temperatures are not usually under automatic control.

- Let us consider controlling the freeboard temperatures $T_{\mathrm{F}}$, instead of bed temperatures. The Bristol's matrix is given by

$$\Lambda = \begin{bmatrix} 0 & 1.4734 & -0.4734 \\ 0 & -0.4734 & 1.4734 \\ 1 & 0 & 0 \end{bmatrix} \tag{8.22}$$

The suggestion is still to control the power by fuel feed (note that this is simple to reason using physical arguments, too). For the temperatures and air flows the situation is more complicated. The suggestion is now to use primary air for $O_2$ control and secondary air for the freeboard temperatures; if chosen otherwise the open- and closed-loop gains will have different signs[3]. In practice, freeboard temperatures are not under automatic control.

If the number of input and output variables is not the same, then several Bristol's matrices need to be formed. Assume that there are $O$ output variables and $I$ ($O \leq I$) possible manipulated variables. Then an $O \times O$ matrix of relative gains can be formed for all different combinations of $O$ manipulated variables. All the matrices need to be examined before selecting the $O$ loops with minimal interaction. The rule for the selection of control loops remains the same, *i.e.* the control loops that have relative gains positive and as close to unity as possible are recommended.

The RGA-method indicates how the inputs should be coupled (paired) with the outputs to form loops with the smallest amount of interaction. However, this interaction may not be small enough, even if it is the smallest possible. In this case, decouplers can be applied. These will be considered in the next section.

## 8.2   Decoupling of interactions

The purpose of decouplers is to *cancel the interaction* between the loops. The remaining system can then be considered (and designed) as having no interactions at all. Hence, a multivariable control design problem is converted into a set of SISO control design problems, by introducing artificial decoupling compensations (see, *e.g.*, [90], pp.504–509).

The interactions can be perfectly decoupled only if the process is perfectly known. In practice, a perfect model is rarely available. Thus only a partial decoupling can be obtained, with some (weak) interactions persisting. It may also be that the decouplers are not realizable, or that the degree of decouplers would be too high for a practical implementation. In this case, some realizable form of the decoupler can be considered. For a stable process,

---

[3]The gains from $F_1$ and $F_2$ to $C_F$ are equal, but the gain from $F_1$ to $T_F$ is significantly smaller than that from $F_2$. If $F_2$ is used for $O_2$ control, and $F_1$ for temperature control then each action taken by the $F_2$ would need to be compensated by a (larger) counteraction in $F_1$. Thus the open- and closed-loop gains would have different signs depending on wheather $T_F - F_1$ controller is on or off.

a steady-state decoupler is always realizable. Remember that for a severely interacting system, static decoupling is better than no decoupling at all.

There are a number of different approaches, the most famous being perhaps the Rosenbrock's (inverse) Nyquist array method (see, *e.g.*, [57]), which is a frequency response method seeking to reduce the interaction by using a compensator to first make the system diagonally dominant. In what follows, a simple scheme for designing a discrete-time multivariable PI controller is presented.

## 8.2.1   Multivariable PI-controller

In [74] a multivariable PI controller was suggested. The main idea is to decouple the system both at low and high frequencies. The original derivation was based on a continuous-time state-space model, in what follows the discrete-time case is considered.

Consider a linear time-invariant stable multivariable plant described by the following discrete-time equations

$$\mathbf{x}(k+1) = \mathbf{A}\mathbf{x}(k) + \mathbf{B}\mathbf{u}(k) \qquad (8.23)$$
$$\mathbf{y}(k) = \mathbf{C}\mathbf{x}(k) \qquad (8.24)$$

and controlled by a multivariable PI-controller

$$\Delta\mathbf{u}(k) = \mathbf{K}_P\boldsymbol{\alpha}_P\Delta\mathbf{e}(k) + \mathbf{K}_I\boldsymbol{\alpha}_I\mathbf{e}(k) \qquad (8.25)$$

where

$$\mathbf{e}(k) = \mathbf{w}(k) - \mathbf{y}(k) \qquad (8.26)$$

and $\boldsymbol{\alpha}_P$ and $\boldsymbol{\alpha}_I$ are tuning variables (diagonal matrices) and $\mathbf{w}(k)$ contains the set points. The idea is that the P-part decouples the system at high frequencies, while the I-part decouples the system at low frequencies (steady-state). Let us first consider the P- and I-controllers separately, and then combine them together.

### P-controller

Let us first assume that the system is controlled by a P-controller and that the aim is to drive the error $\mathbf{e}$ to zero as fast as possible. For the high frequencies we can write ($\Delta\mathbf{y}$ is the component with the highest frequency that can be described by the discrete-time model):

$$\begin{align}
\Delta \mathbf{y}\,(k+1) &= \mathbf{C}\Delta \mathbf{x}\,(k+1) \tag{8.27}\\
&= \mathbf{C}\,[\mathbf{x}\,(k+1) - \mathbf{x}\,(k)] \tag{8.28}\\
&= \mathbf{C}\,[\mathbf{Ax}\,(k) + \mathbf{Bu}\,(k) - \mathbf{x}\,(k)] \tag{8.29}\\
&= \mathbf{C}\,(\mathbf{A} - \mathbf{I})\,\mathbf{x}\,(k) + \mathbf{CBu}\,(k) \tag{8.30}\\
&= \mathbf{C}\,(\mathbf{A} - \mathbf{I})\,\mathbf{x}\,(k) + \mathbf{K}_{\mathrm{high}}\mathbf{u}\,(k) \tag{8.31}
\end{align}$$

where

$$\mathbf{K}_{\mathrm{high}} = \mathbf{CB} \tag{8.32}$$

Consider that, at sample instant $k$, $\mathbf{x}$ is initially in a desired steady-state and a step change in the reference signal, $\Delta \mathbf{w}\,(k+1)$, occurs at $k+1$. We then have

$$\begin{align}
\mathbf{e}\,(k+1) &= \Delta \mathbf{w}\,(k+1) \tag{8.33}\\
&= \mathbf{w}\,(k+1) - \mathbf{w}\,(k) \tag{8.34}\\
&= \mathbf{w}\,(k+1) - \mathbf{y}\,(k) \tag{8.35}
\end{align}$$

In order to drive the error $\mathbf{w}\,(k+1) - \mathbf{y}\,(k+1)$ to zero in one control sample (if possible), we need to have

$$\begin{align}
\mathbf{y}\,(k+1) &= \mathbf{w}\,(k+1) \tag{8.36}\\
\Delta \mathbf{y}\,(k+1) &= \Delta \mathbf{w}\,(k+1) \tag{8.37}\\
\mathbf{C}\,(\mathbf{A} - \mathbf{I})\,\mathbf{x}\,(k) + \mathbf{K}_{\mathrm{high}}\mathbf{u}\,(k) &= \mathbf{e}\,(k+1) \tag{8.38}
\end{align}$$

and we can solve for the manipulated variables

$$\begin{align}
\mathbf{u}\,(k) &= \mathbf{K}_{\mathrm{high}}^{-1}\,[\mathbf{e}\,(k+1) - \mathbf{C}\,(\mathbf{A} - \mathbf{I})\,\mathbf{x}\,(k)] \tag{8.39}\\
\Delta \mathbf{u}\,(k) &= \mathbf{K}_{\mathrm{high}}^{-1}\Delta \mathbf{e}\,(k+1) \tag{8.40}
\end{align}$$

where the last equality is obtained using $\Delta \mathbf{x}\,(k) = 0$ since $\mathbf{x}\,(k)$ was a steady-state. Thus, $\mathbf{K}_P = \mathbf{K}_{\mathrm{high}}^{-1}$ in (8.25), if the inverse exists.

## I-controller

Let us now consider the case where the system is controlled by an I-controller. From the system model we obtain the following relationship for a steady-state (by setting $\mathbf{x}\,(k+1) = \mathbf{x}\,(k) = \mathbf{x}_{ss}$):

$$\begin{align}
\mathbf{x}_{ss} &= \mathbf{Ax}_{ss} + \mathbf{Bu}_{ss} \tag{8.41}\\
\mathbf{x}_{ss} &= (\mathbf{I} - \mathbf{A})^{-1}\mathbf{Bu}_{ss} \tag{8.42}
\end{align}$$

and

$$\mathbf{y}_{ss} = \mathbf{Cx}_{ss} \tag{8.43}$$

which gives

$$\mathbf{y}_{ss} = \mathbf{C}\left(\mathbf{I} - \mathbf{A}\right)^{-1}\mathbf{Bu}_{ss} \tag{8.44}$$

$$\mathbf{y}_{ss} = \mathbf{K}_{ss}\mathbf{u}_{ss} \tag{8.45}$$

where

$$\mathbf{K}_{ss} = \mathbf{C}\left(\mathbf{I} - \mathbf{A}\right)^{-1}\mathbf{B} \tag{8.46}$$

In order to drive the steady-state error (for a step change in the reference signal) to zero, we need to have

$$\mathbf{y}_{ss,\text{new}} = \mathbf{w} \tag{8.47}$$

$$\mathbf{K}_{ss}\mathbf{u}_{ss,\text{new}} = \Delta\mathbf{w} + \mathbf{y}\left(k\right) \tag{8.48}$$

$$\mathbf{K}_{ss}\left(\mathbf{u}_{ss,\text{old}} + \Delta\mathbf{u}_{ss,\text{new}}\right) = \Delta\mathbf{w} + \mathbf{y}_{ss,\text{old}} \tag{8.49}$$

$$\mathbf{K}_{ss}\Delta\mathbf{u}_{ss,\text{new}} = \Delta\mathbf{w} \tag{8.50}$$

The required change at the controller output at $k + 1$ is then

$$\Delta\mathbf{u}\left(k\right) = \mathbf{K}_{ss}^{-1}\mathbf{e}\left(k + 1\right) \tag{8.51}$$

Thus, $\mathbf{K}_I = \mathbf{K}_{ss}^{-1}$ in (8.25), if the inverse exists.

**PI-controller**

The PI controller can now be constructed by combining the tuning for P and I controllers. The controller was given by (8.25)

$$\Delta\mathbf{u}\left(k\right) = \mathbf{K}_P\alpha_P\Delta\mathbf{e}\left(k\right) + \mathbf{K}_I\alpha_I\mathbf{e}\left(k\right) \tag{8.52}$$

where

$$\mathbf{e}\left(k\right) = \mathbf{w}\left(k\right) - \mathbf{y}\left(k\right) \tag{8.53}$$

$$\Delta\mathbf{e}\left(k\right) = \mathbf{e}\left(k\right) - \mathbf{e}\left(k - 1\right) \tag{8.54}$$

and

$$\mathbf{K}_P = \left[\mathbf{CB}\right]^{-1} \tag{8.55}$$

$$\mathbf{K}_I = \left[\mathbf{C}\left(\mathbf{I} - \mathbf{A}\right)^{-1}\mathbf{B}\right]^{-1} \tag{8.56}$$

$\mathbf{K}_P$ and $\mathbf{K}_I$ provide decoupling at high and low frequencies. The tuning of the controller is conducted by adjusting the $\boldsymbol{\alpha}_P$ and $\boldsymbol{\alpha}_I$, starting with small positive values $(0 < \alpha_{P,n} \ll 1, 0 < \alpha_{I,n} \ll 1)$:

$$\boldsymbol{\alpha}_P = \begin{bmatrix} \alpha_{P,1} & & & 0 \\ & \alpha_{P,2} & & \\ & & \ddots & \\ 0 & & & \alpha_{P,N} \end{bmatrix} \tag{8.57}$$

$$\boldsymbol{\alpha}_I = \begin{bmatrix} \alpha_{I,1} & & & 0 \\ & \alpha_{I,2} & & \\ & & \ddots & \\ 0 & & & \alpha_{I,N} \end{bmatrix} \tag{8.58}$$

Setting $\alpha_{P,n} = 0$ results in an I-controller only; similarly $\alpha_{I,n} = 0$ results in pure P-control. With $\alpha_{P,n} = 1$, an aggressive P-control is obtained, which tries to drive the error to zero in one sample. With $\alpha_{I,n} = 1$, the controller output at $k+1$ is set to the value which provides the new steady-state (mean-level control). In the presence of noisy measurements and modelling errors, these can provide instability to the closed loop system, and unrealizable control signals. Therefore, smoother control is usually desired, at the cost of closed-loop performance.

## 8.3 Multivariable predictive control

In Chapter 7, the generalized predictive control (GPC) for SISO systems was considered. In this section, we will extend the concepts of SISO GPC to the control of MIMO processes (see [45][96]).

### 8.3.1 State-space model

A MIMO system can be conveniently described by a state-space model. Let us consider a multivariable input–output polynomial model of the form

$$\mathbf{F}\left(q^{-1}\right) \mathbf{y}\left(k\right) = \mathbf{B}\left(q^{-1}\right) \mathbf{v}\left(k\right) + \mathbf{C}\left(q^{-1}\right) \mathbf{e}\left(k\right) \tag{8.59}$$

where the polynomial matrices $\mathbf{A}$, $\mathbf{B}$ and $\mathbf{C}$ are given by

$$\mathbf{F}\left(q^{-1}\right) = \mathbf{I} + \mathbf{F}_1 q^{-1} + \cdots + \mathbf{F}_N q^{-N} \tag{8.60}$$

$$\mathbf{B}\left(q^{-1}\right) = \mathbf{B}_1 q^{-1} + \cdots + \mathbf{B}_N q^{-N} \tag{8.61}$$

$$C\left(q^{-1}\right) = I + C_1 q^{-1} + \cdots + C_N q^{-N} \qquad (8.62)$$

The output and noise vectors **y** and **e**, respectively, are of size $O \times 1$:

$$\begin{aligned}
\mathbf{y}\left(k\right) &= \left[y_1\left(k\right), y_2\left(k\right), \cdots, y_O\left(k\right)\right]^T & (8.63) \\
\mathbf{e}\left(k\right) &= \left[e_1\left(k\right), e_2\left(k\right), \cdots, e_O\left(k\right)\right]^T & (8.64)
\end{aligned}$$

Consequently, matrices $F\left(q^{-1}\right)$ and $C\left(q^{-1}\right)$ are of size $O \times O$, (as well as matrices $\mathbf{F}_n$ and $\mathbf{C}_n$). Input vector **v** is of size $I \times 1$:

$$\mathbf{v}\left(k\right) = \left[v_1\left(k\right), v_2\left(k\right), \cdots, v_I\left(k\right)\right]^T \qquad (8.65)$$

matrix $B\left(q^{-1}\right)$ is of size $O \times I$ (as well as matrices $\mathbf{B}_n$). Without loss of generality we assume that all polynomials are of order $N$.

The above polynomial model can be represented in a (canonical observable) state-space form:

$$\begin{aligned}
\mathbf{x}\left(k+1\right) &= \mathbf{A}\mathbf{x}\left(k\right) + \mathbf{B}\mathbf{v}\left(k\right) + \mathbf{G}\mathbf{e}\left(k\right) & (8.66) \\
\mathbf{y}\left(k\right) &= \mathbf{C}\mathbf{x}\left(k\right) + \mathbf{e}\left(k\right) & (8.67)
\end{aligned}$$

Please note that **A**, **B**, **C** and **G** are matrices in the state-space model, whereas $F\left(q^{-1}\right)$, $B\left(q^{-1}\right)$ and $C\left(q^{-1}\right)$ are polynomial matrices. The matrices **A**, **B**, **G** and **C** are given by

$$\mathbf{A} = \begin{bmatrix} -F_1 & I & 0 & \cdots & 0 \\ -F_2 & 0 & I & & 0 \\ \vdots & & \vdots & \ddots & \\ -F_{N-1} & 0 & 0 & & I \\ -F_N & 0 & 0 & \cdots & 0 \end{bmatrix} \qquad (8.68)$$

$$\mathbf{B} = \begin{bmatrix} B_1 & B_2 & \cdots & B_{N-1} & B_N \end{bmatrix}^T \qquad (8.69)$$

$$\mathbf{G} = \begin{bmatrix} C_1 - F_1 & C_2 - F_2 & \cdots & C_{N-1} - F_{N-1} & C_N - F_N \end{bmatrix}^T \ (8.70)$$

$$\mathbf{C} = \begin{bmatrix} I & 0 & \cdots & 0 \end{bmatrix} \qquad (8.71)$$

The matrices will now have the following sizes: $[\mathbf{A}] = NO \times NO$, $[\mathbf{B}] = NO \times I$, $[\mathbf{G}] = NO \times O$ and $[\mathbf{C}] = O \times NO$.

**Example 42 (Representation of a $2 \times 2$ system)** For a $2 \times 2$ P-canonical system with common denominators we have (8.59)

$$
\left[ \begin{array}{cc} 1 + f_{1,1,1}q^{-1} + \ldots + f_{N,1,1}q^{-N} & 0 \\ 0 & 1 + f_{1,2,2}q^{-1} + \ldots + f_{N,2,2}q^{-N} \end{array} \right] \times
$$

$$
\left[ \begin{array}{c} y_1(k) \\ y_2(k) \end{array} \right] \tag{8.72}
$$

$$
= \left[ \begin{array}{cc} b_{1,1,1}q^{-1} + \ldots + b_{N,1,1}q^{-N} & b_{1,1,2}q^{-1} + \ldots + b_{N,1,2}q^{-N} \\ b_{1,2,1}q^{-1} + \ldots + b_{N,2,1}q^{-N} & b_{1,2,2}q^{-1} + \ldots + b_{N,2,2}q^{-N} \end{array} \right] \left[ \begin{array}{c} v_1(k) \\ v_2(k) \end{array} \right]
$$

$$
+ \left[ \begin{array}{cc} c_{1,1,1}q^{-1} + \ldots + c_{N,1,1}q^{-N} & 0 \\ 0 & c_{1,2,2}q^{-1} + \ldots + c_{N,2,2}q^{-N} \end{array} \right] \left[ \begin{array}{c} e_1(k) \\ e_2(k) \end{array} \right]
$$

With this notation, the elements $(i,j)$ of the matrices $\mathbf{F}_n$ consist then of the scalar coefficients $f_{n,i,j}$, matrices $\mathbf{B}$ and $\mathbf{C}$ are constructed in a similar way. We then have a state-space representation (8.66)–(8.67) with matrices $\mathbf{A}$, $\mathbf{B}$, $\mathbf{C}$ and $\mathbf{G}$ given by:

$$
\mathbf{A} = \left[ \begin{array}{cccc} \left[ \begin{array}{cc} -f_{1,1,1} & 0 \\ 0 & -f_{1,2,2} \end{array} \right] & \left[ \begin{array}{cc} 1 & 0 \\ 0 & 1 \end{array} \right] & \cdots & \left[ \begin{array}{cc} 0 & 0 \\ 0 & 0 \end{array} \right] \\ \vdots & \vdots & \ddots & \vdots \\ \left[ \begin{array}{cc} -f_{N-1,1,1} & 0 \\ 0 & -f_{N-1,2,2} \end{array} \right] & \left[ \begin{array}{cc} 0 & 0 \\ 0 & 0 \end{array} \right] & \cdots & \left[ \begin{array}{cc} 1 & 0 \\ 0 & 1 \end{array} \right] \\ \left[ \begin{array}{cc} -f_{N,1,1} & 0 \\ 0 & -f_{N,2,2} \end{array} \right] & \left[ \begin{array}{cc} 0 & 0 \\ 0 & 0 \end{array} \right] & \cdots & \left[ \begin{array}{cc} 0 & 0 \\ 0 & 0 \end{array} \right] \end{array} \right] \tag{8.73}
$$

$$
\mathbf{B} = \left[ \begin{array}{c} \left[ \begin{array}{cc} b_{1,1,1} & b_{1,1,2} \\ b_{1,2,1} & b_{1,2,2} \end{array} \right] \\ \vdots \\ \left[ \begin{array}{cc} b_{N-1,1,1} & b_{N-1,1,2} \\ b_{N-1,2,1} & b_{N-1,2,2} \end{array} \right] \\ \left[ \begin{array}{cc} b_{N,1,1} & b_{N,1,2} \\ b_{N,2,1} & b_{N,2,2} \end{array} \right] \end{array} \right] \tag{8.74}
$$

$$
\mathbf{G} = \left[ \begin{array}{c} \left[ \begin{array}{cc} c_{1,1,1} - f_{1,1,1} & 0 \\ 0 & c_{1,2,2} - f_{1,2,2} \end{array} \right] \\ \vdots \\ \left[ \begin{array}{cc} c_{N-1,1,1} - f_{N-1,1,1} & 0 \\ 0 & c_{N-1,2,2} - f_{N-1,2,2} \end{array} \right] \\ \left[ \begin{array}{cc} c_{N,1,1} - f_{N,1,1} & 0 \\ 0 & c_{N,2,2} - f_{N,2,2} \end{array} \right] \end{array} \right] \tag{8.75}
$$

$$\mathbf{C} = \left[ \begin{bmatrix} 1 & 0 \\ 0 & 1 \end{bmatrix} \begin{bmatrix} 0 & 0 \\ 0 & 0 \end{bmatrix} \cdots \begin{bmatrix} 0 & 0 \\ 0 & 0 \end{bmatrix} \right] \qquad (8.76)$$

where all the elements of the matrices are scalars.

### 8.3.2  $i$-step ahead predictions

The optimal predictions at sample instant $k + i$ will be

$$\begin{aligned}
\widehat{\mathbf{y}}\,(k+i) = {} & \sum_{j=1}^{i} \mathbf{C}\mathbf{A}^{i-j}\mathbf{B}\mathbf{v}\,(k+j-1) \qquad (8.77) \\
& + \mathbf{C}\mathbf{A}^{i-1}\,[\mathbf{A} - \mathbf{G}\mathbf{C}]\,\mathbf{x}\,(k) \\
& + \mathbf{C}\mathbf{A}^{i-1}\mathbf{G}\mathbf{y}\,(k)
\end{aligned}$$

Let us use the following condensed notation:

$$\begin{aligned}
\widehat{\mathbf{Y}}\,(k+1) &= \left[ \widehat{\mathbf{y}}^{T}\,(k+1), \cdots, \widehat{\mathbf{y}}^{T}\,(k+H_{\mathrm{p}}) \right]^{T} & (8.78) \\
\mathbf{V}\,(k) &= \left[ \mathbf{v}^{T}\,(k), \cdots, \mathbf{v}^{T}\,(k+H_{\mathrm{p}}-1) \right]^{T} & (8.79)
\end{aligned}$$

that is

$$\widehat{\mathbf{Y}}\,(k+1) = \begin{bmatrix} \widehat{\mathbf{y}}\,(k+1) \\ \vdots \\ \widehat{\mathbf{y}}\,(k+H_{\mathrm{p}}) \end{bmatrix} = \begin{bmatrix} \begin{bmatrix} \widehat{y}_1\,(k+1) \\ \widehat{y}_2\,(k+1) \\ \vdots \\ \widehat{y}_O\,(k+1) \end{bmatrix} \\ \begin{bmatrix} \widehat{y}_1\,(k+2) \\ \widehat{y}_2\,(k+2) \\ \vdots \\ \widehat{y}_O\,(k+2) \end{bmatrix} \\ \vdots \\ \begin{bmatrix} \widehat{y}_1\,(k+H_{\mathrm{p}}) \\ \widehat{y}_2\,(k+H_{\mathrm{p}}) \\ \vdots \\ \widehat{y}_O\,(k+H_{\mathrm{p}}) \end{bmatrix} \end{bmatrix} \qquad (8.80)$$

$$V(k) = \begin{bmatrix} \mathbf{v}(k) \\ \vdots \\ \mathbf{v}(k + H_{\mathrm{p}} - 1) \end{bmatrix} = \begin{bmatrix} \begin{bmatrix} v_1(k) \\ v_2(k) \\ \vdots \\ v_I(k) \end{bmatrix} \\ \begin{bmatrix} v_1(k+1) \\ v_2(k+1) \\ \vdots \\ v_I(k+1) \end{bmatrix} \\ \vdots \\ \begin{bmatrix} v_1(k + H_{\mathrm{p}} - 1) \\ v_2(k + H_{\mathrm{p}} - 1) \\ \vdots \\ v_I(k + H_{\mathrm{p}} - 1) \end{bmatrix} \end{bmatrix} \tag{8.81}$$

We have then a global predictive model for the $H_{\mathrm{p}}$ future time instants $k+1$, $k+2, \cdots, k + H_{\mathrm{p}}$:

$$\widehat{\mathbf{Y}}(k+1) = \mathbf{K}_{\mathrm{CAGC}} \mathbf{x}(k) + \mathbf{K}_{\mathrm{CAB}} \mathbf{V}(k) + \mathbf{K}_{\mathrm{CAG}} \mathbf{y}(k) \tag{8.82}$$

where

$$\mathbf{K}_{\mathrm{CAGC}} = \left[\mathbf{C}\left[\mathbf{A} - \mathbf{GC}\right], \cdots, \mathbf{CA}^{H_{\mathrm{p}}-1}\left[\mathbf{A} - \mathbf{GC}\right]\right]^T \tag{8.83}$$

$$\mathbf{K}_{\mathrm{CAB}} = \begin{bmatrix} \mathbf{CB} & \cdots & 0 \\ \vdots & \ddots & \vdots \\ \mathbf{CA}^{H_{\mathrm{p}}-1}\mathbf{B} & \cdots & \mathbf{CB} \end{bmatrix} \tag{8.84}$$

$$\mathbf{K}_{\mathrm{CAG}} = \left[\mathbf{CG}, \cdots, \mathbf{CA}^{H_{\mathrm{p}}-1}\mathbf{G}\right]^T \tag{8.85}$$

## 8.3.3   Cost function

Let us consider the following cost function

$$\begin{aligned} J &= \left[\mathbf{W}(k+1) - \widehat{\mathbf{Y}}(k+1)\right] \mathbf{Q} \left[\mathbf{W}(k+1) - \widehat{\mathbf{Y}}(k+1)\right]^T \\ &\quad + \mathbf{V}^T(k) \mathbf{R} \mathbf{V}(k) \end{aligned} \tag{8.86}$$

where

$$\mathbf{W}(k+1) = \left[\mathbf{w}^T(k+1), \cdots, \mathbf{w}^T(k + H_{\mathrm{p}})\right]^T \tag{8.87}$$

that is

$$
\mathbf{W}\left(k+1\right) = 
\begin{bmatrix} \mathbf{w}\left(k+1\right) \\ \vdots \\ \mathbf{w}\left(k+H_{\mathrm{p}}\right) \end{bmatrix} = 
\begin{bmatrix} 
\begin{bmatrix} w_1\left(k+1\right) \\ w_2\left(k+1\right) \\ \vdots \\ w_O\left(k+1\right) \end{bmatrix} \\
\begin{bmatrix} w_1\left(k+2\right) \\ w_2\left(k+2\right) \\ \vdots \\ w_O\left(k+2\right) \end{bmatrix} \\
\vdots \\
\begin{bmatrix} w_1\left(k+H_{\mathrm{p}}\right) \\ w_2\left(k+H_{\mathrm{p}}\right) \\ \vdots \\ w_O\left(k+H_{\mathrm{p}}\right) \end{bmatrix}
\end{bmatrix}
\tag{8.88}
$$

The optimal control sequence is given by

$$
\begin{aligned}
\mathbf{V}\left(k\right) &= \left[\mathbf{K}_{\mathrm{CAB}}^{T}\mathbf{Q}\mathbf{K}_{\mathrm{CAB}} + \mathbf{R}\right]^{-1} \\
&\quad \times \mathbf{K}_{\mathrm{CAB}}^{T}\mathbf{Q}\left[\mathbf{W}\left(k+1\right) - \mathbf{K}_{\mathrm{CAGC}}\mathbf{x}\left(k\right) - \mathbf{K}_{\mathrm{CAG}}\mathbf{y}\left(k\right)\right]
\end{aligned}
\tag{8.89}
$$

Comparing with the SISO case, we see that the equations have the same form. Only the sizes of the vectors and matrices are different since an $O$-output $I$-input system is considered instead of a SISO system. Instead of a scalar prediction, the predictions $\widehat{\mathbf{y}}$, (8.77), and targets $\mathbf{w}$ are given by an $O \times 1$ vector. Likewise, instead of a scalar input and noise, the system inputs are now given by a $I \times 1$ vector $\mathbf{v}$ as well as the noise $\mathbf{e}$. The collection of all future predictions $\widehat{\mathbf{Y}}$ as well as future targets $\mathbf{W}$ are now long vectors of length $OH_{\mathrm{p}}$, system inputs $\mathbf{V}$ are contained in a vector of length $IH_{\mathrm{p}}$. Similarly, the elements of the gain matrices (8.83)–(8.85) have been composed by piling the future predictions on top of each other. However, from a technical point of view, the future predictions, controller outputs and targets are still given in column vectors just as in the SISO case. Therefore, the solution also has a similar form and we will give no proof for it.

## 8.3.4   Remarks

The implementation of the minimum horizons $H_{\mathrm{m}}$, prediction horizons $H_{\mathrm{p}}$ and the control horizons $H_{\mathrm{c}}$ can be done in the same manner as in the SISO case, by 'cutting' the matrices. For simplicity, in the above derivation same *horizons* were assumed for all inputs and outputs. Note, however, that when different horizons are used, $H_{m,o} \neq H_{m,j}$; $H_{p,o} \neq H_{p,j}$ and/or $H_{c,i} \neq H_{c,j}$ for input $i \neq j$ and/or output $o \neq j$, the 'removed' elements in the matrices need to be replaced by zeros. If $\mathbf{R}$, (8.86), is a nonzero matrix, this results in numerical problems. To avoid this, all the rows and columns of $\mathbf{K}_{CAB}$ containing only zero elements need to be removed, as well as the corresponding rows and columns of $\mathbf{Q}$ and $\mathbf{R}$, $\mathbf{K}_{CAGC}$, $\mathbf{K}_{CAG}$ and $\mathbf{W}$.

In the same way as in the SISO case, a *fixed gain observer* is given by

$$\hat{\mathbf{x}}(k) = [\mathbf{A} - \mathbf{GC}]\hat{\mathbf{x}}(k-1) + \mathbf{B}\mathbf{v}(k-1) + \mathbf{Gy}(k-1) \qquad (8.90)$$

This is the estimate where the Kalman filter tends to, when time approaches infinity, *i.e.* the asymptotic estimate obtained under the condition that the eigenvalues of the matrix $(\mathbf{A} - \mathbf{GC})$ are less than one.

## 8.3.5 Simulation example

Let us illustrate the multivariable predictive control on the FBC process. From Appendix B we obtain a model for the relations between combustion power $P$ and flue gas oxygen $C_F$ (controlled variables) and fuel feed $Q_C$ and secondary air $F_2$ (manipulated variables), using a sampling time of 10 seconds. Linearizing, discretizing, and converting to a state-space form, we have a state-space description in the form

$$\mathbf{x}(k+1) = \mathbf{Ax}(k) + \mathbf{B}\Delta\mathbf{u}(k) \qquad (8.91)$$
$$\mathbf{y}(k) = \mathbf{Cx}(k) \qquad (8.92)$$

where

$$\mathbf{y} \leftarrow \begin{bmatrix} C_F \\ P \end{bmatrix}, \mathbf{u} \leftarrow \begin{bmatrix} Q_C \\ F_2 \end{bmatrix} \qquad (8.93)$$

Let us design a multivariable GPC controller with the following setting: $H_c = \begin{bmatrix} 3 & 3 \end{bmatrix}$, and $H_p = \begin{bmatrix} 90 & 90 \end{bmatrix}$ corresponding to a 15 min prediction horizon. The weighting matrix $\mathbf{Q}$ was determined such that the varying interval (different scales) of the corresponding input and output variables was taken into account, resulting in $q_1 = 278$ and $q_2 = 0.01$ where $q_1$ are the diagonal elements of $\mathbf{Q}$ corresponding to $C_F$ and $q_2$ the elements corresponding to $P$. The control weighting was set to zeros. An integral noise model was assumed for both outputs ( $C_{i,i}(q^{-1}) = A_{i,i}(q^{-1})$ $i = 1, 2$; $C_{i,j}(q^{-1}) = 0$ for $i \neq j$).

The plant was simulated using the differential equation model (Appendix B). In addition, an unmeasured disturbance (25% drop in fuel power) effected the process at $t = 55$ min. Simulation results are shown in Fig. 8.1.

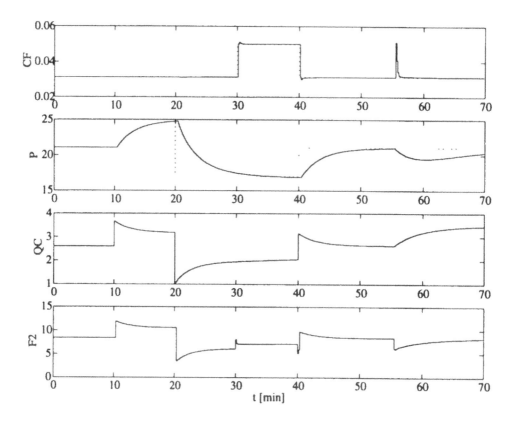

Figure 8.1: Multivariable GPC of an FBC. The upper plots show the process outputs: flue gas oxygen $C_F$ $\left[\frac{Nm^3}{Nm^3}\right]$ and combustion power $P$ $[MW]$. The dashed line indicates the targets. The lower plots show the manipulated variables: fuel feed $Q_C$ $\left[\frac{kg}{s}\right]$ and secondary air $F_2$ $\left[\frac{Nm^3}{s}\right]$. At $t = 55$ min an unmeasured disturbance affects the process.

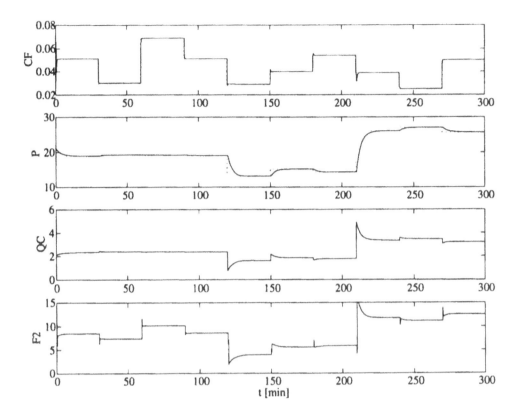

Figure 8.2: Multivariable GPC of an FBC. See legend of Fig. 8.1 for notation. Measured steady states (see Appendix B) were used as target values.

In a second simulation, the steady-states measured from the true plant (see Appendix B) were used as reference targets. Fig. 8.2 illustrates these simulations.

# Chapter 9

# Time-varying and Non-linear Systems

In this chapter, some aspects of the control of time-varying and non-linear systems are considered. The field of control of non-linear systems is wide. The aim of this chapter is to give the interested reader, with the help of illustrative examples, a flavor of the problems encountered and the solutions available. First, a brief introduction to adaptive control is given and the two main approaches of gain scheduling and indirect adaptive control are presented. In non-linear control, the Wiener and Hammerstein systems are of particular interest, as the non-linear control problem can be reformulated such that linear control design methods can be applied. A general approach for Wiener and Hammerstein systems, based on the availability of an inverse of the static part, is introduced, and illustrated *via* a simulation example using Wiener GPC for the control of a pH neutralization process. Then a special case of second order Hammerstein systems is considered. This chapter is concluded by a presentation of a 'pure' non-linear predictive control approach, using SNN and optimization under constraints. The control method is illustrated using two examples concerned with a fermentor and a tubular reactor.

## 9.1 Adaptive control

Let us use the following definition for *adaptive control* systems [38], p. 362, as a starting point:

**Definition 14 (Adaptive control)** Adaptive control systems adapt (adjust) their behavior to the changing properties of controlled processes and their signals.

Basically, there are two main motivations for adaptive control:

- *non-linear* processes, and

- *time-varying* processes.

In real life, all industrial processes exhibit non-linear time-varying behavior. Adaptive control may need to be considered for a process that *changes with time*, or when the process is non-linear to the extent that *one set of control system parameters is not sufficient* to adequately describe the process over its operating region [85].

A major assumption in conventional control design is that the underlying processes are *linear time-invariant* (LTI) dynamical systems. Control design is almost always based on linear descriptions of the process to be controlled, or on the assumption of linearity of the process in its operating region. This is due to the relative easiness of the identification of linear models, as well as to the availability of analytical results in the derivation of control laws based on linear process descriptions. A compete non-linear theory does not exist, and linear design methods work rather well even when applied to non-linear processes.

## Non-linear processes

All industrial processes, however, are inherently non-linear. Non-linearities may be due to constraints, saturations or hysteresis in the process variables (such as upper and lower bounds of the position of an actuator, or an overflow in a tank.) Typically these non-linearities occur at the boundaries of the operation areas of the process. Non-linearities may also be present during the normal operation of the process due to the non-linearity of the process phenomena, for example transport phenomena (transfer heat by conduction or by radiation, *etc.*). These non-linearities are typically smooth and close-to-linear, which justifies the use of linear approximations.

A certain linear model (say, model A) may be valid in the neighborhood if its operating point (a), or in a part of the operating area, and a controller may be designed based on the model description. When the operating point is changed, the model may no longer match with the process. Instead, another linear model (model B) describes well the behavior of the process in the new operating point (point b). Consequently, the controller needs to be redesigned (using the model B) in order to maintain satisfactory behavior of the controlled process.

## Time-varying processes

A major part of conventional system theory is based upon the assumption that the systems have constant coefficients. This assumption of time-invariance is fundamental to conventional design procedures. In real life, however, all industrial processes exhibit time-varying behavior. The properties of the process and/or its signals change with time due to component wearing, changes in process instrumentation, updates in process equipment, failures, *etc.* When changes in the process are significant, the controller needs to be re-designed in order to maintain satisfactory behavior of the controlled process.

With time-varying processes, the model parameters need to be updated on-line. On-line identification may be performed continuously, so that the model is updated at each sample instant. Alternatively, the model may be updated at certain times 'when necessary'. The necessity for identification may be indicated from outside of the system (*e.g.*, by a process operator), or sought out by the adaptive control system itself (*e.g.*, triggered by passing a certain value of an index of performance). (Note how linear control based on on-line identified linear models can be applied to a wide variety of processes, including non-linear time-varying plants.)

An *adaptive system* is able to adjust to the current environment: to gain information about the current environment, and to use it. An adaptive system is memoryless in the sense that it is not able to store this information for later use; all new information replaces the old one and only the current information is available. A *learning system*, instead, is able to recognize the current environment and to recall previously learned associated information from its memory. A learning system is an adaptive system with memory. Learning then means that the system adapts to its current environment and stores this information to be recalled later. Thus, one may expect that a learning system improves its behavior with time; an adaptive system merely adjusts to its current environment.

## 9.1.1 Types of adaptive control

Methods of adaptive control are commonly categorized into three classes:

- gain scheduling,

- indirect adaptive control, and

- direct adaptive control.

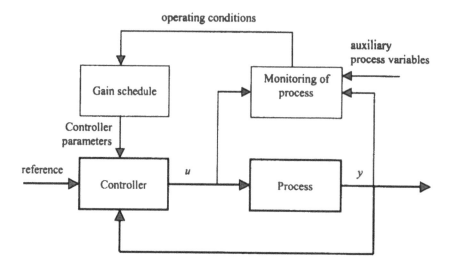

Figure 9.1: Gain scheduling.

## Gain scheduling

In *gain scheduling*, Fig. 9.1, the controller parameters are computed beforehand for each operating region. The computation of controller parameters may be based on a known non-linear model linearized at each operating point, or on linear models identified for each operating region. The model parameters (or, rather, the pre-computed controller parameters) are then tabulated. A *scheduling variable* permits to select which parameter values to use, the tabulated information is then applied for control. Since there is no feedback from the closed loop signals to the controller, gain scheduling is feed forward adaptation.

In gain scheduling, the process operating conditions are monitored, possibly using some auxiliary process variables. Based on the observed operating conditions, pre-computed controller parameters are selected using the 'gain schedule', and then used in process control. The controller is switched between the pre-computed settings, as the operating parameters vary. The name 'gain scheduling' is due to a historical background, since the scheme was originally used to accommodate for changes in the process gain only.

The design of gain scheduling can be seen as consisting of two steps:

- finding suitable scheduling variables, and

- designing of the controller at a number of operating conditions.

The main task in the design is to find suitable scheduling variables. This is normally done based on physical knowledge of the system. In process control, the production rate can often be chosen as a scheduling variable, since time constants and time delays are often inversely proportional to production rate. When the scheduling variables have been found, the controller is designed at a number of operating conditions, and the controller parameters are stored for each specific operating region. The stability and performance of the system are typically evaluated by simulation, with particular attention given to the transition between different operating conditions.

The application of gain scheduling is usually a straightforward process. An advantage of gain scheduling is that the controller parameters can be changed very quickly in response to process changes, since there is no estimation involved. The lack of estimation also brings about the main drawback of the system: there is no feedback to compensate for an incorrect schedule. Gain scheduling is feed foward adaptation (or open-loop adaptation): There is no feedback from the performance of the closed-loop to the controller parameters.

## Indirect adaptive control

*Indirect adaptive controllers* [5][38] try to attain an optimal control performance, subject to the design criterion of the controller and to the obtainable information on the process. The indirect adaptive control scheme is illustrated in Fig. 9.2. Three stages can be distinguished:

- the *identification* of the process (in closed loop);

- the *controller design*; and

- the adjustment of the controller.

Conceptually, an indirect adaptive control scheme is simple to develop. A control design procedure is taken that is based on the use of a process model; the chosen controller design procedure is automated; and the procedure is applied every time a new process model has been identified. Thus there exists a large number of different adaptive indirect (self-tuning) controllers, based on different identification procedures and control laws. The model parameters are estimated in real time, the estimates are then used as if they were equal to the true ones (certainty equivalence principle) and the uncertainties of the estimates are, in general, not concerned.

The indirect adaptive control has been shown to yield good results in practice. Unfortunately, analysis of adaptive control systems is difficult due to the interaction between the controller design and the parameter estimation. This

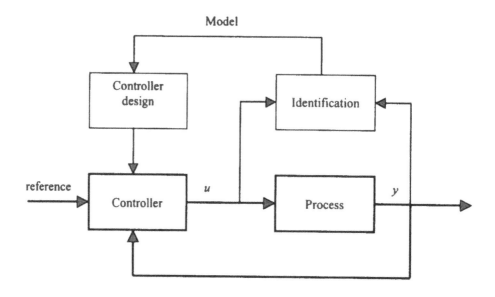

Figure 9.2: Indirect adaptive self-tuning control.

interaction can, however, play a key role in determining the convergence and stability of the adaptive control. Often, this problem is handled by looking at the states of the adaptive control as separated into two categories which change at different rates. This introduces the idea of two time scales: the fast scale is the ordinary feed back, and the slower one is for updating the controller parameters.

In *direct adaptive control* methods, the controller parameters are directly identified based on data (without first identifying a model of the process).

## 9.1.2   Simulation example

Let us look at a simulation example of the performance of an adaptive version of the generalized predictive control. In GPC, an explicit process model is required, *i.e.* a model structure needs to be selected [delay $d$ and orders of model polynomials $A\left(q^{-1}\right)$ and $B\left(q^{-1}\right)$]. In addition, the coefficients of $A$ and $B$ need to be determined. In an adaptive version of GPC, the model parameters are updated using on-line measurement information. The updated process model is then instantly used in minimizing the GPC cost function. Thus, a typical indirect adaptive controller is obtained.

## Process

Consider a linear process (from [13]) described by its Laplace transform:

$$\frac{y(s)}{u(s)} = \frac{1}{1 + 10s + 40s^2} \tag{9.1}$$

The process can be written as a discrete-time model:

$$y(k) = \frac{0.0114 + 0.0106q^{-1}}{(1 - q^{-1})(1 - 1.7567q^{-1} + 0.7788q^{-2})}\Delta u(k-1) \tag{9.2}$$

In the following simulations, a sudden change at sampling instant $k = 100$ in the process is considered, so that the process gain is reduced to one-fifth $(B(q^{-1}) = 0.0023 + 0.0021q^{-1})$ of the initial gain.

## GPC with a fixed process model

The initial process model was assumed to be known (correct structure and parameters). The parameters of the GPC were set to $H_m = 1$ (minimum output horizon), $H_p = 10$ (prediction horizon), $H_c = 1$ (control horizon) and $r = 0$ (control weighting). Fig. 9.3 depicts the resulting mean-level control with fixed parameters. Clearly, after $k = 100$, the mean-level type of control performance is not obtained anymore. Instead, a significant overshoot appears and the settling time increases.

## Adaptive GPC

In a second simulation, the parameters of the process model were updated using RLS with exponential forgetting (forgetting factor $\lambda = 0.99$). The evolution of the parameters in $B(q^{-1})$ during estimation is illustrated in Fig. 9.4. The updated model was then used in the GPC computations. The result of the control is shown in Fig. 9.5. The original design specifications are fulfilled even if the process changes with time.

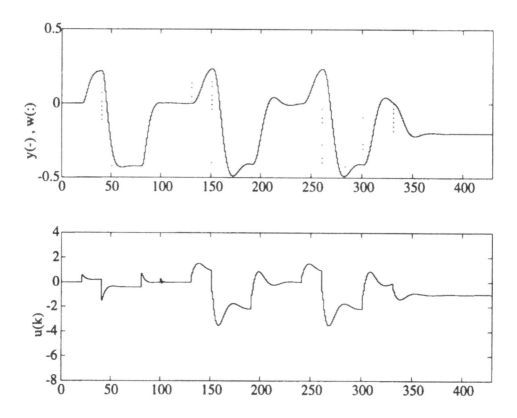

Figure 9.3: GPC using a process model with constant parameters.   At $k = 100$, the process gain is decreased abrubtly to one-fifth of the original. The performance of the designed mean-level GPC deteriorates at sampling instants $k > 100$ with large overshoots and a longer settling time.

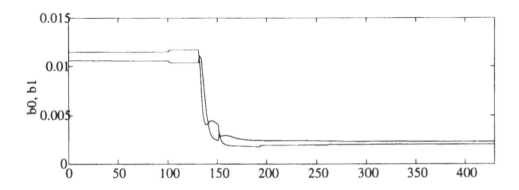

Figure 9.4: The coefficients in B polynomial are correctly estimated. A relatively slow estimator was designed, with equivalent memory horizon equal to 100 samples.

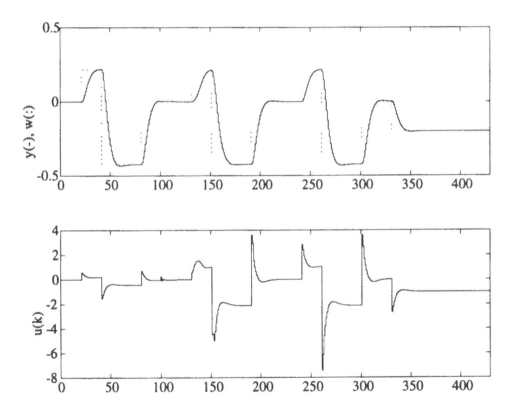

Figure 9.5: With adaptive GPC, the performace of the GPC remains as designed for the original process, even when the process gain changes (at $k = 100$).

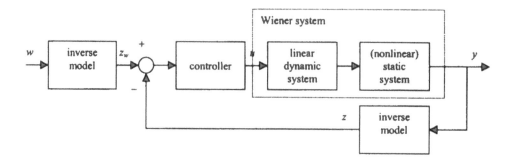

Figure 9.6: Schematic drawing of the control of a Wiener system based on the inverse of the model of the static nonlinear part.

## 9.2    Control of Hammerstein and Wiener systems

For a given process, the predictive control problem can be tackled in many ways. This section provides a very simple and interesting approach for the design of control strategies on the basis of Hammerstein and Wiener models.

The control based on non-linear models such as the Hammerstein or Wiener models can be simplified and reduced to the design of controllers for linear systems. Figures 9.6–9.7 illustrate the control structures for Wiener and Hammerstein systems. Provided that the inverse of the non-linear static part exists and is available[71], the remaining control problem is linear. Thus any linear control design method (such as the GPC, for example) can be applied in a straightforward way.

What is required is the model of the inverse static system. In many cases, it is simplest to identify the inverse model directly from input–output data (process inverse). Alternatively, the static (forward) non-linearity can be identified and its inverse mapping then identified (model inverse). If a forward static model is available, the inverse may also be solved 'on-line' by iterating (on-line solution of model inverse). In some cases it may also be possible to have the inverse model from other sources, such as first principle-based process models, *etc.*

We will next illustrate this *large scale linearization* approach with a simulated Wiener GPC example.

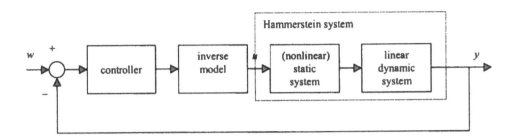

Figure 9.7: Schematic drawing of the control of a Hammerstein system based on the inverse of the model of the static nonlinear part.

## 9.2.1 Simulation example

A MISO Wiener model for the pH neutralization process is identified, and generalized predictive control (GPC) applied for the control of the process. The control of this process has been studied in a number of papers, *e.g.,* [61][68][28].

### Process and data

Acid ($q_1$), buffer ($q_2$) and base ($q_3$) streams are mixed in a tank and the effluent pH is measured. The process model [61] consists of three non-linear ordinary differential equations and a non-linear output equation for the pH ($pH_4$):

$$\frac{dh}{dt} = \frac{1}{A}\left(q_1 + q_2 + q_3 - C_v\left(h + z\right)^n\right) \tag{9.3}$$

$$\frac{dW_{a_4}}{dt} = \frac{1}{Ah}\left[\left(W_{a_1} - W_{a_4}\right)q_1 + \left(W_{a_2} - W_{a_4}\right)q_2 + \left(W_{a_3} - W_{a_4}\right)q_3\right] \tag{9.4}$$

$$\frac{dW_{b_4}}{dt} = \frac{1}{Ah}\left[\left(W_{b_1} - W_{b_4}\right)q_1 + \left(W_{b_2} - W_{b_4}\right)q_2 + \left(W_{b_3} - W_{b_4}\right)q_3\right] \tag{9.5}$$

$$0 = W_{a_4} + 10^{pH_4 - 14} + W_{b_4}\frac{1 + 2 \times 10^{pH_4 - pK_2}}{1 + 10^{pK_1 - pH_4} + 10^{pH_4 - pK_2}} - 10^{pH_4} \tag{9.6}$$

where $h$ is the liquid level in the tank, and $W_{a_4}$ and $W_{b_4}$ are the reaction invariants of the effluent stream. Table 9.1 gives the nominal values used in the simulations.

| variable | value |
|---|---|
| tank area $A$ | 207 cm$^2$ |
| valve coefficient $C_v$ | 8.75 |
| log of equilibrium constant $pK_1$ | 6.35 |
| log of equilibrium constant $pK_2$ | 10.25 |
| reaction invariant $W_{a_1}$ | $3 \times 10^{-3}$ M |
| reaction invariant $W_{a_2}$ | $-3 \times 10^{-2}$ M |
| reaction invariant $W_{a_3}$ | $-3.05 \times 10^{-3}$ M |
| reaction invariant $W_{b_1}$ | 0 M |
| reaction invariant $W_{b_2}$ | $3 \times 10^{-2}$ M |
| reaction invariant $W_{b_3}$ | $5 \times 10^{-5}$ M |
| time delay $\theta$ | 0 min |
| acid flowrate $q_1$ | 16.6 ml/s |
| buffer flowrate $q_2$ | 0.55 ml/s |
| base flowrate $q_3$ | 15.6 ml/s |
| liquid level in tank $h$ | 14 cm |
| effluent pH $pH_4$ | 7.0 |
| vertical distance between outlet and bottom of tank $z$ | 0 cm |
| valve exponent $n$ | 0.5 |
| reaction invariant $W_{a_4}$ | $-4.32 \times 10^{-4}$ M |
| reaction invariant $W_{b_4}$ | $5.28 \times 10^{-4}$ M |

Table 9.1: Estimated parameters of the linear dynamic block.

| | $b_0$ | $b_1^*$ | $a_1$ | $a_2$ |
|---|---|---|---|---|
| $\dfrac{B_1}{A_1}$ | 0.2763 | (−0.0757) | −0.6737 | −0.1257 |
| $\dfrac{B_2}{A_2}$ | 0.2230 | (0.2799) | 0.0509 | −0.5480 |
| $\dfrac{B_3}{A_3}$ | 0.2329 | (−0.0653) | −0.8200 | −0.0123 |

Table 9.2: Identified parameters of the linear dynamic transfer polynomials.

Training data of 500 samples was generated by simulating the model (9.3)–(9.6). The input signal consisted of pseudo-random sequences for each input, with a maximum amplitude of 50% of the nominal value. A test set of 500 data patterns was generated in a similar way.

## Model structure and parameter estimation

A Wiener model was identified using a SNN of 5 hidden nodes for the static part, the linear dynamics were identified using $n_{B_1} = n_{B_2} = n_{B_3} = 1$, $n_{A_1} = n_{A_2} = n_{A_3} = 2$ , and $d_1 = d_2 = d_3 = 1$. Model inputs consisted of the three input flows to the tank, $q_1$, $q_2$ and $q_3$, the system output was the pH at the outlet. For reference purposes, also a Wiener model with a linear static part was identified ($y = \mathbf{K}^T \mathbf{z}$). Parameters were estimated using the Levenberg–Marquardt method.

## Results on identification

The performance of the identified model is illustrated in Figs. 9.8–9.10. Fig. 9.8 shows the performance on training set. The Wiener model output follows closely the output of the true plant. Simulation on test set, Fig. 9.9, reveals that the mapping is not perfect. This is also indicated by the root-mean-squared errors on training set ($RMSE = 0.1192$) and test sets ($RMSE = 0.2576$). However, a reasonably accurate description of the pH process was obtained. The static mapping is illustrated in Fig. 9.10, showing the titration curves for each input flow, when other flows have their nominal values. Table 9.2 shows the estimated parameters of the transfer polynomials.

## Control design

The objective was to control the pH ($pH_4$) in the tank by manipulating the base flow rate ($q_3$). In order to fulfill the objective, a GPC controller was designed for the plant.

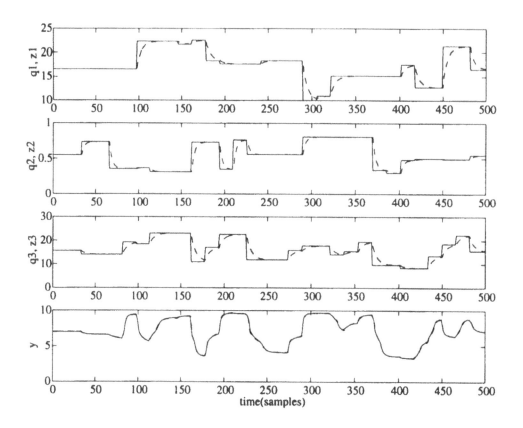

Figure 9.8: Training data for the pH neutralization model. The system inputs are shown in the three upper plots. The system output is the lower plot (solid line). The predicted output (dashed line) follows closely the training data; dashed lines on upper plots show the corresponding intermediate variables.

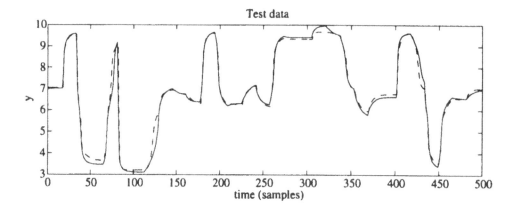

Figure 9.9: Performance on test data.

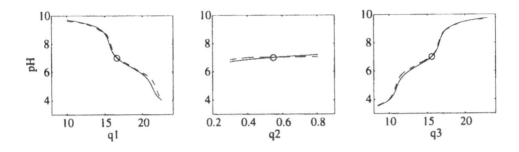

Figure 9.10: True (solid line) and identified (dashed line) static mappings.

For Wiener and Hammerstein systems, a linear control design can be accomplished if the inverse of the non-linear static part is available. In the Wiener system, when the process output and the target are mapped through an inverse non-linear mapping of the static part, the remaining system is a linear one (see Fig. 9.6). In the simulations, the inverse problem was solved on-line: The process model and a Gauss–Newton search were used in order to find a $z_3$ (SNN input) such that $\widehat{y}(k)$ (SNN output) equals $y(k)$ (measured output)[1]. The control problem was then based on the error between the desired output $z_w(k)$ and the intermediate signal $z_3$ so that the desired performance characteristics were fulfilled.

The GPC cost function is of the form

$$ J = \sum_{i=H_m}^{H_p} \left[ z(k+i) - z_w(k+i) \right]^2 + r \sum_{i=0}^{H_c} \left[ \Delta u(k+i) \right]^2 \qquad (9.7) $$

where $z_w(k+i)$ are the desired future responses. In the case of Wiener systems, the $z_w$ and $z$ can be obtained from the desired and measured process outputs $w$ and $y$, by solving the inverse of the static mapping.

## Control simulations

First, a dead-beat controller was designed for the system: $H_m = \deg B + d + 1 = 3$, $H_c = \deg A + 1 = 3$, $H_p \geq \deg A + \deg B + 1$ ($H_p = 8$) and $r = 0$. The ideal response is shown in Fig. 9.11, obtained using the identified Wiener model both in the controller and as the simulated plant. The upper part of Fig. 9.11 shows the control signal $u$ and the intermediate signal $z$. The

---

[1]Note that in this case the (global) inverse does not need to exist, as the local solution (depending on the initial value of the search) was found. For the considered process, the global inverse would exist and could be identified.

lower part shows the target $w$ and the plant output $y$. The desired target sequence consisted of four step changes to the process (at $t = \{375, 750, 1125, 1500\}$ seconds). At $t = \{1875, 2250, 2625\}$ seconds three disturbances affect the process (unmeasured $\pm 10\%$ changes in $q_1$). As shown by the simulation responses, the ideal dead beat response is fast and accurate. Note, that the steady-state gain of the $u$–$z$ controller is one. However, the magnitude and rate of the control input signal makes it unrealistic for many real process control applications. (Note how in the simulation the input $q_3$ was restricted to be non-negative which effects the control at $t = 375$, Fig. 9.11.) A correct estimate of the I/O behavior of the process also needs to be available.

In order to get an implementable control signal, it is common in GPC to decrease the control horizon $H_c$ (often $H_c = 1$). In mean-level control ($H_c = 1$, $H_p$ large) the closed loop will have open loop plant dynamics; with smaller $H_p$ a tighter control is obtained. Alternatively (not excluding), a realizable controller can be obtained by introducing a non-zero value for the parameter $r$. Fig. 9.12 shows the simulation when using the true process, (9.3)–(9.6), dead beat control settings, and $r = 0.1$. This parameter setting results in a relatively smooth control input, and a small overshoot. The small deviations from unit static gain are due to the process–model mismatch. Due to the integral action in GPC, there is no steady-state error.

For comparison, a GPC controller based on the linear pH model was experimented. The simulations are shown in Fig. 9.13. Let us assume that the desired performance was a small overshoot and smooth control actions, as in Fig. 9.12. Then the system response at $pH = 7$ is as desired. At $pH = 9$, the system is overdamped, however, and at $pH = 5$ the system is underdamped. Clearly, the changes in the gain of the system affect the control, and the closed-loop performance changes depending on the operating point. These design difficulties were avoided in the Wiener control approach.

**Results**

In the example, identification and control of a pH neutralization process were considered. First, a MISO Wiener model was identified for the process. Using the identified model, a GPC controller was designed. Simulations showed good results.

Simplicity of non-linear process control is one of the main motivations for Wiener and Hammerstein systems. Provided that the inverse of the static part is available, linear control design methods (*e.g.*, pole placement) can be directly applied for the control of the linear subsystem. In the example we showed that an explicit inverse model is not always necessary. (Note that in some cases a global inverse may not exist). Although fixed parameter models

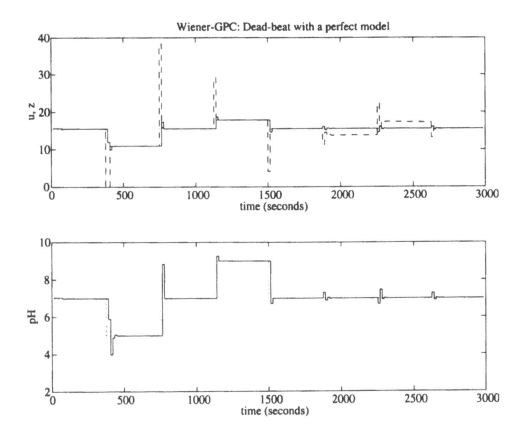

Figure 9.11: Ideal dead beat control. Upper part: controller output $u$ (dashed line) and 'measured' intermediate signal $z$. Lower part: plant output $y$ (solid line) and target $w$ (dotted line).

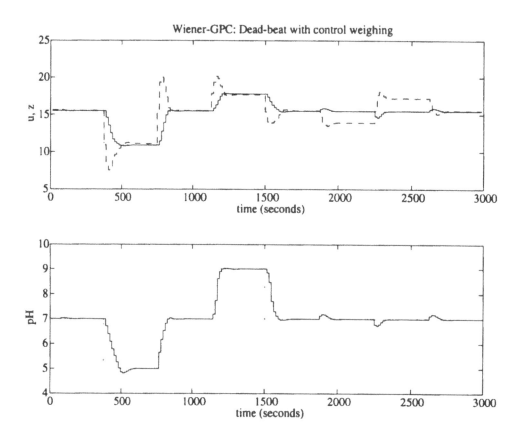

Figure 9.12:  Simulation of Wiener-GPC control of the pH neutralization process.

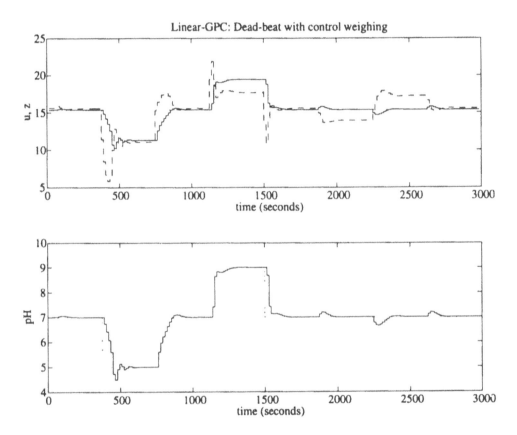

Figure 9.13: Simulation of linear GPC control of the pH neutralization process.

were applied in the example, adaptive control applications are straightforward to conduct, provided that robustness of the closed-loop learning system can be guaranteed. From this point of view, the Wiener and Hammerstein approaches do not pose any additional difficulties.

## 9.2.2   Second order Hammerstein systems

In this sub-section, we will consider the special case of predictive control for a second order Hammerstein system.

### Second order Hammerstein model

In order for a control system to function properly, it should be unduly insensitive to inaccuracies in the process model. We shall be concerned with a class of SISO discrete time Hammerstein models

$$A\left(q^{-1}\right)y\left(k\right) = \sum_{p=1}^{P} B_p\left(q^{-1}\right)u^p\left(k-1\right) + \frac{\xi(k)}{\Delta\left(q^{-1}\right)} \tag{9.8}$$

where $B_p\left(q^{-1}\right)$ is a polynomial of degree $n_{B_p}$. This model belongs to the following class

$$A\left(q^{-1}\right)y\left(k\right) = B\left(q^{-1}\right)\mathrm{f}\left(u\left(k\right)\right) + \frac{\xi(k)}{\Delta\left(q^{-1}\right)} \tag{9.9}$$

where $\mathrm{f}(\cdot)$ is a non-linear function.

Let us introduce the following auxiliary (pseudo) input [42][43]

$$x\left(k\right) = \sum_{p=1}^{P} B_p\left(q^{-1}\right)u^p\left(k\right). \tag{9.10}$$

The process model (9.8) will be rewritten as follows:

$$A\left(q^{-1}\right)y\left(k\right) = q^{-1}x\left(k\right) + \frac{\xi(k)}{\Delta\left(q^{-1}\right)} \tag{9.11}$$

Let us derive a predictive controller for the special case $P = 2$:

$$
\begin{aligned}
y\left(k\right) &= \frac{1}{A\left(q^{-1}\right)}\left[B_1\left(q^{-1}\right)u\left(k-1\right) + B_2\left(q^{-1}\right)u^2\left(k-1\right)\right] \\
&\quad + \frac{1}{A\left(q^{-1}\right)\Delta\left(q^{-1}\right)}\xi(k)
\end{aligned} \tag{9.12}
$$

## Prediction

In order to separate the available information from the unavailable (separate past and future noise), let us consider the following polynomial identity (see Chapter 3, Section 3.3.6)

$$\frac{1}{A\left(q^{-1}\right)\Delta\left(q^{-1}\right)} = E_j\left(q^{-1}\right) + q^{-j}\frac{F_j\left(q^{-1}\right)}{A\left(q^{-1}\right)\Delta\left(q^{-1}\right)} \tag{9.13}$$

from which we have

$$E_j\left(q^{-1}\right)A\left(q^{-1}\right)\Delta\left(q^{-1}\right) = 1 - q^{-j}F_j\left(q^{-1}\right) \tag{9.14}$$

Multiplying both sides of the model (9.12) by $q^j E_j\left(q^{-1}\right)A\left(q^{-1}\right)\Delta\left(q^{-1}\right)$ and substituting (9.14) leads to

$$
\begin{aligned}
y\left(k+j\right) =\ & G_{1,j}\left(q^{-1}\right)\Delta\left(q^{-1}\right)u\left(k+j-1\right) \\
& +G_{2,j}\left(q^{-1}\right)\Delta\left(q^{-1}\right)u^2\left(k+j-1\right) \\
& +F_j\left(q^{-1}\right)y\left(k\right) + E_j\left(q^{-1}\right)\xi(k+j)
\end{aligned}
\tag{9.15}
$$

where

$$
\begin{aligned}
G_{i,j}\left(q^{-1}\right) =\ & E_j\left(q^{-1}\right)B_i\left(q^{-1}\right) \\
=\ & g_{j,0}^i + g_{j,1}^i q^{-1} + \ldots + g_{j,n_{bi}+j-1}^i q^{n_{bi}+j-1}.
\end{aligned}
\tag{9.16}
$$

Since the degree of the polynomial $E_j\left(q^{-1}\right)$ is $j-1$, the noise components $E_j\left(q^{-1}\right)\xi(k+j)$ are all in the future and since $\xi(k)$ is assumed to be white, the prediction $\hat{y}\left(k+j\right)$ of $y\left(k+j\right)$ in the mean squares sense is given by

$$
\begin{aligned}
\hat{y}\left(k+j\right) =\ & G_{1,j}\left(q^{-1}\right)\Delta u\left(k+j-1\right) \\
& +G_{2,j}\left(q^{-1}\right)\Delta u^2\left(k+j-1\right) \\
& +F_j\left(q^{-1}\right)y\left(k\right).
\end{aligned}
\tag{9.17}
$$

Notice that the prediction $\hat{y}\left(k+j\right)$ depends upon: i) past and present measured outputs; ii) past known control increments; and iii) present and future control increments yet to be determined.

Let us denote by $f\left(k+j\right)$ the component of the prediction $\hat{y}\left(k+j\right)$ composed of signal known (available) at sampling instant $k$. For example, the expressions of $f\left(k+1\right)$ and $f\left(k+m\right)$ are respectively given by

$$
\begin{aligned}
f\left(k+1\right) =\ & \left[G_{1,1}\left(q^{-1}\right) - g_{1,0}^1\right]\Delta u\left(k\right) \\
& + \left[G_{2,1}\left(q^{-1}\right) - g_{1,0}^2\right]\Delta u^2\left(k\right) \\
& +F_1\left(q^{-1}\right)y\left(k\right)
\end{aligned}
\tag{9.18}
$$

$$f(k+m) = \left[ G_{1,m}\left(q^{-1}\right) - g_{m,0}^1 - g_{m,1}^1 q^{-1} - \cdots - g_{m,m-1}^1 q^{-(m-1)} \right] \Delta u(k)$$
$$+ \left[ G_{2,m}\left(q^{-1}\right) - g_{m,0}^2 - g_{m,1}^2 q^{-1} - \cdots - g_{m,m-1}^2 q^{-(m-1)} \right] \Delta u^2(k)$$
$$+ F_m\left(q^{-1}\right) y(k). \tag{9.19}$$

Let us rewrite equation (9.17) for $j = 1, ..., H_p$ in the following matrix form

$$\widehat{\mathbf{Y}} = \mathbf{G}_1 \mathbf{U} + \mathbf{G}_2 \mathbf{U}^2 + \mathbf{F} \tag{9.20}$$

where

$$\widehat{Y} = \begin{bmatrix} \widehat{y}(k+1) \\ \widehat{y}(k+2) \\ \vdots \\ \widehat{y}(k+H_p) \end{bmatrix} \tag{9.21}$$

$$\mathbf{U} = \begin{bmatrix} \Delta u(k) \\ \Delta u(k+1) \\ \vdots \\ \Delta u(k+H_p-1) \end{bmatrix} = \begin{bmatrix} u_0 \\ u_1 \\ \vdots \\ u_{H_p-1} \end{bmatrix} \tag{9.22}$$

$$\mathbf{U}^2 = \begin{bmatrix} \Delta u^2(k) \\ \Delta u^2(k+1) \\ \vdots \\ \Delta u^2(k+H_p-1) \end{bmatrix} \tag{9.23}$$

$$\mathbf{F} = \begin{bmatrix} f(k+1) \\ f(k+2) \\ \vdots \\ f(k+H_p) \end{bmatrix} \tag{9.24}$$

$$\mathbf{G}_i = \begin{bmatrix} g_{1,0}^i & 0 & \cdots & & & & 0 \\ g_{2,1}^i & g_{2,0}^i & 0 & & & & 0 \\ g_{3,2}^i & g_{3,1}^i & g_{3,0}^i & 0 & & & \\ \vdots & & & \ddots & 0 & & \vdots \\ g_{m,m-1}^i & g_{m,m-2}^i & \cdots & g_{m,1}^i & g_{m,0}^i & 0\cdot & \\ \vdots & & & & & \ddots & 0 \\ g_{H_p,H_p-1}^i & g_{H_p,H_p-2}^i & & \cdots & & & g_{H_p,0}^i \end{bmatrix} \tag{9.25}$$

$i = 1, 2$. We denote by $\mathbf{g}_k^i$ the $k^{th}$ column of the matrices $\mathbf{G}_i$ $(i = 1, 2)$

$$\mathbf{G}_i = \begin{bmatrix} \mathbf{g}_0^i & \mathbf{g}_1^i & \cdots & \mathbf{g}_{H_p-1}^i \end{bmatrix} \tag{9.26}$$

## Cost function

In what follows, we shall be concerned with the minimization of the following control objective

$$J = E\left\{\left[\sum_{j=1}^{H_{\mathrm{p}}}[\hat{y}(k+j) - w(k+j)]^2 + \sum_{j=1}^{H_{\mathrm{c}}} r\left[\Delta u(k+j-1)\right]^2\right] | k\right\} \quad (9.27)$$

which with (9.20), yields

$$\begin{aligned} J &= E\left\{(\mathbf{G}_1\mathbf{U} + \mathbf{G}_2\mathbf{U}^2 + \mathbf{F} - \mathbf{W})^T (\mathbf{G}_1\mathbf{U} + \mathbf{G}_2\mathbf{U}^2 + \mathbf{F} - \mathbf{W}) \quad (9.28)\\ &\quad + r\mathbf{U}^T\mathbf{U}\right\} \end{aligned}$$

where

$$\mathbf{W} = [w(k+1) \ \cdots \ w(k+H_{\mathrm{p}})]^T. \quad (9.29)$$

Let us set

$$\begin{aligned} \mathbf{V} &= \mathbf{G}_1\mathbf{U} + \mathbf{G}_2\mathbf{U}^2 + \mathbf{F} - \mathbf{W} \quad (9.30)\\ &= u_0\mathbf{g}_0^1 + u_1\mathbf{g}_1^1 + \dots + u_{H_{\mathrm{p}}-1}\mathbf{g}_{H_{\mathrm{p}}-1}^1\\ &\quad + u_0^2\mathbf{g}_0^2 + u_1^2\mathbf{g}_1^2 + \dots + u_{H_{\mathrm{p}}-1}^2\mathbf{g}_{H_{\mathrm{p}}-1}^2 + \mathbf{F} - W \end{aligned}$$

Then (9.28) is equivalent to

$$J = E\left\{\mathbf{V}^T\mathbf{V} + r\mathbf{U}^T\mathbf{U}\right\} \quad (9.31)$$

## Minimization of cost function

To minimize this quadratic cost function, we have to calculate the gradient of $J$ with respect to the control increments $u_i$ and their squares $u_i^2$ ($i = 1, ..., H_{\mathrm{p}} - 1$).

$$\frac{\partial J}{\partial u_i} = 2\mathbf{V}^T\frac{\partial \mathbf{V}}{\partial u_i} + 2ru_i. \quad (9.32)$$

From (9.30), it follows that the gradient $\frac{\partial \mathbf{V}}{\partial u_i}$ is given by

$$\frac{\partial \mathbf{V}}{\partial u_i} = \mathbf{g}_i^1 + 2u_i\mathbf{g}_i^2 \quad (9.33)$$

which leads to

$$\frac{\partial J}{\partial u_i} = 2\mathbf{V}^T\mathbf{g}_i^1 + 4u_i\mathbf{V}^T\mathbf{g}_i^2 + 2ru_i \quad (9.34)$$

The partial derivative of the criterion $J$ with respect to $\mathbf{U}$

$$\frac{\partial J}{\partial \mathbf{U}} = \left[\begin{array}{cccc} \frac{\partial J}{\partial u_0} & \frac{\partial J}{\partial u_1} & \cdots & \frac{\partial J}{\partial u_{H_p-1}} \end{array}\right] \qquad (9.35)$$

may be written as

$$\begin{aligned}
\frac{\partial J}{\partial \mathbf{U}} =\ & \left[\begin{array}{cccc} 2\mathbf{V}^T\mathbf{g}_0^1 & 2\mathbf{V}^T\mathbf{g}_1^1 & \cdots & 2\mathbf{V}^T\mathbf{g}_{H_p-1}^1 \end{array}\right] \\
& + \left[\begin{array}{cccc} 4u_0\mathbf{V}^T\mathbf{g}_0^2 & 4u_1\mathbf{V}^T\mathbf{g}_1^2 & \cdots & 4u_{H_p-1}\mathbf{V}^T\mathbf{g}_{H_p-1}^2 \end{array}\right] \\
& + \left[\begin{array}{cccc} 2ru_0 & 2ru_{10} & \cdots & 2ru_{H_p-1} \end{array}\right]
\end{aligned} \qquad (9.36)$$

This gradient can be rewritten in a more compact form as

$$\frac{\partial J}{\partial \mathbf{U}} = 2\mathbf{V}^T\mathbf{G}_1 + 4\mathbf{V}^T\mathbf{G}_2\operatorname{diag}(\mathbf{U}) + 2r\mathbf{U}^T \qquad (9.37)$$

The current control law is given as the first element of the vector $\mathbf{U}$, obtained by equating the gradient $\frac{\partial J}{\partial \mathbf{U}}$ to zero, *i.e.*,

$$\frac{\partial J}{\partial \mathbf{U}} = 2\mathbf{V}^T\mathbf{G}_1 + 4\mathbf{V}^T\mathbf{G}_2\operatorname{diag}(\mathbf{U}) + 2r\mathbf{U}^T = 0. \qquad (9.38)$$

The complexity of this expression makes it intractable to calculate the optimal control increment. It is not easy to derive the analytical expression of the current control. In the sequel, we shall present Newton's method for the computation of this current control. For this purpose, let us calculate the elements of the Hessian

$$\frac{\partial^2 J}{\partial \mathbf{U}^2} = \left[\begin{array}{cccc} \frac{\partial^2 J}{\partial u_0^2} & \frac{\partial^2 J}{\partial u_0 u_1} & \cdots & \frac{\partial^2 J}{\partial u_0 u_{H_p-1}} \\ \vdots & & & \vdots \\ \frac{\partial^2 J}{\partial u_{H_p-1}u_0} & \frac{\partial^2 J}{\partial u_{H_p-1}u_1} & \cdots & \frac{\partial^2 J}{\partial u_{H_p-1}^2} \end{array}\right]. \qquad (9.39)$$

From (9.32), we derive

$$\frac{\partial^2 J}{\partial u_i u_j} = \frac{\partial}{\partial u_i}\left(\frac{\partial J}{\partial u_j}\right) = \frac{\partial J}{\partial u_i}\left[2\mathbf{V}^T\frac{\partial \mathbf{V}}{\partial u_j} + 2ru_j\right] \qquad (9.40)$$

$$= \frac{\partial}{\partial u_i}\left[2\mathbf{V}^T\frac{\partial \mathbf{V}}{\partial u_j}\right] + 2r\delta_{i,j} \qquad (9.41)$$

where $\delta_{i,j}$ represents the Kronecker symbol. By (9.33), we obtain

$$\frac{\partial^2 J}{\partial u_i u_j} = \frac{\partial}{\partial u_i}\left[2\mathbf{V}^T\mathbf{g}_j^1 + 4u_j\mathbf{V}^T\mathbf{g}_j^2\right] + 2r\delta_{i,j} \qquad (9.42)$$

$$= 2\left(\frac{\partial \mathbf{V}^T}{\partial u_i}\right)\left[\mathbf{g}_j^1 + 2u_j\mathbf{g}_j^2\right] + 4\delta_{i,j}\mathbf{V}^T\mathbf{g}_j^2 + 2r\delta_{i,j} \qquad (9.43)$$

Again, taking (9.33) into account, we deduce that

$$\frac{\partial^2 J}{\partial u_i u_j} \tag{9.44}$$

$$= 2 \left[ \mathbf{g}_i^1 + 2u_i \mathbf{g}_i^2 \right]^T \left[ \mathbf{g}_j^1 + 2u_j \mathbf{g}_j^2 \right] + 4\delta_{i,j} \mathbf{V}^T \mathbf{g}_j^2 + 2r\delta_{i,j} \tag{9.45}$$

$$= 2 \left[ \left( \mathbf{g}_i^1 \right)^T \mathbf{g}_j^1 + 2u_i \left( \mathbf{g}_i^2 \right)^T \mathbf{g}_j + 2u_i \left( \mathbf{g}_i^1 \right)^T \mathbf{g}_j^2 + 4u_i u_j \left( \mathbf{g}_i^2 \right)^T \mathbf{g}_j^2 \right] \tag{9.46}$$

$$+ 4\delta_{i,j} \mathbf{V}^T \mathbf{g}_j^2 + 2r\delta_{i,j}$$

Finally, the second partial derivative of the criterion $J$ with respect to $\mathbf{U}$ is given by

$$\frac{\partial^2 J}{\partial \mathbf{U}^2} = 2\mathbf{G}_1^T \mathbf{G}_2 + 4\mathrm{diag}\left(\mathbf{U}\right) \left[ \mathbf{G}_2^T \mathbf{G}_1 \right] + 4 \left[ \mathbf{G}_1^T \mathbf{G}_2 \right] \mathrm{diag}\left(\mathbf{U}\right) \tag{9.47}$$

$$+ 8\mathrm{diag}\left(\mathbf{U}\right) \left[ \mathbf{G}_2^T \mathbf{G}_2 \mathrm{diag}\left(\mathbf{U}\right) \right] + 4\mathrm{diag}\left( \mathbf{V}^T \mathbf{G}_2 \right) + 2r\mathbf{I}$$

The optimal control increment can be obtained iteratively on the basis of the Hessian (9.47).

Even if the considered model (second order Hammerstein model) is relatively simple, the development of the long-range predictive control strategy requires many calculations which alter considerably the inherent robustness of long-range predictive controllers.

The next section is dedicated to the development of predictive control strategies for both unconstrained and constrained systems. This development is based on stochastic approximation techniques and sigmoid neural networks. Any non-linear basis function approach can also be used.

# 9.3 Control of non-linear systems

Several studies on the use of SNN (sigmoid neural network) as the basis for model predictive controllers (with finite prediction horizons), have been published [82][77][91] [81]. To avoid the use of locally linearized models of the process to be controlled (the classical GPC approach), and complex optimization techniques, we present a simple solution of deriving long-range predictive control based on SNN [66][2]. The design of this solution is based on the training of two dynamic neural networks. The NOE and the NARX

---

[2]This section is based on K. Najim, A. Rusnak, A Meszaros and M. Fikar. Constrained Long-Range Predictive Control Based on Artificial Neural Networks. *International Journal of Systems Science*, 28(12): 1211–1226, 1997. Reproduced with permission from Taylor & Francis Ltd.

neural networks are respectively used as a multi-step predictor and for calculating the control signal (neural controller). The multi-layer feed forward SNN is trained so as to achieve the control objective. The main idea presented in this study concerns the use of stochastic recursive approximation techniques as learning tool for the design of neural networks controllers to solve both unconstrained and constrained predictive control problems (minimization of a long-range quadratic cost function and preventing violations of process constraints). The control approach described below is general and does not depend on the structure of the control objective and the constraints.

### 9.3.1  Predictive control

The formulation is based on a NARIMAX model, and on the minimization of the conditional expectation of a quadratic function measuring the control effort and the distance between the predicted system output and some predicted reference sequence over the receding horizon, *i.e.*

$$J = E \left\{ \sum_{j=H_m}^{H_p} (w(k+j) - y(k+j))^2 + r \sum_{j=1}^{H_p} (\Delta u(k+j-1))^2 \middle| k \right\} \quad (9.48)$$

where $y$, $\Delta u$, and $w$ are the controlled variable, future control increments, and set point, respectively. $H_m$, and $H_p$, are respectively, the minimum and the maximum prediction horizon. The weighting factor $r$ serves for penalization of the future control increments $\Delta u$.

In what follows, the use of neural networks (see Chapter 5) for prediction and control is considered.

### 9.3.2  Sigmoid neural networks

Consider the problem concerning the design of an algorithm which at time $k$ predicts simultaneously the outcomes of the process $\{y(k)\}$ at time $k$, $k+1$, $k+2$, ..., $k+H_p$ where $H_p$ is the prediction horizon. NARX SNN, using one-step-ahead structure, generally perform poorly over a trajectory (prediction horizon) because errors are amplified when inaccurate network outputs are recycled to the input layer. Recurrent SNNs are more appropriate for application in model predictive control [94], and in this study we have used a NOE SNN as a multi-step predictor. Unlike NARX SNN where information flows only from input layer to the outputs, recurrent SNN include delayed information flow back to preceding layers. An NOE SNN predictor is depicted in Fig. 9.14. The SNN inputs consisted of the previous and current values of process inputs and predicted plant outputs. The values of process outputs come into the NOE SNN only indirectly in the process of training when the future predicted output is compared with the actual process output. The training process of this SNN predictor was carried out using a back propagation through the time algorithm [94].

A multilayer feed forward SNN was used as a controller. The proposed structure of this SNN controller is depicted in Fig. 9.15. The controller inputs consisted of the plant predictions which are provided by the predictor and the desired value of the plant output. The outputs correspond to the present and future increments of the control signal. The weights of this SNN controller were updated directly using a stochastic approximation algorithm, which minimizes the control objective (9.48) subject to constraints (any control objective can be considered in this control approach). These weights are considered as the controller parameters. Since control action is based on the prediction of the plant behavior, offset can occur due to disturbances and model mismatch when the SNN is used as a dynamic model of the controlled process. Therefore the plant output is predicted at each sampling time as follows:

$$y(k) = \mathrm{f}(\mathbf{U}, \boldsymbol{\theta}) + d(k) \qquad (9.49)$$

where $\mathbf{U}$ is the SNN input vector, $\boldsymbol{\theta}$ is a vector of the weights to be optimized, and $d$ is a disturbance. This disturbance (correction) of the prediction is computed by the following equation:

$$d(k + i) = d(k) = y(k) - \widehat{y}(k) \qquad (9.50)$$

where $y$ is the current value of the plant output and $\widehat{y}$ is the prediction of $y$ generated by the SNN predictor. The disturbance is assumed to be constant over the prediction horizon.

The general schematic diagram of the predictive controller is depicted in Fig. 9.16. At each sampling period, the signals are fed into the SNN predictors:

- past and present plant outputs,

- past values of control actions applied to the process, and

- the calculated future control sequence from the last sampling period.

The predictor calculates predictions of the plant outputs over the relevant horizon. These are corrected with the calculated deviation between actual process output and predicted process output at time $k$. Next the predictions are fed, together with the set point value (or sequence of future set point

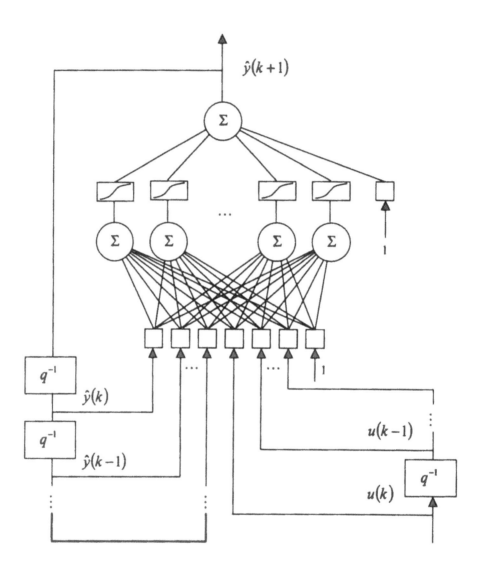

Figure 9.14: Structure of a sigmoid neural network (SNN) predictor.

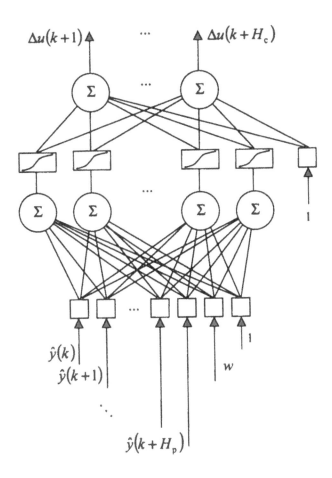

Figure 9.15: Structure of a SNN controller.

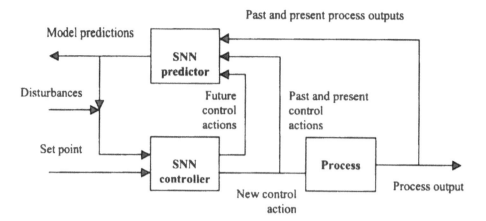

Figure 9.16: Structure of a predictive control system using sigmoid neural networks.

values for programmed set points), into the SNN controller. This minimizes the criterion (9.48), constructs the future control increments, and closes the inner loop. This procedure is repeated until the future control increments converge. The algorithm used for training this neural network controller is described in the following section.

### 9.3.3   Stochastic approximation

In this section our main emphasis will be on stochastic approximation techniques [44]. The synthesis of the neural network controller is formulated as the determination of their associated weights which minimize the unconstrained (constrained) control objective function J, *i.e.*,

$$\theta = \arg \min_{\theta} J(\theta) \qquad (9.51)$$

where $\theta$ is the weight vector. The optimization problem (9.51) can be solved using stochastic approximation techniques.

A key feature of many practical control problems is the presence of constraints on both controlled and manipulated variables. Inequality constraints arise commonly in process control problems due to physical limitations of plant equipment. For example, the control objective may be to minimize some quadratic cost function while satisfying constraints of product quality and avoiding undesirable operating regimes (flooding in a liquid-liquid extraction column, *etc.*).

Let us consider the following constrained optimization problem:

$$\min_{\theta} J_0(\boldsymbol{\theta}) \tag{9.52}$$

under the constraints

$$J_i(\boldsymbol{\theta}) \leq 0, \quad (i = 1, ..., m) \tag{9.53}$$

$J_0(\boldsymbol{\theta})$ is associated with the control objective defined by (9.48). The constraints $J_i(\boldsymbol{\theta}) \leq 0$, ( $i = 1, ...$ , $m$) are usually associated with process physical limitations of valves, reactor volume, *etc.*

Let us introduce the Lagrange function [93] defined by

$$L(\boldsymbol{\theta}, \Phi) = J_0(\boldsymbol{\theta}) + \sum_{j=1}^{m} \phi_j J_j(\boldsymbol{\theta}) \tag{9.54}$$

where $\Phi = \left[ \phi_1, ..., \phi_m \right]^T$ is the Kuhn–Tucker vector. To solve the optimization problem (9.52)–(9.53), an iterative algorithm based on stochastic approximation techniques was proposed by Walk [93]. This algorithm maximizes simultaneously $L(\boldsymbol{\theta}, \Phi)$ with respect to $\Phi$ and minimizes it with respect to $\boldsymbol{\theta}$. It is presented below:

Let the estimates $(\boldsymbol{\theta}_k, \Phi_k)$ be available at time $k$ where $\boldsymbol{\theta}_k$ is a $P$-dimensional random vector, $\Phi_k = (\phi_{k,i})$ $i = 1, ..., m$ is an $m$-dimensional random vector with $\phi_{k,i} \geq 0$ $(k \in N)$ on a probability space $(\theta, \Psi, R)$. Let $a_k, c_k$ $(k \in N)$ be real positive sequences tending to zero and satisfying:

$$\sum a_k^2 c_k^{-2} < \infty, \quad \sum a_k = \infty, \quad \sum a_k c_k < \infty \tag{9.55}$$

The observation noise (the contamination of function values) is modeled by a square integrable real random variables $V_{k,l}^i$ $(i = 0, ..., m; l = 1, ..., P; k \in N)$. The optimization algorithm is given by [93]:

$$\Phi_{k+1} = \Phi_k + a_k \tilde{D}_\phi L(\boldsymbol{\theta}_k, \Phi_k) \tag{9.56}$$

where

$$\tilde{D}_\phi L(\boldsymbol{\theta}_k, \Phi_k)_i = \max \left\{ J_i(\boldsymbol{\theta}_k) - V_k^i, -\frac{\phi_{k,i}}{a_k} \right\}, \quad (i = 1, ..., m) \tag{9.57}$$

and

$$\boldsymbol{\theta}_{k+1} = \boldsymbol{\theta}_k - a_k \tilde{D}_\theta L(\boldsymbol{\theta}_k, \Phi_k) \tag{9.58}$$

where

$$(\tilde{D}_{\theta}L(\theta_k, \Phi_k))_l$$
$$= (2c_k)^{-1}\left[J_0(\theta_k + c_k e_l) - J_0(\theta_k - c_k e_l) - V_{k,l}^0\right] + \qquad (9.59)$$
$$+ \sum_{j=1}^{m} \varphi_{k,j}(2c_k)^{-1}\left[J_j(\theta_k + c_k e_l) - J_j(\theta_k - c_k e_l) - V_{k,l}^j\right]$$

$e_l$ is a $P$-dimensional null-vector with 1 as $l$'th coordinate $(l = 1, ., P)$. With the $\sigma$-algebra $\Psi_k$ defined as follows

$$\Psi_k = \mathcal{F}(\theta_1, \Phi_1, V_1^i, ..., V_{n-1}^i (i = 1, ..., m), \qquad (9.60)$$
$$V_{1,l}^i, ..., V_{k-1,l}^i (i = 0, ..., m; l = 1, ..., P))$$

it is assumed:

$$\forall\, k,l,i \; E\left\{V_{k,l}^i / \Psi_n\right\} = 0 \qquad (9.61)$$

and

$$\forall\, l,i \; + \sum a_k(c_k)^{-2} E\left\{\left(V_{k,l}^i\right)^2\right\} < \infty, \qquad (9.62)$$
$$\forall\, i \; + \sum (a_k)^2 E\left\{\left(V_k^i\right)^2\right\} < \infty \qquad (9.63)$$

Under these assumptions, it has been shown [93] that this algorithm converges almost surely to the optimal solution.

In this algorithm, the components of the gradients of the Lagrange function toward the weights $\theta$ and the Kuhn–Tucker parameters $\Phi$ are estimated by finite differences. The convergence of this algorithm has been proved by Walk [93] as well as a central limit theorem with convergence order $k^{-1/4}$ which is also achieved for the Kiefer–Wolfowitz method to which the considered algorithm reduces if there are no constraints.

To demonstrate the performance and the feasibility of this approach we applied it to control a continuous-flow, stirred biochemical reactor and a fixed bed tubular chemical reactor.

## 9.3.4    Control of a fermenter

Control problems in biotechnological processes have gained increasing interest because of the great number of applications, mainly in the pharmaceutical industry and biological depollution [62]. We considered a model of a continuous-flow, stirred biochemical reactor (fermenter). This model, which describes the growth of *saccharomyces cerevisiae* on glucose with continuous feeding, was adopted from [58]. It is based on a hypothesis in [89]: a limited oxidation capacity leading to formation of ethanol under conditions of oxygen limitation or an excessive glucose concentration.

## Process

The dynamic model is derived from mass balance considerations. It is described by the following equations: Cell mass concentration

$$\frac{dc_x}{dt} = D_1(c_{x,\text{in}} - c_x) + \mu c_x \tag{9.64}$$

glucose (substrate) concentration

$$\frac{dc_s}{dt} = D_1(c_{s,\text{in}} - c_s) - Q_s c_x \tag{9.65}$$

ethanol (product) concentration

$$\frac{dc_e}{dt} = D_1(c_{e,\text{in}} - c_e) + (Q_{e,\text{pr}} - Q_e)c_x \tag{9.66}$$

carbon dioxide concentration

$$\frac{dc_c}{dt} = D_g(c_{c,\text{in}} - c_c) + Q_c c_x \tag{9.67}$$

dissolved oxygen concentration

$$\frac{dc_o}{dt} = D_1(c_{o,\text{in}} - c_o) + Na - Q_o c_x \tag{9.68}$$

gas phase oxygen concentration

$$\frac{dc_g}{dt} = D_g(c_{g,\text{in}} - c_g) - Na\frac{V_1}{V_g} \tag{9.69}$$

where

$$D_1 = \frac{q_1}{V_1} \text{ and } D_g = \frac{q_g}{V_g} \tag{9.70}$$

The mathematical description of the kinetic model mechanisms is given in Table 9.3. The model parameters are given in Table 9.4. The initial conditions for equations (9.64)–(9.69) are given in Table 9.5.

The fermenter model was simulated using the Runge–Kutta method. The bioprocess static behavior is depicted in Fig. 9.17. One can get information about the non-linear behavior of the bioprocess by looking at the variation of the steady-state gain depicted in Fig. 9.17.

The experiments described here illustrate the use of the predictive control algorithm when applied to controlling the dissolved oxygen concentration, $c_o(t)$, using the dilution rate, $D_g(t)$, as the manipulated variable.

| Mechanism | Description |
|---|---|
| Glucose uptake | $Q_s = Q_{s,\max}\frac{c_s}{k_s+c_s}$ |
| Oxidation capacity | $Q_{o,\lim} = Q_{o,\max}\frac{c_o}{k_o+c_o}$ |
| Oxidative ethanol metabolism | $Q_{s,ox} = \min\begin{cases} Q_s \\ Y_{os}Q_{o,\lim} \end{cases}$ |
| Reductive glucose metabolism | $Q_{s,red} = Q_s - Q_{s,ox}$ |
| Ethanol uptake | $Q_e = Q_{e,\max}\frac{c_e}{k_e+c_e}\frac{k_1}{k_1+c_s}$ |
| Oxidative ethanol metabolism | $Q_{e,ox} = \min\begin{cases} Q_e \\ (Q_{o,\lim}-Q_{s,ox}Y_{so})Y_{oe} \end{cases}$ |
| Ethanol production | $Q_{s,pr} = Y_{se}Q_{s,red}$ |
| Growth | $\mu = Y_{sx}^{ox}Q_{s,ox} + Y_{sx}^{red}Q_{s,red} + Y_{ex}Q_e$ |
| Carbon dioxide production | $Q_c = Y_{sc}^{ox}Q_{s,ox} + Y_{sc}^{red}Q_{s,red} + Y_{ec}Q_e$ |
| Oxygen consumption | $Q_o = Y_{so}Q_{s,ox} + Y_{eo}Q_{e,ox}$ |
| Oxygen transfer | $Na = k_{La}(\frac{c_s}{m} - c_o)$ |
| Maximum consumption rates | $\frac{dQ_{i,\max}}{dt} = \frac{1}{t_i}(Q_{i,\max}^p f_{ic} - Q_{i,\max})$ |
| where induction or repression factors are | $f_{sc} = \frac{c_s}{k_n+c_s}$ |
| | $f_{ec} = \frac{c_e}{k_n+c_e}\frac{k_1}{k_1+c_s}\frac{c_o}{k_o+c_o}$ |

Table 9.3: The kinetic model mechanism description.

| | |
|---|---|
| $k_e = 2.2\ 10^{-3}$mol/l | $Y_{ec} = 0.68$ mol/mol |
| $k_1 = 5.6\ 10^{-4}$mol/l | $Y_{eo} = 1.28$ mol/mol |
| $k_m = 1.7\ 10^{-4}$mol/l | $Y_{ex} = 1.32$ C-mol/mol |
| $k_n = 3.6\ 10^{-4}$mol/l | $Y_{sc} = 1.88$ mol/mol |
| $k_o = 3.0\ 10^{-6}$mol/l | $Y_{so} = 2.27$ mol/mol |
| $k_s = 5.6\ 10^{-4}$mol/l | $Y_{sc}^{ox} = 2.35$ mol/mol |
| $k_{La} = 592$ h$^{-1}$ | $Y_{sc}^{red} = 1.89$ mol/mol |
| $Q_{e,\max}^p = 0.13$ mol/(C-mol.h) | $Y_{sx}^{ox} = 3.65$ C-mol/mol |
| $Q_{s,\max}^p = 0.50$ mol/(C-mol.h) | $Y_{sx}^{red} = 0.36$ C-mol/mol |
| $Q_{o,\max}^p = 0.20$ mol/(C-mol.h) | $\tau_e = 2.80$ h |
| $m = 35$ mol/mol | $\tau_o = 1.60$ h |
| $V_l = 0.70$ l | $\tau_s = 2.50$ h |
| $D_g = 1.0$ h$^{-1}$ | $c_{e,in} = c_{c,in} = c_{o,in} = 0.0$ mol/l |
| $c_{s,in} = c_{g,in} = 1.0\ 10^{-3}$mol/l | $\tau_s = 2.50$ h |

Table 9.4: Model parameters. (1 C-mol of biomass has the composition $CH_{1.83}N_{0.17}O_{0.56}$).

| $c_c(0) = 1.5510^{-3}$mol/l | $c_e(0) = 1.010^{-3}$mol/l |
|---|---|
| $c_g(0) = 8.7210^{-5}$mol/l | $c_o(0) = 2.4310^{-6}$mol/l |
| $c_a(0) = 2.4210^{-4}$mol/l | $c_x(0) = 1.010^{-3}$C-mol/l |

Table 9.5: Bioprocess initial conditions.

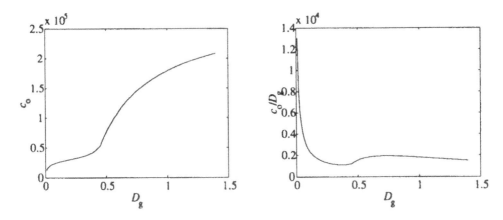

Figure 9.17: Static behavior of dissolved oxygen $c_o$ [mol/l] (left) and static gain (right) as a function of the dilution rate $D_g$ [h$^{-1}$].

The structure of the SNN predictor used was $[6, 5, 1]$: six neurons in the input layer with inputs $[y(k-1), y(k-2), y(k-3), u(k), u(k-1), u(k-3)]$, five neurons in the hidden layer and one neuron in the output layer. The sampling period was set equal to 0.5 hours. The training set contained 600 input–output pairs. The structure of the SNN controller used was $[14, 8, 4]$, and the inputs consisted of the predictions of the process behavior obtained using the SNN predictor.

Apart from predictor and controller parametrization (*i.e.* choice of the number of nodes), there still remain a few design parameters that must be specified *a priori*, *i.e.* the prediction horizon, the control weighting factor, and setting the values of the parameters involved in the stochastic approximation algorithm. The following choices were made:

- the horizons related to the control objective were set equal to $H_m = 1, H_p = 13$.

- the weighting factor $r$ in the control objective was fixed to $r = 0.1$.

- the setting values of the parameters involved in the stochastic approximation algorithm were fixed to $a_k = 0.3$, $c_k = 0.01$.

Notice that $a_k$ must decrease in order to remove the influence of disturbances, according to (9.55). In the noise free case, $a_k$ can be either constant or decreasing sequence that converges to a constant value.

## Control simulations

For the first set of tests, the future reference was considered to be a known square wave, as shown in the upper graph of Fig. 9.18. Figure 9.19 gives an enlargement of Fig. 9.18 for the first 200 hours of the simulation run. The lower graph of Fig. 9.18 represents the control signal $u(k)$ derived from the predictive control calculation. Figure 9.18 and Figure 9.19 show the performance of the control. It can be seen that both steady-state and transient behavior are satisfactory. The variation of the control variable is very smooth. Notice that the set point changes led to the variation of the bioprocess dynamics (steady-state gain change, *etc.*).

The second experiment considers level constraints on the input. The following constraints on the manipulated variable, dilution rate $D_g$, were used: $D_{g,min} \leq D_g(k) \leq D_{g,max}$, $D_{g,min} = 0.4, D_{g,max} = 0.85$. Figure 9.20 shows the evolution of the dissolved oxygen concentration, $c_o(k)$, and the dilution rate, $D_g(k)$. The evolution of the bioprocess output as well as the control variable for the first 200 hours of operation are depicted in Fig. 9.21. These simulation results show that the bioprocess operates well under the constrained control.

In the third experiment, a rate constraint on the input was considered, $|\Delta D_g(k)| \leq 0.1$. Figure 9.22 shows the evolution of the dissolved oxygen concentration $c_o(k)$ and the dilution rate $D_g(k)$. Figure 9.23 gives an enlargement of Fig. 9.22 for the first 200 hours of this simulation run. There are a large number of set point changes. Some of them occur randomly (see Fig. 9.22 for $k \geq 200$ hours). Due to the fulfillment of the constraint, the control signal has no large variations, thus it corresponds to industrial requirements.

In practice, it is impossible to obtain perfect measurements and uniform dissolved oxygen concentrations. We therefore introduced measurement noise $\xi(k)$ with zero mean and variance equal to 0.02. Figure 9.24 shows the dissolved oxygen concentration and the dilution rate. The control action is smooth. These results demonstrate that the presented control algorithm has good regulation and tracking properties. Next, we will consider the control of a chemical reactor.

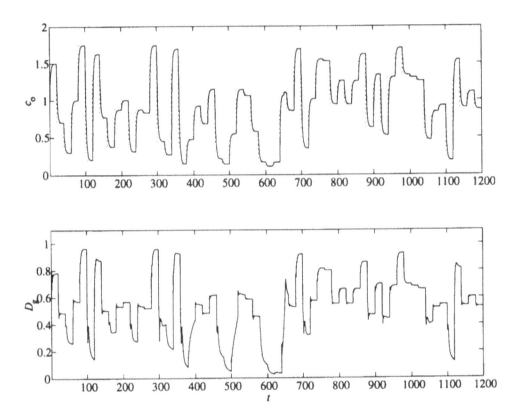

Figure 9.18: Predictive control of a fermenter: unconstrained case. Top: the measured and desired dissolved oxygen, $c_o$, as a function of time [h]. Bottom: the manipulated variable, $D_g$.

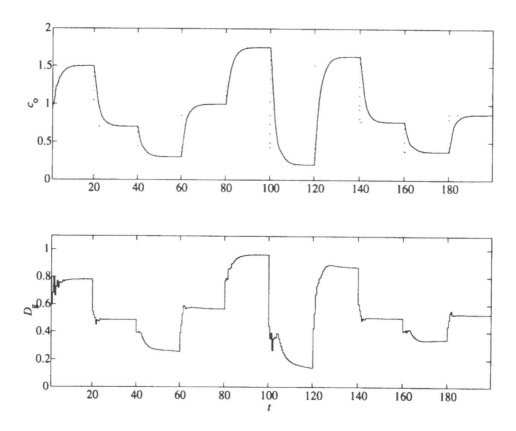

Figure 9.19: Enlargement of Fig. 9.18 for the first 200 hours of operation.

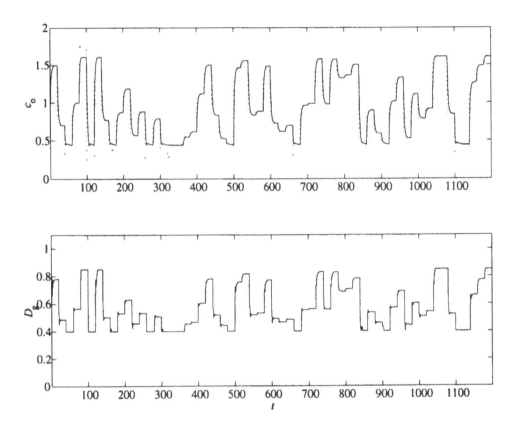

Figure 9.20: Predictive control of a fermenter: level constraint on the manipulated variable, $0.4 \leq D_g \leq 0.85$. Top: the measured and desired dissolved oxygen, $c_o$, as a function of time [h]. Bottom: the manipulated variable, $D_g$.

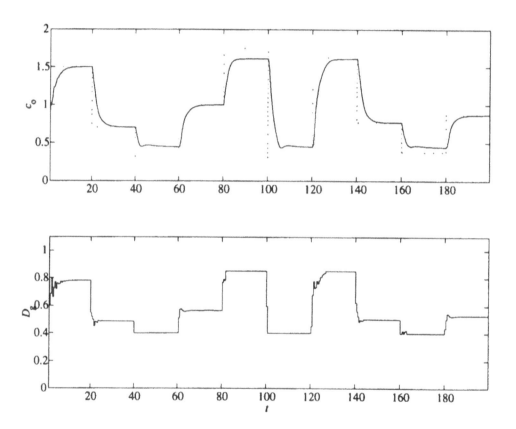

Figure 9.21: Enlargement of Fig. 9.20 for the first 200 hours of operation.

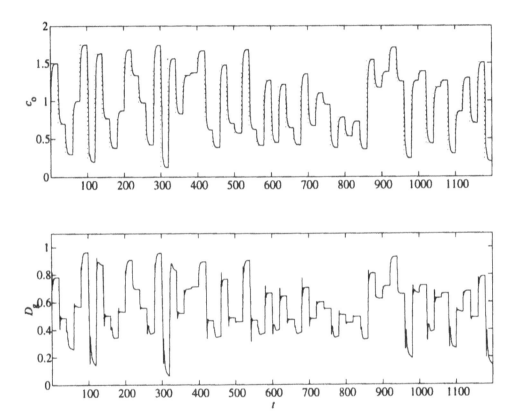

Figure 9.22: Predictive control of a fermenter: rate constraint on the manipulated variable, $|D_g| \leq 0.1$. Top: the measured and desired dissolved oxygen, $c_o$, as a function of time [h]. Bottom: the manipulated variable, $D_g$.

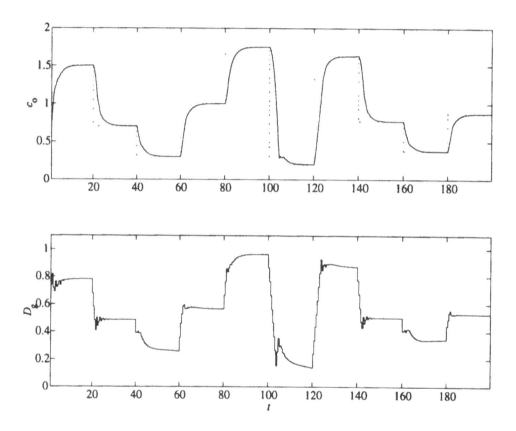

Figure 9.23: Enlargement of Fig. 9.22 for the first 200 hours of operation.

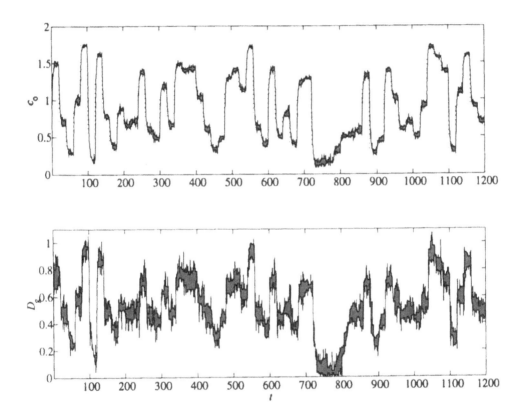

Figure 9.24: Predictive control of a fermenter: noisy measurements. Top: the measured and desired dissolved oxygen, $c_o$, as a function of time [h]. Bottom: the manipulated variable, $D_g$.

## 9.3.5 Control of a tubular reactor

In a second example we consider the temperature control in a tubular chemical reactor. A tubular reactor is a significant and widely used piece of equipment in chemical technology. The object of our study is such a reactor with fixed-bed catalyst and cooling. Efficient control of this type of process is often hampered by its highly non-linear behavior and hazardous operating conditions.

Assuming that $j_1$ reversible exothermic first-order reactions take place in the reactor and some specified simplifying circumstances hold, a structured non-linear mathematical model of the process can be developed [59]. The model was proposed on the basis of both mass and heat balances and its final form is given by a set of non-linear hyperbolic partial differential equations, as follows. Mass balance for the $i$'th component

$$\frac{\partial c_i}{\partial t} + \nu \frac{\partial c_i}{\partial z} = -\sum_{j=1}^{j_1} r_{ij}(c_i, T_k) \qquad (9.71)$$

Energetic balance of reactant mixture

$$\frac{\partial T_g}{\partial t} + \nu \frac{\partial T_g}{\partial z} = A_1(T_k - T_g) - A_2(T_g - T_w) \qquad (9.72)$$

Energetic balance of catalyst

$$\frac{\partial T_k}{\partial t} = B_1 \left[ \sum_{j=1}^{j_1} (-\Delta H_j) r_j - B_2(T_k - T_g) - B_3(T_k - T_w) \right] \qquad (9.73)$$

Energetic balance of the reactor's wall

$$\frac{\partial T_w}{\partial t} = C_1 \left[ C_2(T_k - T_w) + C_3(T_g - T_w) - C_4(T_s - T_g) \right] \qquad (9.74)$$

The initial and boundary conditions are:

$$
\begin{aligned}
c_i(z,0) &= c_{is}(z); \; T_g(z,0) = T_{gs}(z); \; T_k(z,0) = T_{ks}(z) & (9.75) \\
T_w(z,0) &= T_{ws}(z); \; c_i(0,t) = c_{i0}(t); \; T_g(0,t) = T_{g0}(t) & (9.76)
\end{aligned}
$$

Symbols used stand for the following physical quantities: $c_i$ - concentration of the $i$'th component, $t$ - dimensionless time, $z$ - dimensionless spatial variable, $r_{ij}$ - rate of chemical reactions ($i$'th component in $j$'th reaction), $T_g$ - reactant mixture temperature, $T_w$ - wall temperature, $T_k$ - catalyst temperature, $T_c$ - coolant temperature, and coefficients $A_1$, $A_2$, $B_1 - B_3$, $C_1 - C_4$ include the technological parameters of the reactor.

For simulation, two reactions of the following rates were considered:

$$r_1(c, T_k) = 8.7 \ 10^3 \exp\left(-\frac{15200}{1.98T_k}\right) c \qquad (9.77)$$

$$r_2(c, T_k) = 4.57 \ 10^5 \exp\left(-\frac{19800}{1.98T_k}\right) c \qquad (9.78)$$

This case corresponds to the ethylene oxide production in an industrial scale reactor. The parameters involved take the following values:

$$
\begin{aligned}
A_1 &= 51.356307; \ A_2 = 23.796894; \ B_1 = 0.000614; &(9.79)\\
B_1 B_2 &= 2.301454; \ B_1 B_3 = 0.266606; \ C_1 C_2 = 0.080613 &(9.80)\\
C_1 C_3 &= 0.322451; \ C_1 C_4 = 1.048619 &(9.81)
\end{aligned}
$$

For simulation, the following values of initial and boundary conditions were considered:

$$
\begin{aligned}
c_{is}(z) &= 0.015037 \ \text{kmol m}^{-3}; \ T_{g\kappa}(z) = 522.266948 \ \text{K} &(9.82)\\
T_{k\kappa}(z) &= 526.844683 \ \text{K}; \ T_{w\kappa}(z) = 514.216915 \ \text{K} &(9.83)\\
c_{i0}(t) &= 0.015600 \ \text{kmol m}^{-3}; \ T_{g0}(t) = 499.579969 \ \text{K} &(9.84)
\end{aligned}
$$

The coolant temperature was chosen as a control variable. The goal of the control was to maintain a desired profile of the gas mixture average temperature in the reactor. Partial differential equations of the model were solved by dividing the reactor into 10 segments according to the spatial variable.

Experiments included both unconstrained and constrained cases for the control. After some 'pre-experimental' runs we found the appropriate structure for SNN predictor and SNN controller. The structure of the SNN predictor used was $[4, 6, 1]$ with inputs $[y(k), y(k-1), u(k), u(k-1)]$ and output $[y(k+1)]$. The sampling period was set equal to 0.5 min. The training data contained 700 input–output pairs. The structure of the SNN controller used was $[14, 8, 4]$. On the input of the SNN controller, the future predictions of the process behavior were applied and the controller generated the future control actions. The experiments reported here were carried out with the following choices:

- the horizons relating to the control objective were set equal to $H_m = 4$, $H_p = 13$.

- the weighting factor $r$ in the control objective was fixed to $r = 0.1$.

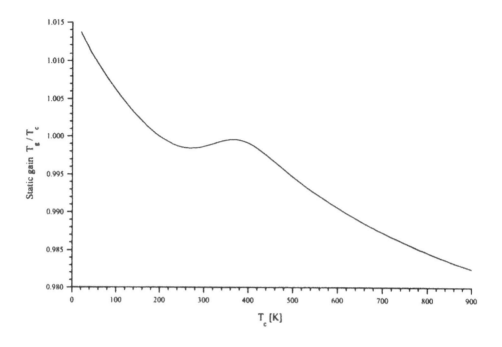

Figure 9.25: Static gain of the chemical reactor [66].

- the values of the parameters involved in the stochastic approximation algorithm were fixed to: $a_k = 0.3$, $c_k = 0.01$.

Figure 9.25 shows the non-linear steady-state (static) gain. The behavior of the controlled reactor in the unconstrained case is depicted in Fig. 9.26. The upper graph shows the reactor output, the middle graph shows the control signal. In the lower graph, the control action increments are depicted.

The behavior of the chemical reactor in the constrained case is illustrated in Fig. 9.27. In this experiment, a constraint of $|\Delta T_c(k)| < 15$ K on the control variable was considered. The description of this figure is similar to that of Fig. 9.26. From the lower graph of Fig. 9.26, it can be seen that the increments of the control signal lie in the interval $[0, 100]$ K. On the other hand, in the constrained case, the control variable increments vary within the interval $[0, 15]$ K and the control signal is smooth.

In this section, we presented some experiments concerning the implementation of the constrained predictive control algorithm based on neural networks. There are a number of other potential applications, since many industrial plants (chemical, mineral, *etc.*) are characterized by non-linear and time-varying behavior, and are subject to several kinds of disturbances.

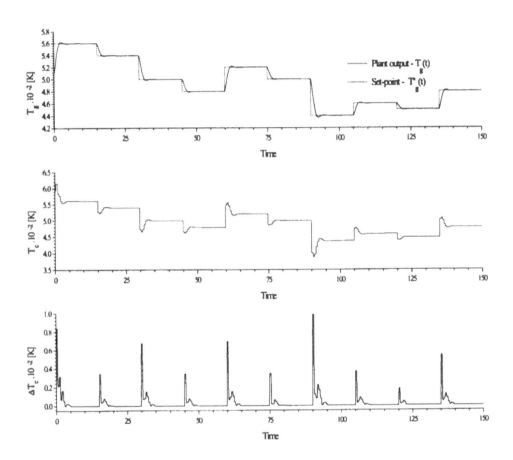

Figure 9.26: The output, desired output and the manipulated variable for predictive control [66].

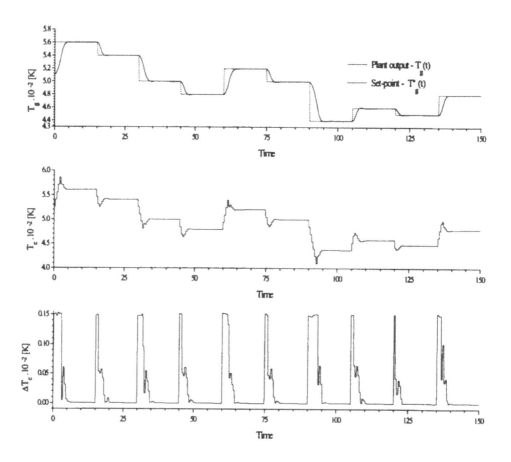

Figure 9.27: The output, desired output and the manipulated variable for the constrained predictive control with constraint on the manipulated variable of the chemical reactor, ($|\Delta T_c(k) \leq 15$ K$|$) [66].

# Part III

# Appendices

# Appendix A

# State-Space Representation

The primary purpose of this appendix is to introduce a number of concepts which are fundamental in the state-space representation and analysis of dynamic systems. We formally define and illustrate the concepts of controllability and observability.

## A.1   State-space description

Let a SISO system (plant, process) be described by a state-space model

$$
\begin{aligned}
\mathbf{x}(k+1) &= \mathbf{A}\mathbf{x}(k) + \mathbf{B}u(k) & \text{(A.1)} \\
y(k) &= \mathbf{C}\mathbf{x}(k) & \text{(A.2)}
\end{aligned}
$$

where

$\mathbf{x}$ is the state vector $(n \times 1)$,

$u$ is the system input (controller output) $(1 \times 1)$

$y$ is the system output (measured) $(1 \times 1)$

$\mathbf{A}$ is the state transition matrix $(n \times n)$

$\mathbf{B}$ is the input transition vector $(n \times 1)$

$\mathbf{C}$ is the state observer vector $(1 \times n)$

**Remark 6 (Characteristic equation)** The characteristic equation is given by

$$
\det(z\mathbf{I} - \mathbf{A}) = 0.
$$

273

**Remark 7 (Number of representations)** For a given system there exists no unique state-space representation. In fact from any state representation we can obtain a new one by using any linear transformation, i.e., $\widetilde{\mathbf{x}}(k) = T\mathbf{x}(k)$ where $T$ is a non-singular matrix.

Let us next introduce two state representations, namely, the control and observer canonical forms.

## A.1.1   Control and observer canonical forms

Consider a system given by a transfer polynomial

$$y\left(k\right) = \frac{B\left(q^{-1}\right)}{A\left(q^{-1}\right)}u\left(k\right) \tag{A.3}$$

where $B\left(q^{-1}\right) = b_1 q^{-1} + ... + b_n q^{-n}$ and $A\left(q^{-1}\right) = 1 + a_1 q^{-1} + ... + a_n q^{-n}$. For notational convenience, without loss of generality, we assume that the polynomials are all of order $n$. The *control canonical form*

$$
\begin{aligned}
\mathbf{x}_c\left(k+1\right) &= \mathbf{A}_c\mathbf{x}_c\left(k\right) + \mathbf{B}_c u\left(k\right) & \text{(A.4)}\\
y\left(k\right) &= \mathbf{C}_c\mathbf{x}_c\left(k\right) & \text{(A.5)}
\end{aligned}
$$

is obtained by substituting

$$
\mathbf{A}_c = 
\begin{bmatrix}
-a_1 & -a_2 & \cdots & -a_{n-1} & -a_n \\
1 & 0 & \cdots & 0 & 0 \\
0 & 1 & & 0 & 0 \\
\vdots & & \ddots & & \vdots \\
0 & 0 & & 1 & 0
\end{bmatrix}
\tag{A.6}
$$

$$
\mathbf{B}_c = 
\begin{bmatrix}
1 \\
0 \\
\vdots \\
0
\end{bmatrix}
\tag{A.7}
$$

$$
\mathbf{C}_c = \begin{bmatrix} b_1 & b_2 & \cdots & b_n \end{bmatrix}
\tag{A.8}
$$

The key idea is that the elements of the first row of $\mathbf{A}_c$ are exactly the coefficients of the characteristic polynomial of the system (the matrix $\mathbf{A}_c$ is known as the *companion matrix* of the polynomial $\lambda^n + a_1\lambda^{n-1} + \cdots + a_n$ (see [64]).

Similarly, the *observer canonical form*

$$\mathbf{x}_o(k+1) = \mathbf{A}_o\mathbf{x}_o(k) + \mathbf{B}_ou(k) \tag{A.9}$$
$$y(k) = \mathbf{C}_o\mathbf{x}_o(k) \tag{A.10}$$

is obtained by substituting

$$\mathbf{A}_o = \begin{bmatrix} -a_1 & 1 & 0 & \cdots & 0 \\ -a_2 & 0 & 1 & & 0 \\ \vdots & \vdots & & \ddots & \\ -a_{n-1} & 0 & 0 & & 1 \\ -a_n & 0 & 0 & \cdots & 0 \end{bmatrix} \tag{A.11}$$

$$\mathbf{B}_o = \begin{bmatrix} b_1 \\ b_2 \\ \vdots \\ b_n \end{bmatrix} \tag{A.12}$$

$$\mathbf{C}_o = \begin{bmatrix} 1 & 0 & \cdots & 0 \end{bmatrix} \tag{A.13}$$

## A.2 Controllability and observability

We will now consider now two fundamental notions concerning dynamics systems that are represented in the state-space form. The first is whether it is possible to transfer (drive, force) a system from a given initial state to any other arbitrary final state. The second is how can we observe (determine) the state of a given system if the only available information consists of input and output measurements. These concepts have been introduced by Kalman as the concepts of *controllability* and *observability*.

**Definition 15 (Controllability)** The system (A.1) is said to be completely state controllable (reachable), or simply controllable, if it is possible to find a control sequence which steers it from any initial state $\mathbf{x}(k_i)$ at any instant $k_i$ to any arbitrary final state $\mathbf{x}(k_f)$ at any instant $k_f > k_i \geq 0$. Otherwise, the system is said to be uncontrollable.

**Definition 16 (Observability)** The system (A.1) is said to be completely state observable, or simply observable, if and only if the complete state of the system can determined over any finite time interval $[k_i, k_f]$ from the available input and output measurements over the time interval $[k_i, k_f]$ with $k_f > k_i \geq 0$. Otherwise, the system is said to be unobservable.

The following theorems state the conditions under which a given system is controllable or observable.

**Theorem 2 (Controllability)** The system (A.1)–(A.2) is controllable if and only if the *controllability matrix* defined by

$$\begin{bmatrix} \mathbf{B} & \mathbf{AB} & \mathbf{A^2B} & \cdots & \mathbf{A^{n-1}B} \end{bmatrix} \tag{A.14}$$

has rank $n$.

**Theorem 3 (Observability)** The system (A.1)–(A.2) is observable if and only if the *observability matrix* defined by

$$\begin{bmatrix} \mathbf{C} \\ \mathbf{CA} \\ \mathbf{CA^2} \\ \vdots \\ \mathbf{CA^{n-1}} \end{bmatrix} \tag{A.15}$$

has rank $n$.

In order to illustrate the usefulness of the controllability notion and the states transformation, we shall consider two applications: a feedback controller based on the pole placement control approach, and the reconstruction of the state variables.

## A.2.1   Pole placement

The design of state-space feedback controllers can be seen as consisting of two independent steps: the design of the control law, and the design of an observer [21]. The final control algorithm will consist of a combination of the control law and the estimator with control-law calculations based on the estimated states.

The state feedback control law is simply a linear combination of the components of the state-space vector

$$u(k) = -\mathbf{K}\mathbf{x}(k) \tag{A.16}$$

where $\mathbf{K} = [k_1, ..., k_n]$. Substituting this to the state equation (A.1) gives

$$\mathbf{x}(k+1) = \mathbf{A}\mathbf{x}(k) - \mathbf{B}\mathbf{K}\mathbf{x}(k) \tag{A.17}$$

and its z-transform is given by

$$(z\mathbf{I} - \mathbf{A} + \mathbf{B}\mathbf{K})\mathbf{x}(z) = 0 \tag{A.18}$$

with a characteristic equation

$$\det(z\mathbf{I} - \mathbf{A} + \mathbf{B}\mathbf{K}) = 0 \tag{A.19}$$

In pole-placement, the control-law design consists of finding the elements of $\mathbf{K}$ so that the roots of (A.19), poles of the closed-loop system, are in the desired locations. The desired characteristic equation is given by

$$P(z) = z^n + \alpha_{n-1}z^{n-1} + ... + \alpha_1 z + \alpha_0 = 0 \tag{A.20}$$

In order to do this, let us consider a linear transformation defined by a matrix $\mathbf{M}$

$$\mathbf{x} = \mathbf{M}\widetilde{\mathbf{x}} \tag{A.21}$$

The system equations then become

$$\begin{aligned}
\widetilde{\mathbf{x}}(k+1) &= \mathbf{M}^{-1}\mathbf{A}\mathbf{M}\widetilde{\mathbf{x}}(k) + \mathbf{M}^{-1}\mathbf{B}u(k) \tag{A.22}\\
y(k) &= \mathbf{C}\mathbf{M}\widetilde{\mathbf{x}}(k)\\
u(k) &= -\mathbf{K}\mathbf{M}\widetilde{\mathbf{x}}(k) \tag{A.23}
\end{aligned}$$

Let us introduce the following notations [64]:

$$\begin{aligned}
\widetilde{\mathbf{A}} &= \mathbf{M}^{-1}\mathbf{A}\mathbf{M} \tag{A.24}\\
\widetilde{\mathbf{B}} &= \mathbf{M}^{-1}\mathbf{B} \tag{A.25}\\
\widetilde{\mathbf{K}} &= \mathbf{K}\mathbf{M} \tag{A.26}
\end{aligned}$$

where $\widetilde{\mathbf{A}}$ et $\widetilde{\mathbf{B}}$ have the following form

$$\widetilde{\mathbf{A}} = \begin{bmatrix} 0 & 1 & 0 & \cdots & 0 \\ 0 & 0 & 1 & & 0 \\ \vdots & & & \ddots & \\ 0 & 0 & 0 & & 1 \\ -a_0 & -a_1 & -a_2 & \cdots & -a_{n-1} \end{bmatrix} ; \ \widetilde{\mathbf{B}} = \begin{bmatrix} 0 \\ 0 \\ \vdots \\ 0 \\ 1 \end{bmatrix} \tag{A.27}$$

The matrix $\tilde{\mathbf{A}}$ is a companion matrix. The characteristic polynomial is given by

$$\det\left(z\mathbf{I} - \tilde{\mathbf{A}}\right) = z^n + a_{n-1}z^{n-1} + \dots + a_1 z + a_0 \qquad (A.28)$$

Let us denote the matrix $\mathbf{M}$ as follows:

$$\mathbf{M} = [\mathbf{m}_1, \mathbf{m}_2, \cdots, \mathbf{m}_n] \qquad (A.29)$$

where $\mathbf{m}_i$ $(i = 1, ..., n)$ represents the columns of the matrix $\mathbf{M}$. We obtain from (A.24)

$$\mathbf{M}\tilde{\mathbf{A}} = \mathbf{AM} \qquad (A.30)$$

that is,

$$
\begin{aligned}
\mathbf{m}_{n-1} - a_{n-1}\mathbf{m}_n &= \mathbf{Am}_n \\
\mathbf{m}_{n-2} - a_{n-2}\mathbf{m}_n &= \mathbf{Am}_{n-1} \\
&\vdots \\
\mathbf{m}_1 - a_1\mathbf{m}_n &= \mathbf{Am}_2 \\
-a_0\mathbf{m}_n &= \mathbf{Am}_1
\end{aligned}
\qquad (A.31)
$$

Hence,

$$
\begin{aligned}
\mathbf{m}_{n-1} &= \left(\mathbf{A} + a_{n-1}\mathbf{I}\right)\mathbf{m}_n \\
\mathbf{m}_{n-2} &= \left(\mathbf{A}^2 + a_{n-1}\mathbf{A} + a_{n-2}\mathbf{I}\right)\mathbf{m}_n \\
&\vdots \\
\mathbf{m}_1 &= \left(\mathbf{A}^{n-1} + a_{n-1}\mathbf{A}^{n-2} + \dots + a_2\mathbf{A} + a_1\mathbf{I}\right)\mathbf{m}_n \\
0 &= \left(\mathbf{A}^n + a_{n-1}\mathbf{A}^{n-1} + \dots + a_1\mathbf{A} + a_0\mathbf{I}\right)\mathbf{m}_n
\end{aligned}
\qquad (A.32)
$$

The equation (A.25)

$$\mathbf{B} = \mathbf{M}\tilde{\mathbf{B}} \qquad (A.33)$$

leads to

$$\mathbf{m}_n = \mathbf{B} \qquad (A.34)$$

From (A.32), we derive

$$
\begin{aligned}
\mathbf{m}_{n-1} &= \left(\mathbf{A} + a_{n-1}\mathbf{I}\right)\mathbf{B} \\
\mathbf{m}_{n-2} &= \left(\mathbf{A}^2 + a_{n-1}\mathbf{A} + a_{n-2}\mathbf{I}\right)\mathbf{B} \\
&\vdots \\
\mathbf{m}_1 &= \left(\mathbf{A}^{n-1} + a_{n-1}\mathbf{A}^{n-2} + \dots + a_2\mathbf{A} + a_1\mathbf{I}\right)\mathbf{B}
\end{aligned}
\qquad (A.35)
$$

The inverse of the matrix $\mathbf{M}$ exists if and only if

$$\text{rank}\left(\mathbf{B}, \mathbf{AB}, \cdots, \mathbf{A}^{n-1}\mathbf{B}\right) = n \qquad (\text{A.36})$$

This corresponds to the controllability condition.

Based on the desired characteristic equation (A.20) we derive

$$\det\left(z\mathbf{I} - \widetilde{\mathbf{A}} + \widetilde{\mathbf{B}}\widetilde{\mathbf{K}}\right) = z^n + \alpha_{n-1}z^{n-1} + \ldots + \alpha_1 z + \alpha_0 \qquad (\text{A.37})$$

We have

$$\widetilde{\mathbf{A}} - \widetilde{\mathbf{B}}\widetilde{\mathbf{K}} = \begin{bmatrix} 0 & 1 & 0 & \cdots & 0 \\ 0 & 0 & 1 & & 0 \\ \vdots & & & \ddots & \\ 0 & 0 & 0 & & 1 \\ -\alpha_0 & -\alpha_1 & -\alpha_2 & \cdots & -\alpha_{n-1} \end{bmatrix} \qquad (\text{A.38})$$

$$\widetilde{\mathbf{B}}\widetilde{\mathbf{K}} = \widetilde{\mathbf{A}} - \left[\widetilde{\mathbf{A}} - \widetilde{\mathbf{B}}\widetilde{\mathbf{K}}\right] = \begin{bmatrix} 0 & 0 & 0 & \cdots & 0 \\ 0 & 0 & 0 & & 0 \\ \vdots & & & & \vdots \\ 0 & 0 & 0 & \cdots & 0 \\ \alpha_0 - a_0 & \alpha_1 - a_1 & \alpha_2 - a_2 & \cdots & \alpha_{n-1} - a_{n-1} \end{bmatrix}$$
$$(\text{A.39})$$

We also have

$$\widetilde{\mathbf{B}}\widetilde{\mathbf{K}} = \begin{bmatrix} 0 \\ \vdots \\ 0 \\ 1 \end{bmatrix} \begin{bmatrix} \widetilde{k}_0 & \widetilde{k}_1 & \cdots & \widetilde{k}_{n-1} \end{bmatrix} \qquad (\text{A.40})$$

It then follows in view of (A.39) and (A.40) that

$$\widetilde{k}_i = \alpha_i - a_i \qquad (\text{A.41})$$

$(i = 0, ..., n-1)$ and $\mathbf{K} = \mathbf{M}^{-1}\widetilde{\mathbf{K}}$.

The ideas behind this control design can also be used in the context of state observation, which will be considered next.

## A.2.2    Observers

The problem of determination of the states of a given system arise in many contexts, *e.g.*, for purposes of control, soft sensors development, diagnosis, *etc.* The state vector can be directly calculated from the available measurements (input-output data). From (A.1)–(A.2), we derive

$$
\begin{aligned}
y(k-n+1) &= \mathbf{C}\mathbf{x}(k-n+1) \\
y(k-n+2) &= \mathbf{C}\mathbf{A}\mathbf{x}(k-n+1) + \mathbf{C}\mathbf{B}u(k-n+1) \\
&\vdots \\
y(k) &= \mathbf{C}\mathbf{A}^{n-1}\mathbf{x}(k-n+1) + \mathbf{C}\mathbf{A}^{n-2}\mathbf{B}u(k-n+1) + \dots \\
&\quad + \mathbf{C}\mathbf{B}u(k-1)
\end{aligned}
\tag{A.42}
$$

These equations can be conveniently arranged in matrix form as, say,

$$
\begin{bmatrix}
y(k-n+1) \\
y(k-n+2) \\
\vdots \\
y(k)
\end{bmatrix}
= \mathbf{Q}\mathbf{x}(k-n+1) + \mathbf{R}
\begin{bmatrix}
u(k-n+1) \\
u(k-n+2) \\
\vdots \\
u(k-1)
\end{bmatrix}
\tag{A.43}
$$

where

$$
\mathbf{Q} =
\begin{bmatrix}
\mathbf{C} \\
\mathbf{C}\mathbf{A} \\
\mathbf{C}\mathbf{A}^2 \\
\vdots \\
\mathbf{C}\mathbf{A}^{n-1}
\end{bmatrix}
\tag{A.44}
$$

This matrix is nothing else than the observability matrix. It is clear that in order to calculate the state $\mathbf{x}(k-n+1)$, this matrix must not be singular (these developments represent the proof of the Theorem 3). In other words, the non-singularity of the matrix $\mathbf{Q}$ is crucial to the problem of observing the states. This state reconstruction approach has the drawback that it may be sensitive to disturbances [5].

Another approach for state reconstruction is based on the use of a dynamic system. Let us consider the following observer for the system states

$$
\hat{\mathbf{x}}(k+1) = \mathbf{A}\hat{\mathbf{x}}(k) + \mathbf{B}u(k) + \mathbf{L}\left[y(k) - \mathbf{C}\hat{\mathbf{x}}(k)\right]
\tag{A.45}
$$

based on the system model and the correction term with gain $\mathbf{L} = [l_1, l_2, ..., l_n]^T$. The estimation error is given by

$$
\tilde{\mathbf{x}}(k+1) = \mathbf{x}(k+1) - \hat{\mathbf{x}}(k+1) = [\mathbf{A} - \mathbf{L}\mathbf{C}]\tilde{\mathbf{x}}(k)
\tag{A.46}
$$

Thus, if the matrix $[\mathbf{A} - \mathbf{LC}]$ represents an asymptotically stable system, $\widetilde{\mathbf{x}}(\cdot)$ will converge to zero for any initial error $\widetilde{\mathbf{x}}(0)$.

For the design of the gain $\mathbf{L}$, we can use the same approach as for the design of the state-feedback control law: The *characteristic equation* associated with the system governing the dynamics of the estimator error is

$$\det(z\mathbf{I} - \mathbf{A} + \mathbf{LC}) \qquad (\text{A.47})$$

and should be identical to the desired estimator characteristic equation. Notice that this characteristic equation is similar to (A.37), and, therefore the mathematical tools used for solving the state reconstruction problem are similar to those employed in the pole placement control design.

# Appendix B

# Fluidized Bed Combustion

In fluidized bed combustion (FBC), the combustion chamber contains a quantity of finely divided particles such as sand or ash. The combustion air entering from below lifts these particles until they form a turbulent bed, which behaves like a boiling fluid. The fuel is added to the bed and the mixed material is kept in constant movement by the combustion air. The heat released as the material burns maintains the bed temperature, and the turbulence keeps the temperature uniform throughout the bed.

The main purpose of an FBC plant is to generate power (energy flux $[W]=[\frac{J}{s}]$). Several powers can be distinguished: The *fuel power* is the power in the fuel (heat value times feed); the *combustion power* is the power released in combustion (dependent on completeness of the combustion and the furnace dynamics). *Boiler power* depends further on the efficiency of the heat exchangers as well as their dynamics. Often, a part of the heat is used for generating electricity in the turbines, while the remaining heat is used for the generation of steam and hot water. We can then distinguish *electrical power* and *thermal power*. Depending on plant constructions these are roughly of order $40\% - 0\%$ (electrical plant), $30\% - 55\%$ (co-generating plant), or $0 - 80\%$ (thermal plant) of the fuel power. In what follows a simplified model of a thermal plant is considered. For a more realistic modeling of the thermal power (steam mass flow), including the drum pressure, see [4].

## B.1 Model of a bubbling fluidized bed

A rough model for a bubbling fluidized bed combustor can be formulated [76] based on mass and energy balances (see also [33][35]). The model divides the furnace into two parts: the bed and the freeboard, see Fig. B.1. Combustion takes place in both: oxygen is consumpted and heat is released and removed.

283

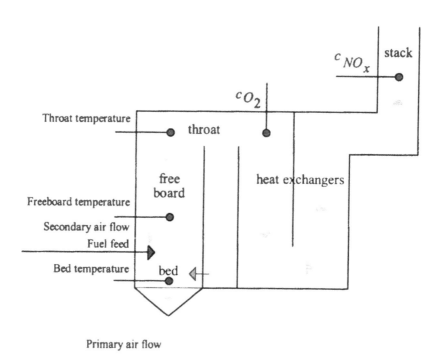

Figure B.1: A schematic drawing of a typical FBC plant. A mixture of inert/sorbent bed material and fuel is fluidized by the primary air. Complete combustion is ensured by the secondary air flow inserted from above the bed. The heat released in combustion is captured by heat exchangers and used for the generation of electricity, steam, or both.

The control inputs of the system are the fuel feed $Q_C$ [$\frac{kg}{s}$], and the primary and secondary air flows $F_1$ and $F_2$ [$\frac{Nm^3}{s}$]. Measurable system outputs are the flue gas $O_2$ [$\frac{Nm^3}{Nm^3}$], and the bed and the freeboard temperatures $T_B$ and $T_F$ [K].

## B.1.1 Bed

The solids (char) in the fuel combust in the bed. When fed to the combustor, the solids are stored in the fuel inventory (the amount of unburned char in the bed). The combustion rate $Q_B$ [$\frac{kg}{s}$] depends on the availability of oxygen in the bed as well as the fuel properties:

$$Q_B(t) = \frac{W_C}{t_C} \frac{C_B(t)}{C_1} \tag{B.1}$$

where $W_C$ is the fuel inventory [kg], and $t_C$ is the (average) char combustion time [s]. $C_B$ and $C_1$ are the oxygen contents [$\frac{Nm^3}{Nm^3}$] in the bed and in the primary air, respectively.

The dynamics of the fuel inventory are given by the difference between the fraction of the fuel feed rate that combusts in the bed $Q_C$ [$\frac{kg}{s}$] and the combustion rate $Q_B(t)$

$$\frac{dW_C(t)}{dt} = (1-V)Q_C(t) - Q_B(t) \tag{B.2}$$

where $V$ is the fraction of volatiles in the fuel [$\frac{kg}{kg}$].

Combustion in the bed consumes oxygen. $O_2$ comes into the bed in the primary air flow $F_1(t)$ [$\frac{Nm^3}{s}$], which is naturally provided by the environment and having an oxygen content $C_1 = 0.21$ [$\frac{Nm^3}{Nm^3}$]. $O_2$ is consumed in the combustion and transported to the freeboard:

$$\frac{dC_B(t)}{dt} = \frac{1}{V_B}[C_1 F_1(t) - X_C Q_B(t) - C_B(t) F_1(t)] \tag{B.3}$$

where $X_C$ [$\frac{Nm^3}{kg}$] is the coefficient describing the amount of $O_2$ consumed by the fuel and $V_B$ [$m^3$] is the volume of the bed.

As a result of combustion, heat is released. The amount of released heat depends on the heat value $H_C$ [$\frac{J}{kg}$] of the solids in the fuel. Heat is removed from the bed by cooling water tubes. The energy balance for the bed temperatures $T_B$ [K] is given by

$$\begin{aligned}
\frac{dT_B(t)}{dt} &= \frac{1}{c_l W_1}\{H_C Q_B(t) - a_{Bt} A_{Bt}[T_B(t) - T_{Bt}] \\
&\quad + c_1 F_1(t) T_1 - c_F F_1(t) T_B(t)\}
\end{aligned} \tag{B.4}$$

where $c_I$ and $W_I$ are the specific heat $[\frac{J}{kgK}]$ and mass [kg] of the bed material (inert sand), $a_{Bt}$ and $A_{Bt}$ are the heat transfer coefficient $[\frac{W}{m^2K}]$ and surface [m$^2$] of the cooling tubes, and $T_{Bt}$ is the temperature [K] of the cooling water. The incoming primary air in temperature $T_1$ [K], with specific heat $c_1$ $[\frac{J}{Nm^3K}]$, conveys some heat into the system. The remaining air, heated in bed temperature, is transported into the freeboard, where $c_F$ $[\frac{J}{Nm^3K}]$ is the specific heat of the flue gases.

## B.1.2   Freeboard

The gaseous components (volatiles) in the fuel are released and transported by the fluidizing air to the freeboard where immediate combustion occurs. The combustion of volatile fraction of the fuel consumes oxygen in the freeboard. Oxygen comes to the freeboard from the bed and with the secondary air flow $F_2$ $[\frac{Nm^3}{s}]$ with the $O_2$ content $C_2$ $[\frac{Nm^3}{Nm^3}]$. The dynamics of the freeboard oxygen content $C_F$ $[\frac{Nm^3}{Nm^3}]$ (flue gas oxygen) are given by

$$\frac{dC_F(t)}{dt} = \frac{1}{V_F}\{C_B(t)F_1(t)+C_2F_2(t) \qquad (B.5)$$
$$-X_V V Q_C(t) - C_F(t)[F_1(t)+F_2(t)]\}$$

where $X_V$ $[\frac{Nm^3}{kg}]$ is the coefficient describing the amount of $O_2$ consumed by the volatiles. $V_F$ [m$^3$] is the freeboard volume.

The volatiles release energy when combusted. Heat is removed from the freeboard by cooling water tubes located at the walls of the furnace. The energy balance for the freeboard temperatures $T_F$ [K] is given by

$$\frac{dT_F(t)}{dt} = \frac{1}{c_F V_F}\{H_V V Q_C(t) - a_{Ft}A_{Ft}[T_F(t)-T_{Ft}] \qquad (B.6)$$
$$c_F F_1(t)T_B(t)+c_2 F_2(t)T_2(t)+c_1[F_1(t)+F_2(t)]T_F(t)\}$$

where $a_{Ft}$ and $A_{Ft}$ are the heat transfer coefficient $[\frac{W}{m^2K}]$ and surface [m$^2$] of the cooling tubes, $T_{Ft}$ is the temperature [K] of the cooling water, $c_2$ is the specific heat $[\frac{J}{Nm^3K}]$ of the secondary air, in temperature $T_2$, and $H_V$ is the heat value of the volatiles $[\frac{J}{kg}]$.

## B.1.3   Power

The combustion power $P_C$ [W] is the rate of energy released in combustion

$$P_C(t) = H_C Q_B(t) + H_V V Q_C(t) \qquad (B.7)$$

and simple first order dynamics were assumed for the thermal power $P$ [W]

$$\frac{dP(t)}{dt} = \frac{1}{\tau_{\text{mix}}}[P_{\text{C}}(t) - P(t)] \tag{B.8}$$

where $\tau_{\text{mix}}$ [s] is a time constant.

## B.1.4 Steady-state

The equations can be solved in steady-state. Bed fuel inventory and bed oxygen content are functions of the fuel feed and the primary air flows:

$$W_{\text{C}} = \frac{C_1 t_{\text{C}}(1-V)F_1 Q_{\text{C}}}{C_1 F_1 - X_{\text{C}}(1-V)Q_{\text{C}}} \tag{B.9}$$

$$C_{\text{B}} = C_1 + \frac{X_{\text{C}}(1-V)Q_{\text{C}}}{F_1} \tag{B.10}$$

Let us solve the equations eliminating variables other than the $Q_{\text{C}}$, $F_1$ and $F_2$. Bed temperatures depend on the fuel power:

$$T_{\text{B}} = \frac{H_{\text{C}}(1-V)Q_{\text{C}} + c_1 F_1 T_1 + a_{\text{Bt}} A_{\text{Bt}} T_{\text{Bt}}}{a_{\text{Bt}} A_{\text{Bt}} + c_{\text{F}} F_1} \tag{B.11}$$

Flue gas oxygen is influenced mainly by the secondary air flow and the fuel feed:

$$C_{\text{F}} = \frac{C_1 F_1 + C_2 F_2 - X_{\text{C}}(1-V)Q_{\text{C}} - X_{\text{V}} V Q_{\text{C}}}{F_1 + F_2} \tag{B.12}$$

Freeboard temperatures depend on the heat released by the volatiles:

$$\begin{aligned}
T_{\text{F}} = \{ & a_{\text{Bt}} A_{\text{Bt}}(c_2 F_2 T_2 + a_{\text{Ft}} A_{\text{Ft}} T_{\text{Ft}} + H_{\text{V}} V Q_{\text{C}}) \\
& + c_{\text{F}} F_1 (c_1 F_1 T_1 + c_2 F_2 T_2) \\
& + c_{\text{F}} F_1 (a_{\text{Bt}} A_{\text{Bt}} T_{\text{Bt}} + a_{\text{Ft}} A_{\text{Ft}} T_{\text{Ft}}) \\
& + c_{\text{F}} F_1 (H_{\text{C}}(1-V)Q_{\text{C}} + H_{\text{V}} V Q_{\text{C}}) \} \\
& \times \frac{1}{[a_{\text{Bt}} A_{\text{Bt}} + c_{\text{F}} F_1][a_{\text{Ft}} A_{\text{Ft}} + c_{\text{F}}(F_1 + F_2)]}
\end{aligned} \tag{B.13}$$

Power depends entirely on the fuel feed and its heat value:

$$P = H_{\text{C}}(1-V)Q_{\text{C}} + H_{\text{V}} V Q_{\text{C}} \tag{B.14}$$

# B.2   Tuning of the model

The above model is very simple. The poor aspect is that it describes only a few of the phenomena involved in a complex process such as FBC combustion. For example, the assumptions on fluidization, combusiton and heat transfer are very elementary. Therefore, the model needs to be tuned in order to match plant measurements. The nice thing is that a model with a simple structure is also simple to tune using standard methods.

Much more detailed models have been constructed for the FBC process. From a practical point of view the calculation times and the lack of accurate measurements restricts the use of these models. It is common in practice that a simple mass or energy balance can not be closed based on plant measurements, due to systematic errors in measurements. Many of the internal parameters related to combustion, fluidization, and heat transfer can be accurately measured only in laboratory conditions, and are not applicable to a real plant. The advanced models can be extremely useful in helping to develp and understand the process. In automatic process control, however, their significance is less.

## B.2.1   Initial values

The tuning of the model was divided into three phases. The *initial values* were found from the literature (heat values and heat transfer coefficients, $O_2$ consumption, plant geometry). These are given in Table B.1.

## B.2.2   Steady-state behavior

The *steady-state* behavior of the model was adjusted first. The following tuning knots were used

$$H_C \leftarrow p_1 H_C \quad V \leftarrow p_2 V \quad X_C \leftarrow p_3 X_C \qquad \text{(B.15)}$$

taking further that $H_V = H_C$; $X_V = X_C$. A cost function was formulated as a weighted sum of squared errors between measured steady state values and the predicted ones:

$$J = \sum w_1 \left( C_F - \widehat{C}_F \left( p_2, p_3 \right) \right) + w_2 \left( T_B - \widehat{T}_B \left( p_1, p_3 \right) \right) \qquad \text{(B.16)}$$
$$+ w_3 \left( T_F - \widehat{T}_F \left( p_1, p_3 \right) \right) + w_4 \left( P - \widehat{P} \left( p_1 \right) \right)$$

$w_i$s were chosen according to the scales of the variables. Using a data set of 11 steady-state points, Table B.2, the values of $p_i$ were estimated to $\mathbf{p} = [0.2701,$

| Bed: | | |
|---|---|---|
| bed material specific heat | $c_I = 800$ | $[\frac{J}{kgK}]$ |
| bed inert material | $W_I = 25000$ | $[kg]$ |
| volume | $V_B = 26.3$ | $[m^3]$ |
| heat transfer coefficient | $\alpha_{Bt} = 210$ | $[\frac{W}{m^2K}]$ |
| heat exchange surface | $A_{Bt} = 26.8$ | $[m^2]$ |
| cooling water temperature | $T_{Bt} = 573$ | $[K]$ |
| **Freeboard:** | | |
| volume | $V = 128.1$ | $[m^3]$ |
| heat transfer coefficient | $\alpha_{Ft} = 210$ | $[\frac{W}{m^2K}]$ |
| heat exchange surface | $A_{Ft} = 130.7$ | $[m^2]$ |
| cooling water temperature | $T_{Bt} = 573$ | $[K]$ |
| **Air flows:** | | |
| primary air $O_2$ content | $C_1 = 0.21$ | $[\frac{Nm^3}{Nm^3}]$ |
| primary air specific heat | $c_1 = 1305$ | $[\frac{J}{m^3K}]$ |
| primary air temperature | $T_1 = 328$ | $[K]$ |
| secondary air $O_2$ content | $C_2 = 0.21$ | $[\frac{Nm^3}{Nm^3}]$ |
| secondary air specific heat | $c_2 = 1305$ | $[\frac{J}{m^3K}]$ |
| secondary air temperature | $T_2 = 328$ | $[K]$ |
| flue gas specific heat | $c_F = 1305$ | $[\frac{J}{m^3K}]$ |
| **Fuel feed:** | | |
| $O_2$ consumed in combustion | $X_C = 1.886$ | $[\frac{Nm^3}{kg}]$ |
| heat value of char | $H_C = 30 \times 10^6$ | $[\frac{J}{kg}]$ |
| mean combustion rate | $t_C = 50$ | $[s]$ |
| fraction of volatiles | $V = 0.75$ | $[\frac{kg}{kg}]$ |
| $O_2$ consumed in combustion | $X_V = 1.225$ | $[\frac{Nm^3}{kg}]$ |
| heat value of volatiles | $H_V = 50 \times 10^6$ | $[\frac{J}{kg}]$ |
| **Other:** | | |
| time constant | $\tau_{mix} = 300$ | $[s]$ |

Table B.1: Constants for the FBC model.

| $Q_{\mathrm{C}}[\frac{\mathrm{kg}}{\mathrm{s}}]$ | $F_1[\frac{\mathrm{Nm}^3}{\mathrm{s}}]$ | $F_2[\frac{\mathrm{Nm}^3}{\mathrm{s}}]$ | $C_{\mathrm{F}}[\%-\mathrm{vol}]$ | $T_{\mathrm{B}}[°\mathrm{C}]$ | $T_F[°\mathrm{C}]$ | $P[\mathrm{MW}]$ |
|------|------|------|------|-----|-----|------|
| 2.2 | 3.5 | 7.9  | 5.1 | 696 | 556 | 19.1 |
| 2.3 | 3.5 | 6.5  | 3.0 | 662 | 607 | 19.3 |
| 2.3 | 3.5 | 9.8  | 6.9 | 696 | 550 | 19.2 |
| 2.3 | 3.5 | 8.0  | 5.1 | 686 | 572 | 19.1 |
| 1.6 | 2.5 | 4.4  | 2.9 | 650 | 581 | 13.1 |
| 1.7 | 2.8 | 5.2  | 4.0 | 668 | 569 | 15.1 |
| 1.7 | 2.8 | 6.4  | 5.4 | 696 | 530 | 14.3 |
| 3.1 | 3.7 | 10.2 | 3.9 | 691 | 646 | 26.0 |
| 3.0 | 3.7 | 8.6  | 2.5 | 681 | 628 | 27.0 |
| 3.0 | 3.7 | 11.0 | 5.0 | 659 | 599 | 25.6 |

Table B.2: Steady-state data from a FBC plant.

$0.9956, 0.4238]^T$ using the Levenberg–Marquardt method. The efficient heat value was found to be only $\sim 30$ % of the heat value of dry fuel. As the fuel feed was taken as the measured kg-input flux to the furnace, and moisture was not taken into account, this is acceptable. For volatiles the 3/4 assumption was reasonable. The $O_2$ consumption coefficient reflects the fact that less than half of the input feed consists of combustible components.

## B.2.3  Dynamics

The *dynamics* were tuned by hand by comparing measurements from step-response experiments and corresponding simulations. First, the *delays* were examined and found negligible (equal to zero) from air flows $F_1$ and $F_2$, and 20 seconds from fuel feed $Q_{\mathrm{C}}$. This was judged reasonable as there is a transport delay from a change in fuel belt conveyor speed to the introduction of a change in the flow to the furnace. In addition, some delay is due to ignition of fuel, which was not taken into account in the model. Thus we have

$$Q_{\mathrm{C}}(t) = Q_{\mathrm{C,actuator}}(t - 20) \tag{B.17}$$

where $Q_{\mathrm{C,actuator}}$ is the measured fuel flow. The transport delay in air flows was insignificant.

The time constants for $C_{\mathrm{F}}$ were found adequate (bed and freeboard volumes). The time constants for temperatures were adjusted by altering the mass of the bed inert material $W_{\mathrm{I}}$ and the freeboard time constant $c_{\mathrm{F}}V_{\mathrm{F}}$ (in $dT_{\mathrm{F}}/dt$ equation (B.6) only); for power these were adjusted by setting $\tau_{\mathrm{mix}}$. For $W_{\mathrm{I}}$ a value of 480 kg was found reasonable; the $c_{\mathrm{F}}V_{\mathrm{F}}$ for $T_{\mathrm{F}}$ was multiplied by 35 in order to have reasonable responses.

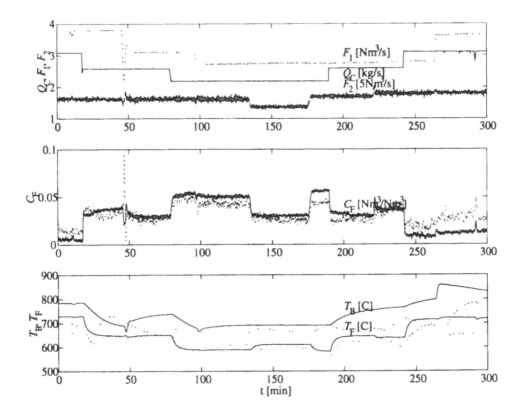

Figure B.2: Response of the tuned FBC model (solid lines) against measurements (dotted lines) for a 25 MW FBC plant.

## B.2.4 Performance of the model

Figures B.2–B.3 illustrate the performance of the model with respect to data measured from a 25 MW semi-circulated FBC plant, for step-like changes in $Q_C$, $F_1$ and $F_2$. Figure B.2 illustrates the steady-state performance. Note that the data used for tuning the steady-states of the model was not taken from these experiments. The main characteristics of the $O_2$ in flue gases as well as freeboard temperatures are captured by the model. For bed temperatures the response is poor.

Figure B.3 shows a smaller section of the same simulation. For $O_2$ the dynamic response is good. For freeboard temperatures the response seems like an acceptable first order approximation of a second order process. Again, the predictability for bed temperatures is poor.

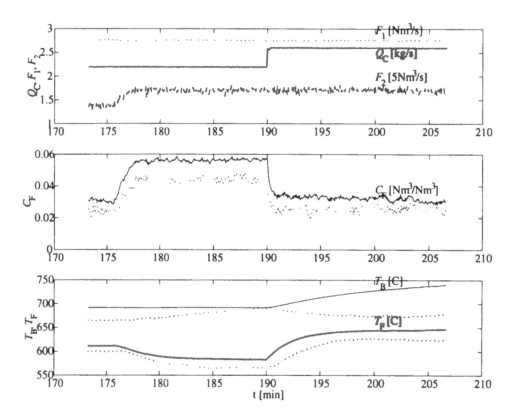

Figure B.3: Response of the tuned FBC model (solid lines) against measurements (dotted lines) for a 25 MW FBC plant.

# B.3  Linearization of the model

The differential equation model (B.1)–(B.6) is a non-linear one, containing bilinear terms such as $W_C C_B$ in (B.1) and $C_B(t) F_1(t)$ in (B.3). This model can be discretized around an operating point.

Let the operating point be given by:

$$\{\overline{Q}_C, \overline{F}_1, \overline{F}_2, \overline{W}_C, \overline{C}_B, \overline{C}_F, \overline{T}_B, \overline{T}_F, \overline{P}\} \tag{B.18}$$

If this is a steady state point, then $\{\overline{W}_C, \overline{C}_B, \overline{C}_F, \overline{T}_B, \overline{T}_F, \overline{P}\}$ can be determined by specifying $\{\overline{Q}_C, \overline{F}_1, \overline{F}_2\}$ and using the steady state equations (B.9)–(B.14). Using a Taylor-series expansion, we obtain the following linearized continuous time model:

$$
\begin{aligned}
\frac{dW_C(t)}{dt} &= (1-V)\left[Q_C(t) - \overline{Q}_C\right] - \frac{\overline{W}_C}{t_C C_1}\left[C_B(t) - \overline{C}_B\right] \\
&\quad - \frac{\overline{C}_B}{t_C C_1}\left[W_C(t) - \overline{W}_C\right] + (1-V)\overline{Q}_C - \frac{\overline{W}_C \overline{C}_B}{t_C C_1} \quad \text{(B.19)}
\end{aligned}
$$

$$
\begin{aligned}
\frac{dC_B(t)}{dt} &= \frac{(C_1 - \overline{C}_B)}{V_B}\left[F_1(t) - \overline{F}_1\right] - \frac{\overline{C}_B X_C}{V_B t_C C_1}\left[W_C(t) - \overline{W}_C\right] \\
&\quad + \frac{(-\frac{\overline{W}_C X_C}{t_C C_1} - \overline{F}_1)}{V_B}\left[C_B(t) - \overline{C}_B\right] \quad \text{(B.20)} \\
&\quad + \frac{C_1 \overline{F}_1 - \overline{C}_B \overline{F}_1 - \frac{\overline{W}_C \overline{C}_B X_C}{t_C C_1}}{V_B}
\end{aligned}
$$

$$
\begin{aligned}
\frac{dC_F(t)}{dt} &= -\frac{V X_V}{V_F}\left[Q_C(t) - \overline{Q}_C\right] + \frac{\overline{C}_B - \overline{C}_F}{V_F}\left[F_1(t) - \overline{F}_1\right] \\
&\quad + \frac{C_2 - \overline{C}_F}{V_F}\left[F_2(t) - \overline{F}_2\right] + \frac{\overline{F}_1}{V_F}\left[C_B(t) - \overline{C}_B\right] \quad \text{(B.21)} \\
&\quad - \frac{\overline{F}_1 + \overline{F}_2}{V_F}\left[C_F(t) - \overline{C}_F\right] \\
&\quad + \frac{\overline{C}_B \overline{F}_1 - V\overline{Q}_C X_V + C_2 \overline{F}_2 - \overline{C}_F(\overline{F}_1 + \overline{F}_2)}{V_F}
\end{aligned}
$$

$$\frac{dT_B(t)}{dt} = \frac{c_1 T_1 - c_F \overline{T}_B}{c_1 W_1}(F_1(t) - \overline{F}_1) + \frac{H_C \overline{C}_B}{c_1 W_1 t_C C_1}(W_C(t) - \overline{W}_C)$$
$$+ \frac{H_C \overline{W}_C}{c_1 W_1 t_C C_1}(C_B(t) - \overline{C}_B) \qquad (B.22)$$
$$+ \frac{-a_{Bt} A_{Bt} - c_F \overline{F}_1}{c_1 W_1}(T_B(t) - \overline{T}_B)$$
$$+ \frac{(\frac{H_C \overline{W}_C \overline{C}_B}{t_C C_1} - a_{Bt} A_{Bt}(\overline{T}_B - T_{Bt}) + c_1 \overline{F}_1 T_1 - c_F \overline{F}_1 \overline{T}_B)}{c_1 W_1}$$

$$\frac{dT_F(t)}{dt} = \frac{H_V V}{c_F V_F \tau_F}[Q_C(t) - \overline{Q}_C] + \frac{c_F \overline{T}_B - c_F \overline{T}_F}{c_F V_F \tau_F}[F_1(t) - \overline{F}_1]$$
$$+ \frac{c_2 T_2 - c_F \overline{T}_F}{c_F V_F \tau_F}[F_2(t) - \overline{F}_2] + \frac{\overline{F}_1}{V_F \tau_F}[T_B(t) - \overline{T}_B] \quad (B.23)$$
$$+ \frac{-c_F(\overline{F}_1 + \overline{F}_2) - a_{Ft} A_{Ft}}{c_F V_F \tau_F}[T_F(t) - \overline{T}_F]$$
$$+ \frac{H_V V \overline{Q}_C - c_F(\overline{F}_1 + \overline{F}_2)\overline{T}_F - a_{Ft} A_{Ft}(\overline{T}_F - T_{Ft})}{c_F V_F \tau_F}$$
$$+ \frac{c_2 \overline{F}_2 T_2 + c_F \overline{F}_1 \overline{T}_B}{c_F V_F \tau_F}$$

$$\frac{dP(t)}{dt} = \frac{H_V V}{\tau_{mix}}[Q_C(t) - \overline{Q}_C] + \frac{H_C \overline{C}_B}{\tau_{mix} t_C C_1}[W_C(t) - \overline{W}_C]$$
$$+ \frac{H_C \overline{W}_C}{\tau_{mix} t_C C_1}[C_B(t) - \overline{C}_B] - \frac{1}{\tau_{mix}}[P(t) - \overline{P}] \qquad (B.24)$$
$$+ \frac{\frac{H_C \overline{W}_C \overline{C}_B}{t_C C_1} - \overline{P} + H_V V \overline{Q}_C}{\tau_{mix}}$$

Equations (B.19)–(B.24) can be expressed as a linear state-space model around an operating point

$$\frac{d\mathbf{x}(t)}{dt} = \mathbf{A}_c \mathbf{x}(t) + \mathbf{B}_c \mathbf{u}(t) \qquad (B.25)$$
$$\mathbf{y}(t) = \mathbf{C}_c \mathbf{x}(t) \qquad (B.26)$$

where the vectors $\mathbf{x}$, $\mathbf{u}$ and $\mathbf{y}$ are given by deviations from the point of linearization

$$\mathbf{x}(t) = \begin{bmatrix} W_C(t) - \overline{W}_C \\ C_B(t) - \overline{C}_B \\ C_F(t) - \overline{C}_F \\ T_B(t) - \overline{T}_B \\ T_F(t) - \overline{T}_F \\ P(t) - \overline{P} \end{bmatrix} ; \mathbf{u}(t) = \begin{bmatrix} Q_C(t) - \overline{Q}_C \\ F_1(t) - \overline{F}_1 \\ F_2(t) - \overline{F}_2 \end{bmatrix} \qquad (B.27)$$

The coefficients of the matrices $\mathbf{A}_c$ and $\mathbf{B}_c$ are obtained from (B.19)–(B.24), $a_{1,1} = -\frac{\overline{C}_B}{t_C \overline{C}_1}$, $b_{1,1} = (1 - V)$, *etc.* Usually the flue gas $O_2$, bed and freeboard temperatures and power outtake are measured:

$$\mathbf{y}(t) = \begin{bmatrix} C_F(t) - \overline{C}_F \\ T_B(t) - \overline{T}_B \\ T_F(t) - \overline{T}_F \\ P(t) - \overline{P} \end{bmatrix} \text{ and } \mathbf{C}_c = \begin{bmatrix} 0 & 0 & 1 & 0 & 0 & 0 \\ 0 & 0 & 0 & 1 & 0 & 0 \\ 0 & 0 & 0 & 0 & 1 & 0 \\ 0 & 0 & 0 & 0 & 0 & 1 \end{bmatrix} \qquad (B.28)$$

For many practical purposes, the model needs to be discretized. The approximation of a continous-time state-space model by a discrete-time model is straightforward (see, *e.g.*, [64]), and results in

$$\begin{aligned} \mathbf{x}(k+1) &= \mathbf{A}\mathbf{x}(k) + \mathbf{B}\mathbf{u}(k) & (B.29) \\ \mathbf{y}(k) &= \mathbf{C}\mathbf{x}(k) & (B.30) \end{aligned}$$

where $t = kT_s$ and $T_s$ is the sampling time.

**Example 43 (Linearization)** Let us linearize the model in a steady state given by

$$\overline{Q}_C = 2.6 \frac{kg}{s}, \ \overline{F}_1 = 3.1 \frac{Nm^3}{s}, \ \overline{F}_2 = 8.4 \frac{Nm^3}{s} \qquad (B.31)$$

The remaining states of the operating point are then given by

$$\begin{aligned} \overline{W}_C &= 165kg, \ \overline{C}_B = 0.042, \ \overline{C}_F = 0.031 & (B.32) \\ \overline{T}_B &= 749°C, \ \overline{T}_F = 650°C, \ \overline{P} = 21.1MW & (B.33) \end{aligned}$$

For a continuous time, state-space model we obtain

$$\mathbf{A}_c = \begin{bmatrix} -0.0040 & -15.6819 & 0 & 0 & 0 & 0 \\ -0.0001 & -0.5908 & 0 & 0 & 0 & 0 \\ 0 & 0.0242 & -0.0898 & 0 & 0 & 0 \\ 0.0027 & 10.5892 & 0 & -0.0008 & 0 & 0 \\ 0 & 0 & 0 & 0.0005 & -0.0051 & 0 \\ 0.0001 & 0.4236 & 0 & 0 & 0 & -0.0033 \end{bmatrix}$$
$$(B.34)$$

$$\mathbf{B}_c = \begin{bmatrix} 0.2533 & 0 & 0 \\ 0 & 0.0064 & 0 \\ -0.0046 & 0.0001 & 0.0014 \\ 0 & -0.755 & 0 \\ 0.7238 & 0.0155 & -0.0929 \\ 0.0202 & 0 & 0 \end{bmatrix} \tag{B.35}$$

$$\mathbf{C}_c = \begin{bmatrix} 0 & 0 & 1 & 0 & 0 & 0 \\ 0 & 0 & 0 & 1 & 0 & 0 \\ 0 & 0 & 0 & 0 & 1 & 0 \\ 0 & 0 & 0 & 0 & 0 & 1 \end{bmatrix} \tag{B.36}$$

Using $T_s = 4$ s, a linearized discrete-time state-space model is obtained (this is simple to accomplish numerically using a suitable software like Matlab, for example). For convenience, the model is given in a transfer polynomial form. We have from the fuel feed:

$$\frac{C_F\left(q^{-1}\right)}{Q_C\left(q^{-1}\right)} = \frac{-0.01549q^{-6}\left(1 - 0.9957q^{-1}\right)\left(1 - 0.09308q^{-1}\right)}{\left(1 - 0.6983q^{-1}\right)\left(1 - 0.9968q^{-1}\right)\left(1 - 0.09293q^{-1}\right)} \tag{B.37}$$

$$\frac{T_B\left(q^{-1}\right)}{Q_C\left(q^{-1}\right)} = \frac{0.003362q^{-5}\left(1 - 0.6233q^{-1}\right)\left(1 + 0.5543q^{-1}\right)}{\left(1 - 0.9968q^{-1}\right)^2\left(1 - 0.09293q^{-1}\right)} \tag{B.38}$$

$$\frac{T_F\left(q^{-1}\right)}{Q_C\left(q^{-1}\right)} = \frac{2.866q^{-6}\left(1 - 1.994q^{-1} + 0.9936q^{-2}\right)}{\left(1 - 0.9799q^{-1}\right)\left(1 - 9968q^{-1}\right)^2} \tag{B.39}$$

$$\frac{P\left(q^{-1}\right)}{Q_C\left(q^{-1}\right)} = \frac{0.08027q^{-6}\left(1 - 0.0923q^{-1}\right)\left(1 - 0.9958q^{-1}\right)}{\left(1 - 0.9868q^{-1}\right)\left(1 - 0.9968q^{-1}\right)\left(1 - 0.09293q^{-1}\right)} \tag{B.40}$$

From primary air flow:

$$\frac{C_F\left(q^{-1}\right)}{F_2\left(q^{-1}\right)} = \frac{0.00084931q^{-1}\left(1 - 0.9872q^{-1}\right)\left(1 + 0.2433q^{-1}\right)}{\left(1 - 0.6983q^{-1}\right)\left(1 - 0.9968q^{-1}\right)\left(1 - 0.09293q^{-1}\right)} \tag{B.41}$$

$$\frac{T_B\left(q^{-1}\right)}{F_2\left(q^{-1}\right)} = \frac{-0.020027q^{-1}\left(1 - 1.006q^{-1}\right)\left(1 - 7.9q^{-1}\right)}{\left(1 - 0.9968q^{-1}\right)^2\left(1 - 0.09293q^{-1}\right)} \tag{B.42}$$

$$\frac{T_F\left(q^{-1}\right)}{F_2\left(q^{-1}\right)} = \frac{0.06135q^{-1}\left(1 - 0.095q^{-1}\right)\left(1 - 0.9872q^{-1}\right)\left(1 - 1.002q^{-1}\right)}{\left(1 - 0.9799q^{-1}\right)\left(1 - 0.9968q^{-1}\right)^2\left(1 - 0.09293q^{-1}\right)} \tag{B.43}$$

$$\frac{P\left(q^{-1}\right)}{F_2\left(q^{-1}\right)} = \frac{0.011219q^{-1}\left(1 - q^{-1}\right)\left(1 + 0.4645q^{-1}\right)}{\left(1 - 0.9868q^{-1}\right)\left(1 - 0.9968q^{-1}\right)\left(1 - 0.09293q^{-1}\right)} \tag{B.44}$$

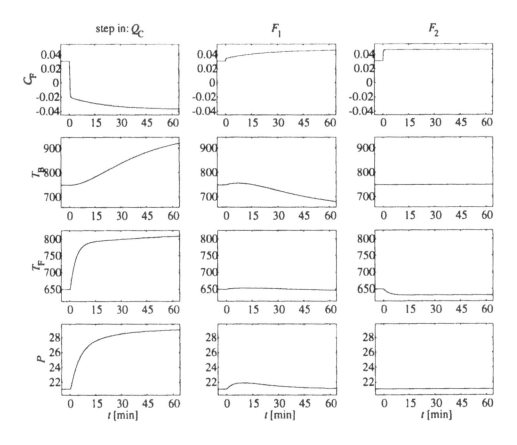

Figure B.4: Step responses of the linearized model, linearized at steady state $Q_C = 2.6$ [kg/s], $F_1 = 3.1$ [Nm$^3$/s], $F_2 = 8.4$ [Nm$^3$/s].

and from the secondary air flow:

$$\frac{C_F(q^{-1})}{F_2(q^{-1})} = \frac{0.0046901q^{-1}}{1 - 0.6983q^{-1}} \tag{B.45}$$

$$\frac{T_B(q^{-1})}{F_2(q^{-1})} = 0 \tag{B.46}$$

$$\frac{T_F(q^{-1})}{F_2(q^{-1})} = \frac{-0.36785q^{-1}}{1 - 0.9799q^{-1}} \tag{B.47}$$

$$\frac{P(q^{-1})}{F_2(q^{-1})} = 0 \tag{B.48}$$

The performance of the linearized model is depicted in Fig. B.4.

# Bibliography

[1] J Andrews. A mathematical model for the continuous culture of micro-organisms using inhibitory substrate. *Biotechnology and Bioengineering*, 10:707–723, 1968.

[2] K Åstrom. *Introduction to Stochastic Control Theory*. Academic Press, New York, 1970.

[3] K Åström. Maximum likelihood and prediction error methods. *Automatica*, 16:551–574, 1980.

[4] K Åstrom and R Bell. Drum-boiler dynamics. *Automatica*, 36:363–378, 2000.

[5] K Åström and K Wittenmark. *Computer Controlled Systems: Theory and Design*. Prentice-Hall Inc., Englewood Cliffs, New Jersey, 1990.

[6] N Baba. A new approach for finding the global minimum of error function of neural networks. *Neural Networks*, 2:367–373, 1989.

[7] R Battiti. First- and second-order methods for learning: Between steepest descent and Newton's method. *Neural Computation*, 4:141–166, 1992.

[8] R Bitmead, M Gevers, and V Wertz. *Adaptive Optimal Control – The Thinking Man's GPC*. Prentice-Hall, New York, 1990.

[9] R Brockett and P Krishnaprasad. A scaling theory for linear systems. *IEEE Transactions on Automatic Control*, 25(2):197–207, 1980.

[10] W Buntine and A Weigend. Computing second derivatives in feedforward networks: A review. *IEEE Transactions on Neural Networks*, 5:480–488, 1994.

[11] R Bush and F Mosteller. *Stochastic Models for Learning*. John Wiley and Sons, New York, 1958.

[12] J Castro and M Delgado. Fuzzy systems with defuzzification are universal approximators. *IEEE Transactions on Systems, Man and Cybernetics*, 26:149–152, 1996.

[13] D Clarke, C Mohtadi, and P Tuffs. Generalized predictive control - part. *Automatica*, 23(2):137–148, 1989.

[14] C Cutler and B Ramaker. Dynamic matrix control: A computer control algorithm. *JAAC*, pages 0–0, 1980.

[15] T Edgar and D Himmelblau. *Optimization of Chemical Processes*. McGraw-Hill, New York, 1989.

[16] J Edmunds. Input and output scaling and reordering for diagonal dominance and block diagonal dominance. *IEE Proceedings - Control Theory and Applications*, 145(6):523–530, 1998.

[17] E Eskinat, S Johnson, and W Luyben. Use of Hammerstein models in identification of nonlinear systems. *AIChE Journal*, 37(2):255–268, 1991.

[18] R L Eubank. *Nonparametric Regression and Spline Smoothing*. Marcel Dekker, New York, 1999.

[19] P Eykhoff. *System Identification: Parameter and State Estimation*. John Wiley and Sons, New York, 1974.

[20] R Fox and L Fan. Stochastic modeling of chemical process systems: Parts I–III. *Chemical Engineering Education*, XXIV:56–59, 88–92, 164–167, 1990.

[21] G Franklin, J Powell, and M Workman. *Digital Control of Dynamic Systems*. Addison Wesley Longman Inc., Menlo Park, U.S.A., 1998.

[22] D Goldberg. *Genetic Algorithms in Search, Optimization, and Machine Learning*. Addison-Wesley, Reding, Massachusetts, 1989.

[23] G Goodwin and K Sin. *Adaptive Filtering, Prediction and Control*. Prentice-Hall, New Jersey, 1984.

[24] M Hagan and M Menhaj. Training feedforward networks with the marquardt algorithm. *IEEE Transactions on Neural Networks*, 5:989–993, 1994.

[25] W Härdle. *Applied Nonparametric Regression*. Cambridge University Press, New York, 1990.

[26] T Hastie and R Tibshirani. *Generalized Additive Models*. Chapman and Hall, London, 1990.

[27] S Haykin. *Neural Networks: A Comprehensive Foundation*. MacMillan, New York, 1994.

[28] M Henson and D Seborg. Adaptive nonlinear control of a pH neutralization process. *IEEE Transactions on Control Systems Technology*, 2(3):169–182, 1994.

[29] J Hertz, A Krogh, and R Palmer. *Introduction to the Theory of Neural Computation*. Addison-Wesley, Redwood City, 1991.

[30] K Hornik, M Stinchcombe, and H White. Multilayer feedforward neural networks are universal approximators. *Neural Networks*, 2:359–366, 1989.

[31] K Hunt, R Haas, and R Murray-Smith. Extending the functional equivalence of radial basis function networks and fuzzy inference systems. *IEEE Transactions on Neural Networks*, 7:776–781, 1996.

[32] H Hyötyniemi. *Self-Organizing Artificial Neural Networks in Dynamic Systems Modeling and Control*. PhD thesis, Helsinki University of Technology, 1994.

[33] E Ikonen. Pedin polttoainekertymänmallintaminen leijukerrospoltossa. Master's thesis, University of Oulu, 1991.

[34] E Ikonen. *Algorithms for Process Modelling Using Fuzzy Neural Networks: A Distributed Logic Processor Approach*. PhD thesis, University of Oulu, 1996.

[35] E Ikonen and U Kortela. Dynamic model of a fluidized bed coal combustor. *Control Engineering Practice*, 2(6):1001–1006, 1994.

[36] E Ikonen and K Najim. Non-linear process modelling based on a Wiener approach. *Journal of Systems and Control Engineering - Proceedings of the Institution of Mechanical Engineers Part I*, to appear.

[37] E Ikonen, K Najim, and U Kortela. Process identification using Hammerstein systems. In *IASTED International Conference on Modelling, Identification and Control (MIC 2000)*, Insbruck, Austria, 2000. IASTED.

[38] R Isermann. *Digital Control Systems*. Springer-Verlag, Heidelberg, 1981.

[39] A Jazwinski. *Stochastic Processes and Filtering Theory*. Academic Press, New York, 1970.

[40] M Johansson. *A Primer on Fuzzy Control (Report)*. Lund Institute of Technology, 1996.

[41] R Johansson. *System Modeling and Identification*. Prentice-Hall, New Jersey, 1993.

[42] E Katende and A Jutan. Nonlinear predictive control of complex processes. *Industrial Engineering Chemistry Research*, 35:3539–3546, 1996.

[43] E Katende, A Jutan, and R Corless. Quadratic nonlinear predictive control. *Industrial Engineering Chemistry Research*, 37(7):2721–2728, 1998.

[44] J Kiefer and J Wolfowitz. Stochastic estimation of the maximum of a regression. *Annals of Mathematics and Statistics*, 23:462–466, 1952.

[45] M Kinnaert. Adaptive generalized predictive controller for MIMO systems. *International Journal of Control*, 50(1):161–172, 1989.

[46] S Kirkpatrick, C Gelatt, and M Vecchi. Optimization by simulated annealing. *Science*, 220:671–680, 1983.

[47] G Klir and T Folger. *Fuzzy Sets, Uncertainty and Information*. Prentice-Hall, 1988.

[48] T Knudsen. Consistency analysis of subspace identification methods based on a linear regression approach. *Automatica*, 37(1):81–89, 2001.

[49] T Kohonen. The self-organising map. *IEEE Proceedings*, 78:1464–1486, 1990.

[50] R Kruse, J Gebhardt, and F Klawonn. *Foundations of Fuzzy Systems*. John Wiley and Sons, Chichester, 1994.

[51] I Landau, R Lozano, and M M'Saad. *Adaptive Control*. Springer Verlag, London, 1997.

[52] P Lee, R Newell, and I Cameron. *Process Management and Control*. http://wwweng2.murdoch.edu.au/m288/resources/textbook/title.htm, 2000.

[53] P Lindskog. *Methods, Algorithms and Tools for System Identification Based on Prior Knowledge.* PhD thesis, Lindköping University, 1996.

[54] L Ljung and T McKelvey. A least squares interpretation of sub-space methods for system identification. In *Proceedings of the 35th Conference on Decision and Control*, pages 335–342. IEEE, 1996.

[55] L Ljung and T Söderström. *Theory and Practice of Recursive Identification.* MIT Press, Cambridge, Massachusetts, 1983.

[56] R Luus and T Jaakola. Optimization by direct search and systematic reduction of the size of the search region. *AIChE Journal*, 19:760–766, 1973.

[57] J Maciejowski. *Multivariable Feedback Design.* Addison-Wesley, Wokingham, England, 1989.

[58] A Mészaros, M Brdys, P Tatjewski, and P Lednicky. Multilayer adaptive control of continuous bioprocesses using optimising control techniques. case study: Baker's yeast culture. *Bioprocess Engineering*, 12:1–9, 1995.

[59] A Mészaros, P Dostal, and J Mikles. Development of tubular chemical reactor models for control purposes. *Chemical Papers*, 48:69–72, 1994.

[60] J Monod. *Recherche sur la Croissance des Cultures Bactériennes.* Herman, Paris, 1942.

[61] E Nahas, M Henson, and D Seborg. Nonlinear internal model control strategy for neural network models. *Computers and Chemical Engineering*, 29(4):1039–1057, 1992.

[62] K Najim. *Process Modeling and Control in Chemical Engineering.* Marcel Dekker Inc., New York, 1988.

[63] K Najim and E Ikonen. Distributed logic processors trained under constraints using stochastic approximation techniques. *IEEE Transations on Systems, Man, and Cybernetics - A*, 29:421–426, 1999.

[64] K Najim and E Ikonen. *Outils Mathématiques pour le Génie des Procédés - Cours et Exercices Corrigés.* Dunod, Paris, 1999.

[65] K Najim, A Poznyak, and E Ikonen. Calculation of residence time for non linear systems. *International Journal of Systems Science*, 27:661–667, 1996.

[66] K Najim, A Rusnak, A Mészaros, and M Fikar. Constrained long-range predictive control based on artificial neural networks. *International Journal of Systems Science*, 28:1211–1226, 1997.

[67] K Narendra and M A L Thathachar. *Learning Automata an Introduction*. Prentice-Hall, Englewood Cliffs, New Jersey, 1989.

[68] S Norquay, A Palazoglu, and J Romagnoli. Application of Wiener model predictive control (WMPC) to a pH neutralization experiment. *IEEE Transactions on Control Systems Technology*, 7(4):437–445, 1999.

[69] A Ordys and D Clarke. A state-space description for GPC controllers. *International Journal of Systems Science*, 24(9):1727–1744, 1993.

[70] J Parkum, J Poulsen, and J Holst. Recursive forgetting algorithms. *International Journal of Control*, 55(1):109–128, 1990.

[71] T Parthasarathy. *On Global Univalence Theorems*. Springer-Verlag, Berlin, 1983.

[72] R Pearson. *Discrete-Time Dynamic Models*. Oxford University Press, Oxford, 1999.

[73] W Pedrycz. *Fuzzy Control and Fuzzy Systems*. John Wiley and Sons, New York, 1989.

[74] J Penttinen and H Koivo. Multivariable tuning regulators for unknown system. *Automatica*, 16:393–398, 1980.

[75] A Poznyak and K Najim. *Learning Automata and Stochastic Optimization*. Springer, Berlin, 1997.

[76] M Pyykkö. *Leijupetikattilan tulipesän säätöjen simulointi*. Tampere University of Technology, Finland, 1989.

[77] J M Quero, E F Camacho, and L G Franquelo. Neural network for constrained predictive control. *IEEE Transactions on Circuits and Systems – I: Fundamental Theory and Applications*, 40:621–626, 1993.

[78] D Ratkowsky. *Nonlinear Regression Modeling: A Unified Practical Approach*. Marcel Dekker Inc., New York, 1983.

[79] J Richalet, A Rault, J Testud, and J Papon. Model predictive heuristic control: Applications to industrial processes. *Automatica*, 14:413–428, 1978.

[80] V Ruoppila, T Sorsa, and H Koivo. Recursive least-squares approach to self-organizing maps. In *IEEE International Conference on Neural Networks, San Francisco*, 1993.

[81] A Rusnak, M Fikar, K Najim, and A Mészaros. Generalized predictive control based on neural networks. *Neural Processing Letters*, 4:107–112, 1996.

[82] J Saint-Donat, N Bhat, and T McAvoy. Neural net based model predictive control. *International Journal of Control*, 54:1453–1468, 1991.

[83] D Sbarbaro, N Filatov, and H Unbehauen. Adaptive predictive controllers based on othonormal series representation. *International journal of adaptive control and signal processing*, 13:621–631, 1999.

[84] R Setiono and L Hui. Use of quasi-Newton method in a feedforward neural network construction algorithm. *IEEE Transactions on Neural Networks*, 6:273–277, 1995.

[85] S Shah and W Cluett. Recursive least squares based estimation schemes for self-tuning control. *Canadian Journal of Chemical Engineering*, 69:89–96, 1991.

[86] J Sjöberg, Q Zhang, L Ljung, A Benveniste, B Delyon, P Glorennec, H Hjalmarsson, and A Juditsky. Nonlinear black-box modelling in system identification: A unified overview. *Automatica*, 31:1691–1724, 1995.

[87] P van der Smagt. Minimisation methods for training feedforward neural networks. *Neural Networks*, 7:1–11, 1994.

[88] R. Soeterboek. *Predictive Control: A Unified Approach*. Prentice Hall International, London, 1992.

[89] B Sonnleitner and O Käppeli. Growth of saccharomyces crevisiae is controlled by its limited respiratiry capacity. Formulation and verification of a hypothesis. *Biotechnology and Bioengineering*, 28:81–88, 1986.

[90] G Stephanopoulos. *Chemical Process Control: An Introduction to Theory and Practice*. Prentice-Hall, New York, 1984.

[91] K O Temeng, P D Schnelle, and T J McAvoy. Model predictive control of an industrial packed bed reactor using neural networks. *Journal of Process Control*, 5:19–27, 1995.

[92] A Visala. *Modeling of Nonlinear Processes Using Wiener-NN Representation and Multiple Models*. PhD thesis, Helsinki University of Technology, 1997.

[93] H Walk. Stochastic iteration for a constrained optimization problem. *Communications in Statistics, Sequential Analysis*, 2:369–385, 1983–1984.

[94] P Werbos. Backpropagation through time: What it does and how to do it. *Proceedings of the IEEE*, 78:1550–1560, 1990.

[95] T Wigren. Recursive prediction error identification using the nonlinear Wiener model. *Automatica*, 29(4):1011–1025, 1993.

[96] C Yue and W Qinglin. A multivariable unified predictive control (UPC) algorithm based on the state space model. In K Seki, editor, *38th SICE Conference '99*, pages 949–952, Morioka, Japan, 1999. The Society of Instrument and Control Engineers, The Society of Instrument and Control Engineers.

[97] K Zenger. *Analysis and Control Design of a Class of Time-Varying Systems*. Report 88, Helsinki University of Technology, Control Engineering Laboratory, 1992.

# Index

Milton Keynes UK
Ingram Content Group UK Ltd.
UKHW052019071024
449327UK00027B/2338

9 780367 396886